Acclaim for Richard Fortey's

Life

"An eclectic waltz among organisms at once extraordinary and extinct. . . . No one can finish *Life* without having accrued considerable knowledge of evolutionary history and a sense of excitement of discovery that can be conveyed only by a professional scientist chipping away at the rock face of his discipline."   —*The New York Times Book Review*

"[Fortey's] prose, like Darwin's, is spare, confident, and unadorned. As his impressive synthesis of evolution unfolds, a distant world is brought to life."   —*The Economist*

"The personal elements are crucial to the story, for Fortey—a senior paleontologist in London's Natural History Museum—has been at the center of paleontological research for more than two decades. . . . A wonderful blend of science and anecdote."   —*Natural History*

"A true history of the world. . . . A humbling narrative, written with a powerful impact. . . . A tour de force."
                        —*St. Louis Post-Dispatch*

"A celebration of learning, a spellbinding look into prehistoric landscapes and a captivating piece of literature."
                        —*The Columbus Dispatch*

Richard Fortey

Life

Richard Fortey is a senior paleontologist at the Natural History Museum in London. He is the author of several books, including *Fossils: A Key to the Past* and *The Hidden Landscape*, which was named the Natural World Book of the Year in 1993. In 1997 he was elected a Fellow of the Royal Society.

Life

Richard Fortey

Life

A NATURAL HISTORY OF
THE FIRST FOUR BILLION
YEARS OF LIFE ON EARTH

Vintage Books

A DIVISION OF RANDOM HOUSE, INC.

NEW YORK

FIRST VINTAGE BOOKS EDITION, SEPTEMBER 1999

Copyright © 1997 by Richard Fortey

All rights reserved under International and Pan-American Copyright
Conventions. Published in the United States by Vintage Books, a
division of Random House, Inc., New York. Originally published in
hardcover in Great Britain as *Life: An Unauthorized Biography* by
HarperCollins Publishers, London, in 1997, and in hardcover in the
United States by Alfred A. Knopf, Inc., New York, in 1998.

Vintage Books and colophon are registered trademarks of
Random House, Inc.

The Library of Congress has cataloged the Knopf edition as follows:
Fortey, Richard A.
Life: a natural history of the first four billion years of life on earth /
Richard Fortey. —1st American ed.
p. cm.
Includes bibliographical references and index.
ISBN 0-375-40119-9
1. Evolution (Biology) 2. Life—Origin. I. Title.
QH366.2.F69   1998
576.8—dc21   97-49466
CIP

Vintage ISBN: 0-375-70261-X

Author photograph © Jerry Bauer
Book design by Cassandra J. Pappas

www.vintagebooks.com

Printed in the United States of America
10 9 8 7 6 5 4 3 2

For Jackie, with my love

Contents

Illustrations

© Astrid & Hanns-Frieder Michler/ Science Photo Library);
e) paramecium showing its cilia (photo © Michael Abbey/
Science Photo Library)

17. Living primitive jellyfish

18. *Dickinsonia*

19. *Spriggina*

20. Skeletal fossils of animals from the early Cambrian
    (photos Stefan Bengtson)

21. Radiolarians

22. Burgess Shale arthropods

23. Trilobite with sophisticated eye

24. Reefs—ancient and modern

25. Conodonts

26. Professor Lindström's reconstruction and the real conodont
    reconstructed by Mark Purnell

27. Graptolites "writing in the rocks"

28. Charles Lapworth

29. Tree ferns in New Zealand (photo © John Mead/Science Photo Library)

30. The Feathers, Ludlow (photo reproduced courtesy of Regal Hotel
    Group plc)

31. Thin section through Devonian Rhynie Chert

32. Willi Hennig (© Springer–Verlag 1977)

33. *Trigonotarbid*—an early and distant relative of the spiders reconstructed
    by Jason Dunlop

34. Part of the skeleton of "Boris" from the late Devonian of Greenland
    (photo courtesy of Jenny Clack)

35. Jawless fish *Spizbergaspis*

36. Spiny trilobite *Ceratarges*

37. Bark of the Carboniferous tree *Lepidodendron*

38. Ornamental column from the Natural History Museum, London

39. Living *Selaginella*

40. Carboniferous cockroach *Aphthoroblattina johnsoni*

41. "Bolsover dragonfly" (photo © The Natural History Museum, London)

42. Carboniferous shark *Stethacanthus* (photo reproduced courtesy of the
    Hunterian Museum, University of Glasgow)

43. Fossil shark teeth

44. Alfred Wegener's reconstruction of the continents at the time of Pangaea

Unless otherwise stated, photographs come from the author's own collection.

Acknowledgements

THIS BOOK would not have been as it is without the trenchant editing of Heather Godwin, nor without the punctilious attentions of Stuart Proffitt. I owe them both a great debt. I sincerely thank Robin Cocks and Neale Monks for acting as scientific readers, and for commenting on sundry facts. Claire Mellish kindly drew two of the diagrams. I particularly thank Chris Stringer, Theya Molleson and Robert Kruszynski for anthropological advice. Many friends and colleagues helped me by providing illustrations and advice (they carry no responsibility if I did not heed the latter). These include: Stefan Bengtson, Per Ahlberg, Cedric Shute, Paul Cornelius, Angela Milner, Bill Schopf, John Richardson, Brian Rosen, Phil Palmer, Jerry Hooker, Giles Miller, Joyce Pope, Andrew Ross, Sally Young, Alison Longbottom, John Whittaker, Clive Jones, Richard Thomas, Peter Forey, Sandy Knapp, Charlotte Jeffery, Jeremy Young, Solene Morris, Andy Gale, Norman MacLeod, Anthony Sutcliffe, Harry Taylor, Bob Bloomfield and Tracey Elliott. I acknowledge the Palaeontological Association for reproduction of a conodont animal. Hamish Francis kindly read the proofs. Finally I thank my family for tolerating my eighteen months' exile in the boxroom in the company of the snake.

R.A.F.

Life

Chapter 1

The Everlasting Sea

SALTERELLA DODGED BETWEEN THE ICEBERGS. While the small boat bucked and tossed, I hung over its side, peering down into the clear Arctic waters. I had not known that there could be such density of life. This frigid sea was a speckled mass of organisms. Tiny copepod crustaceans, looking like so many animated peas, beat their way in their thousands through the surface waters, feeding on plankton that I knew must be there, but which could not be seen without a microscope. There were jellyfish of every size: white, gently pulsing discs as delicate as spun glass; small pink barrage balloons decked with beating cilia, which appeared to be solid—but became gelatinous and impalpable if grasped from the water; an occasional orange monster with tentacles that promised evil stings for fish or mammal. They drifted in their millions, swirling and beating against the dumb tides, concealing purpose in contractions as instinctive as breathing, like protoplasmic lungs dilating and constricting in primitive obedience to the prompting of the currents. Behind the nearest iceberg arctic terns beat and hung in the air, peering down as I was, but with so much more precision, then darting to retrieve some living morsel from the sea. The ice floes were stained pink with their droppings.

Salterella was tackling a stretch of sea, Hinlopenstretet, between the islands of Spitsbergen and Nordauslandet far beyond the Arctic Circle at

80 degrees north. Ice floes had melted in the summer thaw, sculpted by the vagaries of weather into plates or crags, or simulacra of giants. On the water-line they were notched deeply by the sea, lapped by insistent waves, and just occasionally one would teeter into instability, cracking and keeling over with a great resigned splash which sent waves to make our small boat buck and grind against the smaller fragments of ice. It was true: the greater part of an ice floe was always beneath the sea, and you approached too close at your peril. If you looked down, you could see the bluish mass curving down into the deeps, while jellyfish skimmed hidden protuberances with impunity. Little *Salterella* sought the spaces between the floes. Her wooden construc-tion was designed to cope with ice. Winds herded floes into clots that could become almost impenetrable. Then, suddenly, patches of clear water would allow rapid progress, and the bleat of the motor sent little auks and black guillemots fluttering low across the sea to plunder the rich waters elsewhere. In the distance a mysterious coastline lay low on the horizon. Glaciers ran straight down to the sea. Ice cliffs groaned or barked to signal the inexorable creep of sheets of ancient ice. The boat seemed like an interloper.

I was twenty-one and on my first expedition. Cambridge University had a tradition of sending young geologists to Spitsbergen. For a young naturalist it was very heaven. Here there were birds on every side that had only existed as pictures in bird books. The sea, the profligate sea, was a shimmering text-book of zoology. There seemed nothing to interfere with the joy of observa-tion, no end to knowledge, no possibility that any discovery should be less than astounding.

The boat comprised two crew and several scientists, including myself and Geoff. We had already suffered in the old whaling vessel which had carried us from Norway, a switchback ride across the Barents Sea all the way to Spits-bergen. Few on board could face the whale-meat stew. Our expedition leader was the worst sailor of all, having disappeared below decks just after leaving the Norwegian port of Boda, and only reappearing a week later when we reached the base at Longyearbyen.

Geoff and I were to live together for weeks in a small tent, watching our beards grow from speckled patches to whiskers worthy of a Victorian pater-familias. Together, we were in search of ancient fossils. An expedition from the previous year had stopped off to replenish their water supplies from a melt stream running off the great glacier of Valhallfonna in this remote and unwelcoming northern part of the archipelago. To everyone's surprise, the crew had picked up lumps of dark limestone on the beach that teemed with fossils: trilobites and brachiopods and many unrecognizable things besides.

Nobody knew that fossil remains of such animals existed in this part of Spitsbergen. It was all completely new. But there had been no time to investigate that year because the Arctic night was closing in. Perpetual dusk was soon replaced by perpetual night. The few lumps of rock were brought back to Cambridge, and were studied by the great Professor Whittington, who pronounced them very interesting. Thus it was that two students came to be sitting side by side on *Salterella* looking at swarming jellyfish, and in such serendipitous ways lives are decided. It was 1967. "All You Need Is Love" was top of the hit parade, and stayed there for the whole expedition.

Expeditions are curious things. They last for weeks or sometimes months, during which time they acquire a life of their own, a structure, like a drama. Members of the expedition get to play roles, and the most curious aspect of all is that it is impossible to predict in advance quite what those roles will be. People have to get on together; there is simply no choice. Even pathological personalities have to survive the whole affair. There is, of course, the leader—well, there has to be—who has managed many of these things before. Of an evening, he recounts tales of blizzards past that make the present one seem tame. He knows stories of Nordenskjöld and the other great men, who did it all with pemmican and huskies. He legitimizes the whole experience by accommodating the current namby-pamby lot within a great tradition. If you follow in the footsteps of giants, don't you walk taller yourself?

Then there is the expedition joker. He is not necessarily the wittiest man in the party, but he has a knack of igniting humour. Every member of the expedition likes to have him around in the evening. He has a generous gift of appreciating the humour of others, puffing up a glancing remark into hilarity, keeping flagging conversations alive, massaging morale. It is impossible to recall the humour that keeps an expedition afloat. It is concocted by the joker out of nothing and vanishes once more into nothing, but while it is there it seems to be the best thing in the world. The expedition's Practical Man knows how to fix a paraffin stove, or an engine. He can splice a broken guy rope. He can take out splinters, make splints; he can build machines from bits of wire and bottle tops. He is a wonder, as his ham-fisted friends who rely on him never tire of reiterating. I dare say that in ordinary life in suburbia Practical Man may seem a bit of a dullard, but when the outboard engine is failing among the ice floes he has his moment of glory. My own role, a modest one, was that of chef. Our food was nearly all dried: peas, onions, potatoes, rice, oats. Worst of all was the meat bar, 200 grams of dried protein which had to be reconstituted with hot water and which stayed insipid no matter what ingredients you added. Hours of ingenuity went into spinning these ingredients into something spectacular. I tried meat balls, curry, shepherd's pie, patties, pasties and pastries. I bashed them flat, or stuffed them with onion and peas. I married meat bars with oats. I was left undisturbed to follow my arcane trade, which was good news for one incapable of peering at an engine without exhibiting patent confusion. While Practical Man did his vital stuff, the leader led, and the joker cheered up the bystanders, the cook could be quietly abandoned to try to fabricate an onion soufflé with powdered milk, flour, yeast extract and dried shallots.

The oddest role in the expedition is that of the scapegoat. His function is to take the blame for everything that goes wrong. A lost wrench? The scapegoat had it last. A leaking tent? You know who damaged the lining. Unexpected bad weather? Whose turn was it to check the weather forecast? Poor scapegoat. Unlike Practical Man, who can usually be identified in advance, there is no telling who will finish up as scapegoat. However, scapegoats have one thing in common: *they never realize they are the scapegoat*. They tend to be bumptious and self-confident types, convinced equally of their rectitude and their popularity. The scapegoat's function is, however, vital. He personifies mischance. Rather than curse fate, or wonder whether some god is playing tricks on a despised humanity, the scapegoat domesticates and humanizes misfortune. With the scapegoat there, nothing really bad can happen. And if the choice of scapegoat is as it should be, even *he* is unaware of the role he is playing. Peter enjoyed his expedition to Spitsbergen enormously, unaware that he was being blamed for everything from metal fatigue to blizzards. In this way an expedition defines its members. The identification of parts ensures the success of the whole, a formal intimacy is established, and the job gets done.

Geoff and I were eventually dropped off on to the shore of Hinlopenstretet, just the two of us, leaving behind our expedition roles, to find the fossils that had excited the previous year's collectors. It did not take long. In a couple of minutes there was a trilobite showing up all black on a white limestone slab. A few moments later there was another, and then another. The place was prolific! We danced around picking up any piece of rock that attracted our attention. Every rock fragment seemed to have something. This was the delight that animated Howard Carter at the tomb of King Tutankhamun. Nobody had ever seen these creatures before. Our eyes were the first to peer at the primeval rocks, to understand something of the ancient cargo they bore, to wonder at the preservation of extinct creatures on a bleak Arctic shore. In that harsh place there must have been something oddly incongruous about these capering enthusiasts.

But the tent had to be pitched. The shoreline was a beach stranded by the last great Ice Age, covered in shingle. The wind never ceased. Our tough tent was called a Whymper after Colonel Whymper, one of the great expeditionists. We tied the guy ropes on to spars that lay on the beach—logs brought in by the North Atlantic Drift to this island far beyond the habitat of any trees except the tiny arctic birch. Then we buried the spars in the shingle. Any gale would have to rip the tree-trunks out from their graves. Air beds and thick, real eiderdown sleeping bags provided such comfort as was to be had. That afternoon even meat bar *bonne femme* tasted wonderful.

There was no night—we were far too far north for that. But the leader had told us how important it was to keep to a regular pattern of sleep and work. If we failed to do so our minds would spin out of their proper biological rhythms; strange distortions of perception might develop. But sleep did not come easily when just a few yards away lay rocks that had never been explored before, our own personal slice of ancient history. Outside the tent, we could hear the ceaseless suck and rattle of the waves on the strand, the mewing of the gulls and the sharp cries of the terns. We had planted a flag on a pole, which chattered like distant gunfire in the incessant wind. Our minds could play, as we lay there, upon the fearsome polar bear, the *isbjorn*, who could flatten our tent with a single blow, and break a human leg with a single swipe. We even had rifles (of a sort) against his arrival. My feet were always cold when I climbed into the sleeping bag, and I wriggled deliciously in the warm cocoon until a gentle warmth crept slowly into my toes. Then I waited for the unconsciousness that would, finally, steal over me.

When our alarm clock woke us after the obligatory seven hours, we were into our woollen trousers and double anoraks in a rush, and out on to the rocks. Those pieces we had picked up on the beach must have come from rock outcrops beneath the gravel. Within a few minutes we had discovered where these outcrops were. All along the sea's edge there were low ledges of limestone, stacked one of top of the other, dipping down gently towards the water. The movements of the Earth that had long since elevated our ancient rocks had also tipped them gently. Limestone is a sedimentary rock, one that was built beneath ancient seas by the slow accumulation of sediment, making beds each a few inches to a foot or so thick. The top of each bed was a flattish bedding plane, every one the surface of a former sea floor. So the beds of rocks we were admiring were like the successive pages of a book that recorded ancient time, logging time itself in limy mud that further time had hardened and transmuted into rock. On the bedding planes we could see shadows of trilobites, occasionally something clearer—a tail, perhaps. These fossils were the shells of animals that had once lived upon the sea floor, trapped, like Time itself, as part of the narrative in stone. As we looked along the shore we could see the rock beds dipping in ranks into the distance. Ice floes had come to rest against some of the thicker ribs of rock as they struck out into the sea like groynes, and mist concealed still more distant rocks in enticing obscurity. How much time might be buried here along this desolate shore?

And it was all ours! This stone diary had never been read before. There could be almost anything here, just waiting to be split from its rocky pages. We were standing near the top of the thickness of piled strata, so the beds that dipped towards us along the shore were progressively older the further away from us they were. We knew that as we tapped our way along the shore, so we also tapped our way back into geological time, exploring an older and older past, seeing what came before, and before that again. This simple method had built the whole elaborate edifice of geological time, the sum of a thousand narratives in stone-stacked order. Ancient seas had preserved their history in rocks. In time those rocks themselves would be preyed upon by newer seas, eroding history away again. But enough would survive to tell of life by then vanished, of the endless cycles of climate change, and of the hidden poetry of our mutable world.

> Time, like an ever-rolling stream,
> Bears all its sons away . . .

IN THE WEEKS THAT FOLLOWED we broke rock. Every fossil we recovered was logged into its precise place in the historical story. Notebooks were filled, sketches of rock sections were scribbled. Each specimen was wrapped in newspaper and tucked into a canvas collecting bag, and then the bags were collected into the same boxes from which we had taken our food. Out came porridge oats and dried meat, back went fossils. And months later I unwrapped with tremulous hands the little parcels which we had wrapped so laboriously when our fingers had still been stiff with cold.

The tools we used were geological hammers, tough enough to withstand endless battering without splintering, and hand lenses to peer closely at our finds. Some of our finds peered back, because certain trilobites had preserved the oldest eyes—convex, compound eyes, some as large as those of a dragonfly and with myriad polygonal lenses. We looked at one another for the first time, those trilobites and us, though hundreds of millions of years apart, and I understood that there could scarcely be a better metaphor for discovery.

We soon found out that these trilobites had to be Ordovician (that is, about 480 million years old) because a few of them had relatives which we recognized from other sites, and some of the limestone beds also contained the fossils of graptolites, a well-known kind of extinct planktonic animal, which changed with every geological period in ways that students had to memorize. How glad we were then that we had drubbed into our brains the litany of species that spelled out geological time. But even with the little we knew we could see that there were types of fossil in these rocks that had never been seen before. We did not know what to call them, so we gave them nicknames—undergraduate-ish names like Mildred and Fred. Some years later they received the blessing of a scientific name, latinate and slightly pompous as they are supposed to be: *Cloacaspis* or *Svalbardites*. But the nicknames will be remembered, because they were generated in the hectic enthusiasm of those early days.

Sometimes the weather could no longer be ignored. Spitsbergen seems to be home to all the deepest meteorological depressions. One could imagine some Norse saga peopling its wet and inaccessible shores with carbuncular and bad-tempered trolls. Even in the middle of summer it was swept with squalls and driving drizzle. Specially small musk oxen live on the south of the island, sustained by meagre vegetation which would make the tundra seem lush. The struggle for existence was graven on the shores of Hinlopenstretet, too, where bones of stranded whales seemed to provide the only nourishment for an impoverished flora. Tiny arctic poppies huddled in hollows, and purple saxifrages seemed almost indecently colourful in a monochrome world. We were even too far north for mosquitoes (a blessing). The sheer joy of discovery usually kept the cold at bay: but even thick woollen gloves soon became tattered against the sharp rocks, and as our hair and beards grew we came to look more and more like dervishes belonging to some hermetic sect, especially if we jumped up and down to celebrate an unexpected find.

But a blizzard could not be faced out. We crawled into the tent and into our sleeping bags without even taking off our sweaters, and stayed there for several days. The howling wind mercilessly undid the good work of the summer thaw, plastering snow over rocks that had only just begun to yield up their secrets. This was when we discovered why the leader had recommended bringing *War and Peace* and *The Brothers Karamazov*. We got out of bed only to perform the most vital bodily functions, an unkind compromise between biological necessity and fear of frostbite of the private organs in which decorousness had little part to play.

Back in the feathered haven of my sleeping bag I could hear the sea lashing against the shingle. The waves that broke on the cold, grey arctic stones would have sounded the same in the Ordovician or the Jurassic. The sound was audible proof of the endurance of the sea.

The cauldron of life I had admired from *Salterella* was a different life—another set of animals, another ecology—from the life we had been hacking out of the rocks for the first time. There were no living trilobites, nor any Ordovician terns. Yet the sea endured. It was one constant in a fickle biosphere. The sea was cradle to the Ordovician animals which we were discovering: it succoured them while they were alive, and ultimately it provided for their entombment. I visualized the trilobite crawling over a soft mud nearly 500 million years earlier. Perhaps it died while it was moulting, overcome by noxious gases emanating from the sulphurous sediment. The hard carapace remained to record the life it had led, an archive of calcium carbonate, a shell for eternity. More fine mud covered the carapace, and as the millennia passed the beds above were slowly accreted. Occasionally, a tempest would dump more mud in a day than had accumulated for a century or more. Eventually, the pile of sediment hardened as its water was pressed out, and then followed the long annealing of geological time, millions of years, not mere thousands, before it became the dark rock I was now exploring. But still it carried its precious cargo. The chances of my making a connection with the fossil were still remote, because a hundred thousand incidents lay between the life of this humble invertebrate and my own—not least, if the hammer had fallen in a slightly different place the fossil might never have told its story. And consider what intervened between the Ordovician and the blizzard raging outside the tent: three ice ages, the deaths of a dozen continents and the birth of a dozen more, earthquakes and earth movements, mountain ranges thrown up and then eroded deep into their roots, the rise of fish and dinosaurs, the dramatic demise of the latter, bombardments of the Earth from space, and all the tangled skein of life winding around the changing terrestrial world like partners in a *pas de deux*. Chance and the consequences of chance ruled a blow of hammer on rock, arranging an assignation with history.

Fortuitous or not, this trilobite represented one moment in geological time, and within the section of rocks exposed along the Arctic shore a thousand preceding or subsequent moments were recorded. The story of geological time is pieced together from thousands of such fragments; some have been known for 200 years, others were recorded only yesterday. The rocks in Spitsbergen were my own little piece of the story—a doctorate's worth, if you like. The great narrative of geological time is a patchwork, a stitching-together of odd fragments with partial vistas, idiosyncratic, a tale put together by heroes and journeymen. And this narrative has a language of its own, the divisions of geological time that soon trip off the tongue with the ready familiarity of a railway timetable. Cambrian, Ordovician, Silurian, and so on, up the geological column, each of these again subdivided finely and more finely, the better to approximate history. It is an astonishing story, a tale of more than 3,500 million years. Consider what has happened since the death of Napoleon: the interpretations, the parade of historical facts, the controversies; and it will be obvious that a history more than ten million times as long can never be known, even in outline. And when this history has been stitched together from small pieces taken from rock sections where chance has had a part in the glance of every hammer blow, it will be certain that this history is a poor thing, an approximate story peddled by optimists.

The blizzard was over. We could return to our patient battering of the Ordovician rocks. A routine settled in, as regular as any office worker's. The alarm clock would wake us. A ritual curse about the weather and it was time to make the porridge—breakfast was always porridge. We used water from a melt stream, and would throw in raisins to make it more exciting. (A peculiar thing had happened to our taste buds. We loved the porridge to be very sweet—we poured granulated sugar over it until it was crusted. In civilized conditions it would have made us sick, but in this cold waste our bodies were hungry for calories in the most blatant form and the glutinous stuff tasted good.) We gobbled it down. Then to work. If we were working near by we could come back for lunch; if not, we put a few lifeboat biscuits into our pockets, and a bar of chocolate, and that would have to suffice for the day. Unless the weather was particularly miserable we would spend most of our time collecting the new fossil fauna, systematically, from each successive rock bed. Every so often something new or spectacular would turn up, prompting cries of delight, and then the lucky finder would hold it out for the other to admire. But the incessant wind was troubling. It never seemed to stop sweeping over the naked gravel of the vast, raised beach on which we lived. Although the temperature was above freezing the chill factor was dangerous; even with balaclavas and wind-proof jackets, woollen pullovers and woollen combinations that would have seemed old-fashioned to our grandparents the cold still managed to sneak in. At first we wore wellington boots: we often had to trudge for miles through slushy snow, and it was essential to keep the feet dry. But when we stopped walking a profound iciness would creep into the toes, and once it had taken control it was implacable until we climbed into our sleeping bags at the end of the day. It was a big moment when we discovered the *mukluks* in our supply boxes. These were a kind of giant canvas boot lined with felt—an

Eskimo invention. They trapped the heat wonderfully well. Then it was only the fingers that ached with the chill. We were supplied with tins of Three Castles cigarettes made by W. D. & H. O. Wills of Bristol, and I would light up, trembling behind whatever bluff I could find, as much for the brief moment of heat as for the nicotine.

From time to time there would be visitors. On the calmest day of all a walrus paid us a visit, cruising up and down the strait, and diving to grub up his favourite clam, the arctic *Mya*, a good-sized morsel the size of a large mussel. Every time he surfaced he would display his great tusks and blow out a massive, splashy breath, like some old colonel puffing outrage at intrusive strangers. Tiny, albino arctic foxes would appear as if from nowhere. They would eat anything, even our discarded porridge. We were so remote (or they were so hungry) that they seemed to be almost tame, even though their fur was hunted. I suppose their usual diet was the eggs and chicks of birds. They disappeared as quickly as they had come.

There was never a time without birds. They skimmed incessantly along the shore, battling purposefully against the wind: fulmars, effortless and aerodynamically perfect, flirting with the sea surface; terns, more laboured, beating hard sometimes to make progress. Least welcome were the skuas, those unappealing parasites. They lurked on the strand, and then relentlessly pursued some unfortunate gull, harrying and diving at the creature until it regurgitated its meal. The gull's final indignity was to have the contents of its evacuated crop swallowed in mid-air by its tormentor. The rich production of the Arctic sea supported all these animals. They were one with the pulsating jellyfish.

Geoff and I established a kind of working cordiality. With the arbitrariness which is part of an expedition, we had been chosen for our enthusiasm for palaeontology rather than any predictable compatibility. The truth was that we were as different as could be in every regard except for our love of fossils. After the expedition I do not think we ever spoke again. But while we were together we were obliged to share every belch and indiscretion. We evolved strategies. We talked about what we had found, naturally, and what we should collect that day. We talked about what we could do with the meat bars that evening, wondering whether we could perhaps make kebabs if we could only improvise glue out of flour. We speculated about the other members of the expedition. We made comments about the comparative odorousness of our socks. It was an odd, jokey relationship that was proximal without being intimate. I imagine it had the kind of closeness of a moderately successful arranged marriage.

We had chatted only briefly on the long cruise up the Norwegian coast. From the north of Norway we had crossed the Maelstrom together, past the Lofoten Islands to Bear Island, the most forsaken lump of rock in the world. Wild-eyed radio station operators had loomed out of the mist there to take off their supplies, comprising tinned food and whisky in approximately equal parts. A special breed of escapist seeks these places out—men (they are always men) who can only function at the edges of things, in a kind of solitary confinement, men who find the company of others difficult, and move and move again, motivated by a kind of inverse gregariousness, to where they can be almost alone. I was to meet their kind again in the Australian outback. This longing for solitude is apparently a progressive condition. After a while it is no longer enough to man the radio station with others: the committed loner seeks the chance to overwinter in some outstation, absolutely isolated. I met one man who had done just this: he had been confined by cold and darkness at the edge of the world for months. He told me wryly, almost sadly, that he had returned to Norway but found the crowds in Hammerfest, Norway's most northerly town, almost unbearable, and returned to Spitsbergen by the next available boat. Even as he talked to me I could see that his eyes were disengaged, looking out over the pack ice to where seclusion lay.

So Geoff and I developed a relationship which was an expedition relationship. But Geoff was a year my senior—he had completed his finals shortly before we left for the Arctic. Because of this year's difference I was Geoff's assistant, the hired hand—and at first he made the field strategy. But things were to change, because of another stroke of chance that would transform both our lives.

We were finding more and more new species, not just one but dozens. These trilobites were without names, a whole fauna through an uncharted stretch of geological time that had never been known before. As our collections grew, so did the realization that there was a lot of work here, just to make these new animals known. And how did they fit into the Ordovician world? How did they live? Just as, a metre or so away, the sea was profligate with variety, so, it seemed, was the Ordovician one—laid out upon plate after plate of rock before our inquisitive eyes. There were more fossil animals yet: snails and nautili and sponges and curious squiggles that defied our knowledge. Clearly, life in the sea had been a cornucopia of variety hundreds of millions of years before our hammers broke open its hidden plenty.

The excitement of discovery cannot be bought, or faked, or learned from books (although learning always helps). It is an emotion which must have developed from mankind's earliest days as a conscious animal, similar to the feeling when prey had successfully been stalked, or a secret honeycomb located high in a tree. It is one of the most uncomplicated and simple joys, although it soon becomes mired in all that other human business of possessiveness and greed. But the discovery of some beautiful new species laid out on its stony bed provokes a whoop of enthusiasm that can banish frozen fingers from consideration, and make a long day too short. It is not just the feeling that accompanies curiosity satisfied—it is too sharp for that; it arises not from that rational part of the mind that likes to solve crosswords, but from the deep unconscious. It hardly fades with the years. It must lie hidden and unacknowledged beneath the dispassionate prose of a thousand scientific papers, which are, by convention, filleted of emotion. It is the reason why scientists and archaeologists persist in searches which may even be doomed and unacknowledged by their fellows.

The urge to collect is different. It is clearly a deep urge also, because collected objects have been a part of human culture almost from its beginnings. There is even a Grotte du Trilobite, near Les Eyzies in France, a cave in which one of the earliest Europeans secreted a trilobite as a revered relic. Collecting is more than hoarding. Children will collect seashells from the beach, and rigorously sort them into types, by colour and design. They feel it is important, somehow, to get it right and, having done so, to keep the result. This kind of personal museum is part of the way we define ourselves, an archive of self, and is not mere covetousness or "stamp collecting." Children need to classify things in order to get a grasp upon the world. Discrimination and identification have value beyond the obvious separation of edible from poisonous, valuable from worthless, or safe from dangerous. This is a means to gain an appreciation of the richness of the environment and our human place within it. The variety of the world is the product of hundreds of millions of years of evolution, of catastrophes survived, and of ecological expansion. To begin to grasp any of this complexity the first task is to identify and recognize its component parts: for biologists, this means the species of animals and plants, both living and extinct. And to begin to negotiate this astonishing diversity a reference is needed, a sample of one species to compare with the next: in short, a collection. We start to understand our history by seeking to collect and classify.

There will be no end to what can be discovered from the rocks. Every year important new finds are made, and there is no sign that discoveries are becoming rarer. This is scarcely to be wondered at. Consider the millions of species that inhabit the world today. Many of these have still to be named and described. There are even occasional discoveries of new large animals—a previously unknown antelope was discovered in Vietnam only a year before this was written. And there is no end to beetles: we know only a fraction of the living species. A friend of mine, a beetle specialist, has "gassed" the canopies of tropical trees, and catches the animals that fall out in inverted umbrellas. It turns out that very few of the species obtained in this way have ever been seen before. Even with a modest extrapolation it seems that there may be millions of animals yet to be named in the living fauna. For humble animals like nematode worms there are untold hordes of unrecognized species—and these are much harder to define than beetles. Now they are being numbered rather than named, given molecular signatures which can be recognized again even without a name. But all these animals have a history, the millions known and unknown. This history fingers back through hundreds of millions of years, not an evolutionary tree so much as an evolutionary forest, and every branch possibly, or possibly not, recorded in the rocks. It is all a matter of chance.

The dice are loaded in favour of the preservation of some kinds of animals and plants. The most obvious bias favours organisms with hard tissues, such as shell or wood or bone, the common stuff of fossilization. So we would expect molluscs and fish to have a good fossil record, and we might doubt whether jellyfish or earthworms would have much record at all (surprisingly, they do have meagre ones). Because fossils must be preserved in sediments, it seems equally obvious that it will be common to find marine animals fossilized, and perhaps those that lived in lakes and rivers, while it is improbable that the eyries of eagles, or the bones of mountain goats, or the many inhabitants of high plains, will leave anything behind to tell of the history of these habitats. It is possible that Jurassic mountain tops were populated with fabulous, crested dragons that we shall never know.

So little rock is exposed at the surface, where it will catch the eye of the seeker—much more lies buried, hidden beneath the landscape, beneath town centres, golf courses or supermarkets. Occasionally, a lucky blow of a pickaxe, the excavation of new foundations for a house, or the opening of a brick pit will bring something to light that would otherwise have remained concealed. Exposure always partly depends on the toughness of rocks. Hard limestones form crags and escarpments, or sandstones bluffs, where there may be exposure of fossil-bearing surfaces. Soft rocks like clays will be exposed at the coast, but seldom inland, where they tend to form boggy vales in moist climates, or salt pans in dry ones. Then there are other rocks which may once have held rich booty from the past but which have been caught in the great vice of earth movements, buried deep inside mountain chains, squeezed and baked and changed through heat and pressure, and thereby metamorphosed into sparkling schists or mottled marbles; but then their precious cargo from the past will be obliterated. So we see the past through a shattered glass, darkly, nor can we tell which shards will be recovered to give us fragmentary glimpses of what once flourished.

I MUST RETURN TO THE SUBJECT of geological time. All narratives require a scale. In novels, the scale is usually comprehensible in terms of the span of a human life, a few score years, maybe a few generations. This scale is appropriate to us. Our natural longevity is an instinctive yardstick, the measure both of our mortality and the changes wrought by time. We can understand shifts in history within a few generations, empathize with our grandparents, even dimly appreciate the problems of the thirteenth century. This is all but a moment in geological time, an instant which may fall between one hammer blow and the next. The narrative of life requires a scale of thousands to millions of years, acting over a drama of more than 3,000 million years. Geologists are insouciant in the face of these figures. Creation "scientists" simply do not believe them. In truth, it is difficult to grasp such an immensity of time because it is so out of kilter with our own brief lifespan. The temptation is to resort to homely metaphors. Imagine that the history of the world is represented by a clockface, say, then the appearance of "blue green" bacteria in the record happened at about two o'clock, while invertebrates appeared at about ten o'clock, and mankind, like Cinderella suddenly recalling the end of the ball, at about one minute to midnight. I do not know whether such images are useful other than as encyclopaedia illustrations. The domestication of time serves to trivialize a magnitude which should be held in awe. In some ways I prefer something vaguer but truer—at least symbolically. The great raised beach on which Geoff and I laboured in Spitsbergen, with its endless shingle retreating into the mist—the pebble to hand might mark the appearance of *Homo sapiens*. The farthest I could throw it might just reach to the age of the dinosaurs, while beyond that lay further beaches which could be seen more or less clearly, themselves composed of banks of pebbles, and then, into the mist, dimly perceived, more distant beaches, impalpable, remote, the outer reaches of Precambrian time. And, alongside, the sea, the eternal sea, linking pebble with pebble, framing time itself. Then at least we could appreciate the immensity of time, its countless instants, the fossil record perhaps a litter of shells upon the shore, a scattering of jetsam.

There are two scales of time: a relative one and an absolute one. The relative timescale is like the one Geoff and I were piecing together; this rock bed is younger than the one that underlies it, and older beds again lie further down in the rock column. A jigsaw of time is spliced together piece by piece from many such rock sections. Time is stacked in order, period by period. The names we all know—Cretaceous, Jurassic, Cambrian—were labels applied to great segments of this relative timescale long before it was known how many years each one comprised, or how far before the present the Cretaceous lay. The relative scale is still indispensable, and has been growing in precision ever since the great nineteenth-century geologists discerned its important lineaments. Fossil zones serve to define its small slices, a calibration which depends on the fact that extinct species had definite durations, so that the fossil fauna of one time differed from that of another. Some animals—ammonites and trilobites, for example—seemed to change faster than others, and these came to assume particular prominence in discriminating the relative scale. And all this depended on our human capacity for recognition, for placing things in order. Sedimentary rocks are often correlated from one place to another by means of the fossil faunas they contain. Stratigraphy is the science of such correlation. In some cases, rock correlations are quite straightforward. There have been moments in Earth's history when some great event, such as the impact of a meteorite, or a transgression of the sea over land, has laid a clear imprint upon the rocks. Faunas and floras alike bow before such events, which punctuate the fossil record like chapter headings in a story.

But in the majority of cases correlation is a subtle business, cluttered with argument and dispute. Papers are written that bat opinions back and forth. Gradually, a consensus is approached. Martin Rudwick has described how disagreement and personal conflict drove the debate about the Devonian System in the nineteenth century, how correlations sparked furious correspondence and personal enmities, and all for the delineation of one part of the relative timescale. Much of this argument originates in the rather obvious fact that the world is divided into different habitats, each with its own flora and fauna. Now we have no difficulty in recognizing that our arctic faunas are contemporary with tropical ones, freshwater with marine, deep water with shallow, even though they may share few common species. But far in the past these differences conceal contemporaneousness: how are we to distinguish difference in time from difference in space? How do we correlate between rocks of one part of the world and another? We do so by means of stepping stones that bridge separate places. For example, in order to correlate between the rocks laid down in ancient rivers and those laid down at the same time in former seas we are obliged to look for those rocks which accumulated in the contemporary estuaries, where the two environments met and overlapped, and where the calibrating animals of one realm might merge with those from another. The outcome is rather like synchronizing watches between different time zones. Once the connection is made we have only to know the time in one part of the world to be able to deduce what it is in another. In some future geological age the delicate bones of the arctic tern might link together arctic and temperate rocks—even tropical ones—for this far-travelled bird is a messenger for our times, a global ambassador of the Holocene. In this way, little by little, the relative timescale is refined.

The low chance of preservation of many animals and plants; the sheer numbers of species—every one coming with a history of its own; the chequered pattern of ancient environments; and the sundry difficulties in the way of correlating rocks between one part of the world and another, all combine to explain why we are unlikely to reach the stage when rocks will no longer yield news, and where what we know will be close to all there is to know. The history of life is filtered by the very processes that preserve it. "I am soft sift/In an hourglass," as Gerard Manley Hopkins said.

The absolute timescale produces a calibration of the age of rocks utilizing the radioactive decay of unstable isotopes (atomic varieties) of elements, or some other measure that ticks off years or moments. As decay proceeds, sub-atomic particles are emitted, and these can be registered by the crackling of a Geiger counter. This process provides a true clock, yet like all clocks it carries some error. Radioactive elements decay from an unstable state to a stable one; when half the original material has decayed its "half life" has expired. Half lives vary from a few moments to many millions of years, depending on the elements involved. Thus *radiocarbon* dating works best for archaeological material, because this form of the familiar element has a short half life. On the other hand, uranium isotopes provide very slow clocks, leisurely enough to pluck out time from the early phases of the solar system itself. All that is needed is a measure of how much of the original material has decayed, and by a simple calculation based on the rate of decay the time taken to effect the change can be deduced. There is a whole range of different radioactive elements which can provide clocks, like potassium–argon (K–Ar), or rubidium–strontium (Rb–Sr). It is almost a case of designer elements: choose the time period you wish to measure and select your appropriate radioactive decay series. Mass spectrometers, which can sift out atoms of one isotope from those of another, have vastly increased the accuracy to which small quantities of isotopic material can be measured. This is how we know that the Earth is 4,600 million* years old, or that mankind moved into the North American continent across the Bering Straits no more than 15,000 years ago. So absolute numbers of years are spliced on to the older, relative timescale, and even the most ancient history has become a matter of dates. The precision of these dates is continually revised as more and more accurate methods are developed for measuring smaller and smaller amounts of material.

But oddly enough, we seem to find it easier to handle the vastness of geological time by the system of dividing it into named chunks. We are more comfortable thinking in terms of the "Jurassic" than having to remember figures like "155 million years b.p." (before present); we prefer talking about the "end Cretaceous extinction" than recounting the events of "65 million years ago." In a similar way it is easier to understand recorded history if we can attach the label of a king or queen to a time period. Everybody seems to know what the Victorian era means, but how many can give the dates of Queen Victoria's reign? Of course, the real date is important, but the name allows us to grasp the inconceivable stretches of geological time, to turn an endless series of dates into something comparable with a family history. How much easier to toss around the word "Cretaceous" than to grasp the passage of tens of millions of years! And while the names are stable, the dates themselves may change as finer resolution is attained. I have seen the base of the Cambrian move upwards from 600 million years ago to 545 million years ago during my lifetime, and I dare say it will be refined further in the future—but it is still the base of the Cambrian for all that. What has happened is that dates—absolute dates—have been progressively corrected as more dateable rocks have been discovered, and erroneous dating has been eliminated. Measurement has become more accurate; new isotopes have been discovered and used to calibrate time. The hard and fast geological column given in textbooks is truly mutable—only as good as the last date, the latest technical refinement, or the experimenter doing the work.

More important than dates, or names, are the great events that happened. Just as we remember the Great Fire of London, or the Plague, or the age of the Pharaohs or Incas, so the events of prehistory typify their periods; the

---

*In many recent scientific papers and articles 1,000 million is abbreviated to the symbol "Ga"— I shall spell it out in full in this book, if only as a reminder of the magnitude of time concerned.

chronological names serve merely to pigeonhole them in the narrative. Phrases like "the Age of the Dinosaurs" acknowledge a prominent aspect of history without reference to millions of years; this is the stuff that is easily memorized. The Ordovician is distinguished from the Jurassic by a host of particular events, such as an Ice Age, and by having different fossil animals, which could be recognized even without the convenience of a chronology. The trouble is that events without a temporal framework become muddled; they conflate and blend. Who can forget the sight of dinosaurs threatening a voluptuous Raquel Welch in the movie *One Million Years BC*—even though cavemen and dinosaurs were separated by 65 million years?

Even in our relativistic age it is easier to understand a narrative with a beginning, a middle and an end. For the archives of life the beginnings are partly obscured and we do not yet know the end, but we are starting to understand something of the thousands of millions of years in between. The dates are hardening, while the relative calibration established in the last century still serves its purpose to make time familiar, and to give us a language of chronology.

WHEN GEOFF AND I HAD SPENT a month on Hinlopenstretet we were visited again by *Salterella*. By now, we had filled four or five large boxes with our discoveries, and it was clear that we could fill as many more again before we left. The routine of our days had been interrupted by one four-day blizzard, and once by something much more terrifying.

I had been collecting on the shore when I looked up to see the shape of an *isbjorn* rushing towards me from the horizon at the distant headland. It was moving fast. It was clearly hungry, and, worse still, it seemed to be hungry for *me*. I did not trust my marksmanship—in fact, neither did I trust my rifle, which seemed to have been left over from some forgotten Afghan campaign. So I ran. Running in *mukluk*s has all the delicacy of a show elephant performing a foxtrot. But I ran as one possessed, clearing shingle banks at a bound, scuttling over permafrost polygons, until my breath came in short, panicky gasps. I had to stop. My time to enter the food chain had come. Then I risked a glance over my shoulder. I saw that my *isbjorn* pursuer was a hallucination—no more than the top of an ice floe moving behind the headland. The floe had been propelled by a favourable wind, and its topmost crag presented a passable profile of a bear. For a few minutes it seemed as if the sharp keening of the terns was mocking laughter.

Salterella brought the post, the only delivery we had. There was something oddly unreal about reading domestic details from rural England, peaceful Wiltshire, while sitting outside your Whymper at 80 degrees north. The cat had died, poor thing. My sister had gone to university. There were sparrows in the thatch. The crew of *Salterella* told us of the progress of the other members of the expedition, the mappers to the south, the geophysicists inland. These were the real tough guys, pulling man sledges over the ice cap, husky and human combined. These were the only scientists who ate up their entire porridge ration; I guessed that they would have eaten the oat packets themselves to get more calories.

Then there was news of the Tripos. These are the end of year exams of the University of Cambridge for the Bachelor of Arts degree (and you get a BA, scientist or not). The name derives from a three-legged stool on which candidates once sat, and like so many things Cantab., the name was retained as a knowing anachronism. We had left for Spitsbergen soon after taking our exams, and because Geoff was a year ahead of me he had been examined for his final degree. We must have been bouncing across the Maelstrom when the results were posted outside Senate House. Now at last they had caught up with us. With some nervousness we opened the envelopes which had been forwarded from our colleges. Geoff, in his finals, had obtained a lower second class degree; but at the end of my second year I had got a first.

The significance of this took a little while to sink in, but when it did it devastated poor Geoff. He realized that to do an advanced degree like a PhD—even thirty years ago when things were less stringent—you needed to obtain at least an upper second, and preferably a first. It was immediately apparent that he was not, after all, likely to be a research student working on the wonderful new collections. On the other hand, his field assistant, the expedition cook . . . well, if he had obtained a first at the end of his second year, it was more than likely he would do well in his finals. It would, after all, be I who would name all the new wonders from Hinlopenstretet.

Such moments decide lives. A different scrawl on a piece of paper, and both of us would probably have had different histories. Had our results been reversed it is unlikely that I would have been able to follow a life dedicated to pursuing the distant past, and Geoff might now be sitting in my chair at the Natural History Museum in London. Small turnings often lead to new routes, and only retrospect imposes a feeling of inevitability about these changes of direction. It is hard to unpick the effects of pure chance from matters more predictable and determined. Did I deserve my luck? As it was, Geoff went off to Belize after our return, to join the geological survey there, and I never saw him again.

Salterella left that evening. The Cap'n of the small vessel was cheerfully unaware of the serious consequences of his visit. He was more concerned with escaping from a clutch of ice floes that were to seal the two of us in for several weeks. Pack ice moves *en masse*, driven inexorably by the wind. Although there were exceptionally clear seas that year, this far north the major ice fields could be swept together by the perverse winds that scoured Hinlopenstretet, pushing together a dense pack too thick to be penetrated by a small boat like *Salterella*; and so it proved. It was colder, too. Gradually, as the end of the field season approaches in Spitsbergen, so the boreal light is slowly bled from the world. A creeping dimness pervades the day, so that close scrutiny of dark rocks becomes difficult, and you find yourself turning the specimen around towards a brighter quarter that is no longer there. As you prepare to leave the island for familiar and comfortable temperate latitudes a ghastly pallor informs the landscape, the distant ice caps merge with the fading sky, and if a weak sun appears at all it hugs the horizon like an undecided sunset. The shadows it casts are so long that you can project an attenuated self across 100 metres; a spindly stick insect in vaguely anthropoid disguise.

As we sat in our tent that evening Geoff consumed his whole whisky ration, neat, and with some deliberation. He had to get his resentment of our reversed roles out of his system. He became very drunk, and called me all kinds of names. There was certainly no other person or object at which he could direct his anger. Eventually I left the tent, and Geoff followed. Grabbing a piece of driftwood, he pursued me up the long beach, staggering from side to side and muttering, waving the wood in an approximate way, as if he were chasing a hornet. He was far too drunk to pose much of a threat. It was droll: two small figures marooned within a waste of ice on one side and an endless strand of shingle on the other, capering up the beach through a thin drizzle, the one in front loping away while a tottering figure behind flailed about with a sea-worn plank. After a while the expletives petered out. And eventually there was no place to go except back to the tent, and sleep.

In the morning Geoff was rather sheepish. But it was true: our roles had changed. From now on I was collecting for myself, devising the methods, setting the pace. I wanted to leave nothing behind. It says much for Geoff that after a short while he accepted the new conventions. The job had to be done; it was part of the ethos of the expedition. And, after all, in that remote waste, work was all there was. As if to reward us, on one perfect day the gloom cleared, the implacable wind dropped, and the sun bathed us in good light for twenty-four hours. It was so hot that we had to take off our shirts. Even the cries of the arctic skua took on a benign tone that day, lemonade crystals made a delicious refreshment, and meat bars tasted of protein. Days like that hardly ever happen on Spitsbergen. The joy of such moments healed any differences between us. Like W. B. Yeats,

> My body of a sudden blazed;
> And twenty minutes more or less
> It seemed, so great my happiness,
> That I was blessèd and could bless.

For the final week or two we desperately tried to sample the earliest parts of the rock section. By now, all our gloves were in tatters, shredded by hacking at the obstinate limestone. The incessant wind had whipped our hair into gorgon-like curls, matted beyond the reach of any comb. Our beards were thick and curly, a kind of natural pelt. Our woollen combinations, lived in day and night, had become an indefinable shade of grey. When we left, they were thrown over the side of the boat with some ceremony. The garments sank steadily, and I sometimes wonder what generations of jellyfish they nourished. Before we left, we had one last visit—from the Susselman. The Susselman is the Norwegian governor of Spitsbergen, and he travels the island during the summer in an official yacht or schooner, a magnificent sailing boat with brass trimmings maintained in the kind of style that always (perhaps only) seems to accompany a diplomatic function. The arrival of a fairy godmother on Hinlopenstretet could not have been more surprising. We were invited on board for canapés and sweetmeats. There could have been nothing more incongruous than two hirsute and malodorous young men being served canapés by a uniformed servant at 80 degrees north. The Susselman must have regarded his previous visit to the Russian governor of Barentsburg as a comparatively cultural occasion, even though it was well known that a Soviet posting to Barentsburg was a punishment for drunkenness at best, venality at worst, and usually some interesting combination of the two. He must have been glad when we boarded our dory and left for the bleak shore once again.

In truth, we, too, were glad when at last we were picked up by *Salterella* for the journey back to Ny Ålesund towards the south of the island, happy to get out from under canvas, and with our precious cargo of fossils stored in the hold. Like love affairs, expeditions usually continue just a little too long. The end is always a little rueful, and tinged with impatience.

On our way home we stopped at Biskaerhuken at the northern tip of Spitsbergen, where it was still possible to pick up items of Nazi cutlery from the beach (the place had a strategic importance in the Second World War). A few metres away from the hut at Biskaerhuken there were graves of whalers, who had hunted these seas more than a century before the Allies had fought Hitler here. Simple wooden boxes were covered with gravel, and the corners of the makeshift coffins could be seen emerging here and there from their rubbly cover. We were evidently just the last in a long tradition of visitors to the Arctic. In our own way, we had plundered, too, but our visit was for the enrichment of knowledge rather than monetary gain. It was difficult to imagine how we could have explained our motivation to a nineteenth-century whaler, or even to a Nazi U-boat commander. We had discovered the evidence for a small piece of the narrative of the history of life, a unique piece, and our own. It was to be one small contribution to the greater history that can never be fully discovered. We had rifled a small file from rocks that had been set for eternity.

THIS ACCOUNT OF THE EXPEDITION to Spitsbergen in the late 1960s serves as a metaphor for the themes that pervade this book. The expedition had its reverses and surprises; chance had a part to play, both in discovery and in our own story; and on one occasion my personal fate was changed by the merest detail on a piece of paper. The important steps in discovery and history are often down to such details, the pivotal event rubbing shoulders with the mundane. Discovery is intimately intertwined with the discoverer. There are plenty of accounts in encyclopaedias and textbooks that treat the narrative of life as if it were a scenario for a documentary movie. Event seems to tread inexorably upon event, fact upon fact, in a chronology recited by rote. The real business somehow seems to be absent. Neither the awe the story should command is there, nor its curiously human dimension, as when one discoverer wrestles with his rival to uncover history, or a stop for fresh water on Hinlopenstretet reveals hidden treasure. My own account, which I call an unauthorized history, will pick a more idiosyncratic way through thousands of millions of years of life. No event of real moment will be omitted, but it is impossible to be compendious: there is simply too much history, and the story will be shaped as much by what has been left out as by what has been included. Isaac Newton famously described his sampling of phenomena from the physical universe as a kind of beachcombing, whereby he could pick up only the brightest shells that caught his eye from an infinite litter on the strand. Like pebbles on the beach of Hinlopenstretet, history, too, is a succession of endless details, and there is an infinite choice whether to pick this one or that. And where my own experiences with people or places will serve to bring the process of investigation alive then I shall make diversions, the better to illuminate the way forward. Scientists are supposed to eliminate their personal voice, which no doubt works admirably for technical journals, but such spurious objectivity jettisons an awareness of much of what makes the process of discovery exciting, interesting, and informed with the whole inventory of our frailties and virtues.

I must mention the bugbear of teleology—ascribing purpose to all evolutionary activity, as if the whole of creation were striving for perfectibility. It is tempting. The nature of adaptation *is* improvement of design—or at least of fitness. It is attractive to think of animals or plants willing themselves to get better and better, more adapted, as if life were some kind of capitalist enterprise with an eye to ever-increasing sales and productivity. A contrary view attributes almost everything in evolution to the operation of heartless chance, through which life blunders like a blind man in a battlefield. Luck controls survival rather than genetic virtue or perfection of adaptation. Battered by fortune, controlled by the toss of countless dice, fortunate species survive a filter of fate. There is no real conflict here. I see no point in becoming obsessed by narrative objectivity in a fruitless attempt to worm out the human from descriptions of history. Like it or not, human motivation will creep into descriptions of the natural history of animals and plants; it is impossible to consider the biological world without colouring it with our own humanity. We know that the human expression of emotion is mirrored in the snarls or grimaces of animals, and that the genetic bond that unites all mammals runs to more than the sequences of chemical bases in the DNA molecule. There is a commonality to life, and it is not a failing to acknowledge it. Equally, there are cases where chance is in control, as random as the impact of a meteorite, a sudden reverse, like the arrival of a letter bearing unexpected and momentous news.

All stories need a chronology. Geological time is paradoxical and difficult. The further back in time we go the more obscure are the events, the less certain the narrative. To approach the story of life in the most logical way, we might start at the present day and work backwards. We know a great deal about the last few thousand years, and thanks to radiocarbon dating and dendrochronology, we can date events precisely. By the time we get to the Pleistocene, about 1 million years ago, we would only be able to date many events to within a few thousand years. Further back again, to the time of the dinosaurs—say, 120 million years ago—and the accuracy would have shrunk to a few tens of thousands or perhaps hundreds of thousands of years; and back to the time of the trilobites—say, 400 million years ago—maybe half a million years might be the best margin to which we could correlate events. But life stretched back further, much further, until at 3,500 million years ago it has been debated whether the traces of life are life at all. At this distant time, the possibility for aligning an event in one part of the world with that in another might be askew by some millions of years. And the further back in time the less the fossil record has been preserved. As rocks have been recycled time after time—through erosion—less and less of the oldest crust of the Earth survives, and often this durable ancient crust has been baked or squeezed in the vast mill of mountain chains until little of its original character remains. The past is continually erased, and the record of the most distant time survives only by a chain of minor miracles. Some commonplace detail is most likely to endure; the few regular soldiers who walked unscathed from the Battle of the Somme in the First World War were probably more lucky than skilful. Like those distant bluffs that we observed when we first landed on the shore in Spitsbergen, the remotest past is dimly perceived and full of mystery, enticing but very obscure.

In spite of the attractions of working backwards through time—from the familiar to the arcane, from the clearly seen detail to the slimmest speculation—I prefer to begin near the beginning and proceed towards the present time. Stories often work best in chronological order, and this is particularly important in stories like mine with tangled plots, odd subplots, and characters who disappear, never to return. The history of life is more convoluted than any novel by Charles Dickens, and the satisfactory ending is far less guaranteed. So I must begin with what is most ancient but least known, and end with what is almost historical fact, a geological yesterday. The story will come progressively into focus as the present day is approached. Our own memories of our earliest years are also hazy and unfocused; maybe a great event will be recorded vividly but without detail. Events from our youth, and the formative experiences of adulthood, comprise a robust chronology, recalled with the same intensity that still suffuses the memories of my first expedition. Events of yesterday or the day before can be looked up in a diary. So the chronology of geological time parallels the pattern of memory, a faculty which is peculiarly developed in our own species. It is just that the timescale is expanded 45 million times from that of our own puny lifespan.

Even so, it is an incomplete story: for it comprises only the middle part. The early history of the Universe—especially its first few seconds—are the province of astronomers and theoretical physicists, and the language of their history is written in complex mathematical equations. At that genesis, the forces that shaped the story of our Earth, including gravitation, magnetism and electricity, were married in an extraordinary unity that, once shattered, will not, it is claimed, recombine until the end of time itself. The great outward explosion of time and matter at the "big bang" was a factory of creation, making heavier elements from lighter ones in a series of nuclear reactions the like of which have never occurred since. The Universe literally made itself, nearly 13 billion years ago. Matter exploded outwards in a diaspora which is still only dimly understood, forming strings of galaxies, each one of which ultimately came to be composed of millions of stars. And in the corner of one of these galaxies, the Milky Way, clouds of interstellar dust and gases were drawn together by gravity, and thus the sun was born. There are millions of similar stars, for there was nothing unique about the series of events that led to the Sun's birth, but as yet we know of no other star that gave life. When this matter congealed, conditions were right for the onset of the thermonuclear reactions that still fire the furnace at the heart of the Sun's warmth. This is the warmth that nourishes life. An intuitive appreciation of this fact inspires Sun worship by native peoples on almost every continent. They believe that the observation of ritual alone sustains the daily or seasonal reappearance of the Sun, without which life would cease. And they are right: if the warmth were switched off, even for a few seconds, it would be instant death for all living things.

This early history of the solar system predates life, although if it had happened one jot differently, there would have been no living cells. But this is not our story.

At the other end of time there is written history, and all the paraphernalia of human civilization. Here the province of the palaeontologist gives way to that of the archaeologist and historian. But there is an ill-defined area between. The domestication of animals was a momentous event in human evolution, and may have happened on the interface between prehistory and history. Recently, many of the crucial thresholds in the story of humankind are being recognized as having happened earlier than was once thought, just as the origins of our human species are more antique than was realized twenty years ago. If anything, the compass of palaeontology is growing. But like palaeontologists, archaeologists dig, albeit in soil that overlies rock formations, delving into the waste of vanished generations, the shards and ordure, the forgotten fragments of abandoned buildings. Other than ritual burials, the stuff of archaeology is mostly the bits that people no longer wanted and threw away, or buildings neglected until they fell down. Archaeologists, too, often have to make do with fragments of durable material as a means to reconstruct a fuller narrative, and they are also constrained by the principles of stratigraphy as they dig downwards, layer by layer, deeper into the past.

The narrative that concerns me lies in the vast tract of time after the Sun blazed into heat, fired by its hydrogen furnaces, and before humans started making pots, building ceremonial centres, and recording the details of their daily transactions on pottery slabs. This is quite enough to be getting on with.

Chapter 2

Dust to Life

T HE EARTH WAS BORN FROM DEBRIS that circled the nascent Sun. It was a planet spun from dust and rock, and one of the smaller masses that were trapped in the thrall of the Sun's attraction. The debris comprising the belt of asteroids testifies to this time of creation, being a circlet that never congealed into a planetary ball. The other planets show what the Earth might have been like if just one or two circumstances of history had been different. Above all, it would have been dead.

It was once believed that the process of accretion might have been rather gentle, the Earth woven from dust in the manner that candy floss is spun from sugar. But now it is acknowledged that the Earth was conceived and grew violently, a chaos of impact and fragmentation and annealing. All was instability. Meteorites continually plunged into the surface of the growing planet. Because they impacted with such force their energy was spent in melting—even vaporizing—the rocky surface, and they themselves would have fragmented and melted, contributing their substance to the growing Earth. It was a kind of mad sculpture, plastered on in chunks. One especially violent impact may have spun off the Moon, destined to remain our lifeless satellite for eternity. Certainly the Moon records, almost like a fossil, a period long obliterated by time and erosion from its planetary neighbour.

So the spinning ball of the young Earth would now have been melted and cauterized by the feverish concatenation of impacts. Elements would have been shuffled and recombined as new minerals in a frantic alchemy of creation. At the same time, the *inside* of the accumulating planet was heating up. This was because the radioactive decay of unstable isotopes of uranium and other elements provided an internal stoking of the fires that still heat the interior of our planet. The balance between the size of the Earth and the heat supplied from the living fires has to be exquisite: a smaller planet might have "burned out," while a larger one might never have achieved the right temperature at the surface to foster life. The globe spins on its axis, which ensures that not only one face is presented to the implacable Sun. All sides bask in its blessings, but none bake excessively: we are toasted to a turn. Likewise the distance of the Earth from the Sun is finely adjusted—not so close that we are burned, nor yet so far that the rays from the Sun would be too dim to allow the chemical reactions upon which life depends. The placement of the Earth in the firmament, and its pivoting in the solar system, are fine-tuned to make life a possibility. If life is just a matter of chance, then the dice were loaded in its favour.

Meteorites visit us from space with messages from the past of the solar system. They are rocky fragments of history frozen in time, vagrants from the primeval days. Most of them burn away to nothing on entering Earth's oxygenated atmosphere—they flare out, as shooting stars. I have seen them in deserts, in Australia and the Arabian Peninsula, suddenly appearing at intervals of only a few minutes, trailing lines in the sky. They look like fine scratches etched on the great inverted bowl of the sky, yet they are as ephemeral as fireworks; indeed, that is just what they are. In the early days of the Earth they were more numerous, larger, and more frequent. They were not then forces of destruction, but part of creation. Because they brought their elements in with them, meteorites were like stony packages delivered to the accreting and violent young planet. As they melted on impact they donated their gifts to the nascent world. So, by looking at meteorites and comets we can learn about the chemical elements which were thrown into our primitive planet.

Curiously enough, the common elements included carbon.

Carbon lies at the centre of life, its ubiquitous and indispensable ingredient. Carbon atoms link together in chains, and bind with other atoms, to make the whole array of organic chemicals that constitute life itself, from DNA to toenails. Only one other atom is as versatile as carbon, and that is silicon, which comprises the essential ingredient of many rock-forming minerals. It, too, can hold hands with its neighbours through large molecules.

Silicon chip technology exploits its properties, and it is not a coincidence that silicon intelligence is portrayed as the only possible rival to that of our own carbon-based brain. You do not have to be a fanatical reductionist to understand that the soul of life is carbonaceous and the soul of rock siliceous.

There is a class of meteorites called carbonaceous chondrites in which carbon is especially abundant. They are curious little objects, not uncommon, which can often be held in the palm of one hand. They are ball-like, but with a distinctive ridge around them, rather like a priest's tonsure. Their rounded side is where they were polished and melted on their passage through the atmosphere: they have been shaped by heat and melting. Their carbon content includes not only the element itself, but also carbon compounds, some related to substances known to be important in the genesis and sustenance of life itself. These might well have provided some of the "seeds" of life. Comets, too, are rich in carbon and its compounds, and doubtless were also hurled into the Earth at a late stage in its formation. Ultimately, carbon, and life itself, were rooted in the elemental creation that followed the early expansion of the Universe. Only the mode of delivery to our planet is controversial. But there can be little room for doubt that the impact of countless meteorites and comets, particularly at a time when the Earth's atmosphere had not formed sufficiently to burn them up, played a major part in preparing the world for life. It is equally clear that in that searing genesis life could not spring up spontaneously, for these donors also served to sterilize the surface of the planet even as they donated the ingredients of creation. As described in the Bible, it was like a refiner's fire.

Of the 3,000 or so meteorites in the collections of museums around the world, not all are chondrites. There are also meteorites composed of metallic elements, especially nickel and iron, which may weigh several tons. These provide some analogy to the core of the Earth, a core that grew after the coherence of the planet was established. They are extraordinarily heavy objects, which have always attracted attention. Stony meteorites may themselves be fragments of early phases of planetary disintegration, a legacy from the time of constructive chaos. Larger meteorites are capable of great destruction. They, too, deluged upon the early Earth. And lest this seems an aeon so vanished into antiquity as to be arcane and irrelevant, it is important to stress that their influence on the subsequent history of evolution has been profound. I shall return to this when discussing the demise of the dinosaurs.

The Moon still records the legacy of this time, its own history having been frozen thousands of millions of years ago. Lacking an atmosphere, its story has not been revised countless times as has that of its natural sister and

neighbour Earth. Our own past has been obliterated or modified again and
again, caught in the endless cycles of sea, wind and ice. But on the Moon
craters endure. Younger craters still preserve radial scars that testify to violent
impacts, as blatant as bullet wounds. But the greater part of the cratered
architecture of the Moon was formed between 4,600 and 3,500 million years
ago, and has survived to bear witness to those remote times of creation long
forgotten in the fabric of the Earth. The astronauts who wandered, heavily
elegant, over the Moon's ancient, cratered surface collected rocks—it was one
of their most important tasks. These rocks yielded the radioactive dates that
provided proof of the antiquity of the solar system, and a confirmation of
the age of meteorites. Nor was the Moon alone in its ravaged surface, for
other planets and moons have also been frozen since their tumultuous genesis. Spacecraft have allowed us to glimpse the print of the past upon remote
planets in the kind of explicit detail astronomers once only dreamed about.
Mariner 10 showed Mercury's surface dotted with craters, stark and dead.
Voyager found six new moons circling Neptune; these moons, too, are pock-
marked by ancient impacts, and sterile beyond imagination. Life made no
start on these discouraging masses of rock. No spark ignited the first organic
molecules into cells, nor did the complex dance of carbon proceed beyond
the first few steps. That may have happened uniquely on Earth, although as
this book neared completion NASA reported evidence that there may have
been another false start, a genesis that did not prosper, upon our planetary
neighbour, Mars.

The creation of life happened because Earth had a gaseous atmosphere,
and water. During the violent genesis of the planet whatever primitive atmosphere that might have been present would have been stripped away, just as
the heat at the surface would have prevented the first, mysterious alchemy
leading to the appearance of life. Later, as the Earth's solid crust developed,
the gases that made up the atmosphere, and much of the water, steamed out
from volcanoes and vents and a million fumaroles. As the body of the Earth
sorted itself into its constituent layers, the more volatile compounds and
gases found their way to the surface. This process is usually described as "out-
gassing" by those who speculate on the early history of our planet. You might
say that our atmosphere, and the possibility of life itself, was the consequence
of a vast, terrestrial flatulence risen from the bowels of the Earth. The atmosphere was retained by the attraction of the Earth's gravity, and heated and
processed by the Sun's powerful rays. Our distance from the Sun was appropriate for reactions that led to the natural synthesis of the building blocks of
life from carbon, water, nitrogen and a handful of other elements which were

in plentiful supply. The trick of creation was not the exoticism of the ingredients, but the ways in which they could be uniquely combined on Earth. This was the age of chemistry.

We see very little of this process in the rocks that remain on the Earth's surface. The oldest rocks preserved are from Isua, Greenland, which are dated at 3,800 million years old. Why these particular rocks should have survived so long is an interesting conundrum. Most of the rocks of the Earth's ancient crust have been recycled, caught at one time or another in the inexorable vice of tectonic plates. Heat and burial would then transform them, and erosion ultimately break them down again into their constituent minerals, to be embroiled in the next great cycle of sedimentation and change. The crust in Greenland stabilized early. But then, perhaps, its survival through the thousands of millions of years that followed may have been a matter of good fortune. These rocks are rather like a survivor not merely of one campaign or battle, but of a series of wars. A grizzled old soldier may be convinced of some particular virtue to account for his survival when his compatriots have fallen, but it might be no more than luck—bullets and bombs cull their victims without regard for virtue. But the Greenland rocks are baked and altered sediments which must themselves have been derived from rocks still older, weathered away under the primeval rain and winds that swept the ancient Earth, proof of a yet earlier history. And it shows that water, condensed from the steam that roared from volcanoes, was present in quantity—even as shallow seas—from early times. From what we know of the dating of the origin of the solar system, it is simple arithmetic to deduce that the still earlier history must approach 1,000 million years. Of this history there are only dim intimations. It is the real dark age of the Earth.

The early age of physics and chemistry predated life, and is still the province of speculation; so is the crucial question of how life began. This is nothing less than the transformation of matter itself: to forge the indifferent elements into vital systems that can regenerate themselves. The search for the secret of this transformation is still far from complete, and rendering its myriad steps comprehensible is like trying to summarize what is known of human anatomy on a postcard: the shape might be broadly correct but the detail is inevitably approximate.

The idea of transformation is a fundamental one—not only the notion of perfectibility of the human soul, which lies close to the heart of most religions. The transformation of matter has a tradition which extends from the Middle Ages up to our own time of atomic reactions. The alchemists believed in the mutability of the elements: the Philosopher's Stone could bring about transformation of base metals into the Noble Metal—gold. The more sophisticated alchemical philosophers, like Paracelsus (1493–1541), regarded the physical process of transmutation of the elements as a metaphor for the journey of the soul, thus brewing together the physical and the metaphysical. The transformation into gold was also the mirror for medical treatment, the banishment of imperfections in the body's balance. As Robert Browning wrote in his poem on Paracelsus:

> . . . a tincture
> Of force to flush old age with youth, or breed
> Gold, or imprison moonbeams till they change
> to opal shafts . . .

From our own perspective, it is too easy to see these dabbling chemists as a set of charlatans preying on the cupidity and gullibility of the ignorant. But there is reason to suppose that most of those virtuosi who practised alchemy firmly believed in the quest for the Philosopher's Stone, just as they incidentally added real discoveries on the nature of chemical reactions in their fruitless quest. It is an irony that the transmutation of elements was ultimately proven by the developments of "real" science—indeed, the radioactive decay process that effects this transformation underwrites the whole narrative of this book (gold, though, nobly remains aloof from this mutable atomic jostling). Public demonstrations of the Philosopher's Stone not surprisingly failed to convince—or if they did, were proved to be the result of jiggery-pokery. None the less, a conviction of the possibilities of changing the crass and inert into the Noble—a higher form—affords a kind of conceptual underpinning to the important business of the origin of life itself, and, ultimately, the origin of consciousness. Many people still believe in a chain or hierarchy that runs from dull and insentient matter through assertive life, to mind, and thence to God. For all our sophistication, this vision would be familiar to our metaphysically inclined predecessors. As we shall see, the real transformation was even more extraordinary and profound than that visualized by the alchemists.

As yet nobody has produced the new Philosopher's Stone—the means to bring about an artificial synthesis of life. But the explanations that are advanced (although the authors never admit it in so many words) are predicated on the idea that the origin of life is explicable in purely physical terms, and this clearly carries with it the implication that the process is ultimately reproducible. Not yet, of course, they say; not until we know more! But would this disclaimer have been so different from the cry of the alchemist immersed in his mercuric solutions, alembics and retorts, when the public urged a demonstration of the transformation of lead into gold?

On 1 February 1871 Charles Darwin wrote to his friend Joseph Hooker: "If (and oh, what a big if) we could conceive in some warm little pond with all sorts of ammonia and phosphoric salts—light, heat, electricity present, that a protein compound was chemically formed, ready to undergo still more complex changes, at the present day such matter would be instantly devoured, or absorbed, which would not have been the case before living creatures were formed." This was a clear indication of his hope that life could be manufactured from what came to be known in the popular press as the "primeval soup": a kind of nourishing broth from which a living cell could emerge, ready made, as the common ancestor of all things. Darwin himself was circumspect about further public speculation about the origin of life. But the image of cookery has endured, and it is an attractive one. For cookery needs a list of ingredients, and a recipe, and then a transformation by heat to produce the finished dish. It is a comfortably domestic route to what the German zoologist Ernst Haeckel called in 1866 *"radix communis organismorum"* (the common root of organisms): by means of the appropriate cookery the appropriate organic chemicals could be coaxed into life. As early as the 1920s the Russian biologist A. Oparin was able to specify at least some of the ingredients and some of the cookery necessary to produce life in this way. It might even have seemed then that the synthesis of life was almost within grasp. The simple organic compounds necessary for life would have formed naturally upon the early Earth. These carbon compounds would have had the capacity to join together further into chains (or polymerize), and thus to hoist themselves by their own bootstraps into the first molecules that were able to reproduce. All that was needed to fuel the process was a supply of energy—for all life absorbs energy to fuel its reproduction. The chemical cauldron had hundreds of millions of years to bubble and simmer creatively, but once the breakthrough had been made—once in a lottery of hundreds of millions of chemical reactions—then it would have been self-perpetuating. Thus feeding would have come into the world and, with it, life.

But then, if this extraordinary natural splicing of carbon-bearing molecules could happen once, could it not have happened more than once—or even many times? The problem with making a trick seem easy is that anyone can learn it. After all, experiments with organic chemicals are repeatable in the laboratory; unlike the cures of Paracelsus, the scientific method deals in repeatability. If this first concoction of life were not miraculous then it could have been merely routine. This being the case, life could have sprung up independently in one or another organism; plants and animals could have started from a different source; maybe, after all, we are not distant cousins of the mushroom, or nephews of the tapeworm. To those who might like to distance themselves from grisly parasites and diseases this might seem a desirable hypothesis. But we cannot escape the brotherhood of lower invertebrates so easily. For our molecules inform on our history, and implicate us in the past crimes of our descent.

Vastly more is known about the fine structure of life now than was the case when A. Oparin and J. B. S. Haldane deduced possible mechanisms for concocting life from the dull clay. This makes life at the same time more comprehensible, and also more astonishing. The chemistry of the cell has been prised apart. There is not only the mechanism of heredity by way of nucleic acids, DNA and RNA, but also a host of enzymes and other biochemicals which do the chemical business of the cell and the body. The structures are slowly being unpicked, atom by atom. Genes in the arcane double helix of DNA are being described every week as I write, and it is these genes that provide a code for the proteins which Darwin hoped to find in his small pond of creation. All these biologically created molecules are infinitely complex constructions fabricated from simple materials. As represented on the printed page, they resemble blueprints of the models that children build from sticks and balls. This is a thoroughly mechanistic portrayal. Meccano is a construction system that uses struts and bars and nuts and bolts and little more, but working models of the Eiffel Tower can be made from it. Life could be seen as a kind of auto-constructing Meccano of inconceivable complexity—but, at bottom, still Meccano, which can be deconstructed into a box or two of tricks and a manual.

I find it difficult to visualize life in these terms, even though this kind of reductionist view has claimed many (perhaps all) of the great victories in its understanding. What is abundantly clear is that *all* life—from bacterium to elephant—shares common characteristics at the level of molecules. There is a common thread that runs through the whole of biological existence. Individual genes on the ribosomal RNA are common to all life, and these are complex structures. It is hugely improbable that such genetic similarities arose by chance. These molecules run through life in the same way as the musical theme runs through the last movement of Brahms's Fourth Symphony. There is a set of variations which superficially sound very different but which are underpinned by a deeper similarity that binds the whole. The beauty of the structure depends upon the individuality of the passing music, and also upon the coherence of the construction. That vital spark from inanimate matter to animate life happened once and only once, and all living existence depends on that moment. We are one tribe with bacteria that live in hot springs, parasitic barnacles, vampire bats and cauliflowers. We all share a common ancestor.

This astonishing fact carries an obvious corollary. The genesis of life was not easy; rather, it must have been an exquisite gamble in the face of thermodynamic odds, which work against self-replicating systems—those which *acquire*, rather than lose, energy. If it were not so difficult, there should be evidence that some creatures were not born of the same clay as others. The primeval fingerprints of creation should still be ingrained in the molecules of every tissue of their bodies, and they should be different fingerprints from those identifying the rest of life. But this is not so. Descent seems to have been from one common source. But then the nub of the argument is reached: if life is such a remarkable thing, how, then, can we ever hope to duplicate such a singularity that happened so desperately against the odds? And if this difficulty is admitted, do we not also place ourselves for ever beyond the grasp of the new Philosopher's Stone?

The idea of the fundamental commonality of all life is not new. Once the notion of descent through evolution entered intellectual awareness the path was clear back to Darwin's small pond. George Eliot published *Middlemarch* in 1872—not long after Haeckel had postulated his "common root" to all life. The physician Dr. Lydgate was Eliot's embodiment of the optimistic, rational humanist, a believer in enlightenment and the benefits to the common man (which he was not) arising from the pursuit of knowledge. In particular, he sought the common thread to life—the fundamental, or "primitive tissue." It was a search for the "ultimate facts in the living organism, marking the limit of anatomical analysis; but it was open to another mind to say, have not these structures some common basis from which they all have started, as your sarsnet, gauze, net, satin and velvet from the raw cocoon? . . . showing the very grain of things, and revising all former explanations" (Chapter 15). What else is the hunt for the origin of life but a search for "the very grain of things"?

Dr. Lydgate's level of observation was not the atom, but rather the tissue, or something just beneath it at the cellular level. Sixty years earlier, Baron Frankenstein's creation was a monster fabricated of whole organs. There was thereafter a progressive focus on more and more fundamental levels of biological organization. The twentieth century brought the focus to a new level of magnification—beyond that of the light microscope. No doubt, as a sophisticated and intelligent woman in the new mould, George Eliot would have had demonstrations of what wonders could be seen through those elegant brass instruments, the microscopes which had transformed the study of anatomy. Ernst Haeckel himself was a supreme virtuoso of the light microscope. But he could not see the fundamental molecules on which the structures he knew so well ultimately depended.

There is an interesting parallel between this progressive precision of biological focus and that shift in the physical sciences that sought to delve beneath the system of the elements to the atomic reality beneath. The most fundamental structures of life and the atoms themselves finally met when the spiral structure of nucleic acids—the famous double helix—was discovered by Crick and Watson in 1953. Then, and only then, could mankind assume the role of a kind of playful deity and attempt simulacra of creation, by trying to recreate Darwin's "small pond" and the conditions under which it could be brought to life.

The first attempts at brewing a cauldron of life were made somewhat earlier in the 1950s, in a famous series of experiments carried out by S. L. Miller and H. C. Urey: ever since, experiments like these have been known as Miller–Urey reactions. An atmosphere was assumed: it comprised ammonia and methane and steam—but no free oxygen, because a "reducing atmosphere" like the one chosen compared with what was known of the atmosphere of Jupiter at the time. When electric discharges—a surrogate for lightning—were passed through this atmosphere, a rich brew of organic chemicals accumulated. Although nothing like a living organism was created, the chemical broth was richer by far than could have been anticipated. Among these chemicals were quite large yields of amino acids—the fundamental units of nucleic acid construction. The dreams of Oparin had been fulfilled! Life could potentially be cooked by the simplest of means: just as the combination of a few eggs and flour could produce the wondrous complexity of the soufflé, so the animating power of electric sparks could turn the commonplace into the complex, stimulating carbon, water and nitrogen to join together in the right order to create molecules for life. Not since Mary Shelley had dreamed of harnessing the power of lightning to animate Dr. Frankenstein's golem had the natural fire of the heavens been recruited for such a creative role. The boundary between the animate and inanimate was blurred and, no matter what arguments about genesis have arisen since, the barrier has never been erected again.

The outcome of this earthly synthesis depends on the conditions and gases that might have been present in the atmosphere during the first days.

There are those who assert that there was a very small amount of oxygen right from the start—less than 0.1 per cent of present levels. Almost certainly, the original experiments underestimated the importance of the two gaseous oxides of carbon—carbon monoxide and carbon dioxide—in the founding atmosphere. It is likely that cyanide (HCN) was also present. But all scientists agree that the original atmosphere would have been profoundly poisonous to any oxygen-breathing animals, like ourselves. Cyanide is the most poisonous substance of all to our metabolism because it blocks the capacity of haemoglobin (the most active ingredient in our red blood) to carry oxygen. Carbon dioxide is a heavy gas that still erupts from volcanoes in sufficient quantities to cause loss of life. In 1967 carbon dioxide poisoned a whole village in Tanzania, as it crept down a valley into an inhabited hollow. It stifles life most effectively. Very many life-sustaining molecules contain the cyanide radical buried within their organic depths; Professor Oro from Houston has argued that a chemical called cyanamide was a vital mediator in the synthesis of yet more vital molecules. So these poisonous substances are implicated in genesis and also still reside deep in our body chemistry. While the chemists haggle over the details, one thing is clear: from this strange Earth life grew and prospered.

Darwin's other ingredients must have come into the brew. He mentioned phosphorus, which is an active element in the form of phosphate, and a vital mediator in transferring energy through all organisms, a ubiquitous motor of life. In its crude, elemental form phosphorus, too, is dangerously poisonous. When phosphorus was used to make matches, employees in the match factories often suffered a ghastly debilitating and rotting disease known as "phossy jaw." It was only in the embrace of carbon that the energy of these poisonous components was harnessed to creative ends. Phosphates are manufactured inside living cells, and the presence of such compounds in the ancient Isua rocks has been taken by some scientists to show that life was already up and running at the inception of the rock record at 3.8 thousand million years ago. Life without phosphorus is unthinkable.

Then there is the question of clays.

Clays are entirely inorganic, and have been common since water started eroding rock. The minerals that make them have atoms arranged in sheets, loosely held together: this is why clays are slimy and slippery. Between the sheets other atoms may bind—and in this way chemicals can become concentrated so that they react together. Clays thus provide a natural factory for doubling the size and complexity of organic molecules. It is curious to recall that in the biblical account God fashioned Adam from inert clay; science may yet prove that clay truly had a role in that most distant genesis of life.

But even all these molecules (and I have mentioned only a few from a long list), once synthesized, do not guarantee life. To invoke the cooking analogy again: just throwing together eggs and flour in a saucepan does not guarantee a soufflé. The more mechanistic reader might prefer an engineering metaphor: to throw cogs and cylinders and pistons into a box and shake them around does not produce a working engine. Life is not just a matter of chemistry: it is a cooperation between molecules to produce a consequence infinitely greater than the sum of the parts. To be sure, the chemists are writing plausible scenarios for the manufacture of the ingredients—it is just the cooking that is still a problem. Or to use the other analogy: the parts are there but only a few pages of the construction manual. There is, for example, an idea that the first self-replicating world was an "RNA world" that preceded our DNA world, in which protein synthesis became central. It does seem reasonable to assume that there were steps towards the first living cell that were "alive"—but not yet cellular. For it was the creation of the cell that marked the great breakthrough in organization in the history of life. The crucial invention which defined the cell in the most literal sense was a container that confined its component parts (termed organelles) within, a thin barrier between the living and the inert worlds: this is the cell membrane. The first cells were bacteria, a few thousandths of a millimetre in length.

Cellular life is a mass of chemical reactions surrounded by a skin, but it is a skin through which the environment can be heard selectively. Much of the business of life is conducted by different rates of diffusion through cell membranes. Energy enters the cell, and when enough energy has been acquired the cell can split: thus doubling and redoubling, the world is rendered fecund. But how to acquire the skin?

This may not have been as difficult as it might seem. There are some substances which assemble themselves into membranes automatically, and these chemicals are present in the carbonaceous chondrites (meteorites) which delivered organic molecules to the nascent world. Such hydrocarbons have been extracted from their meteoritic hosts, and under experimental conditions they have naturally congealed into membranes surrounding sacs—tiny bubbles known as lipid vesicles. Such skin-formers might have produced a kind of oil slick on the surface of pools and at the edges of those early Precambrian seas that condensed from volcanic steam: think how the surf on the sea today lingers at the water's edge as a drifting foam. I have seen clots blown along the shore like evanescent tumbleweed. Cells may have been thrown up from froth.

BUT WHEN ALL THIS CONSTRUCTION had been done, when all the molecules had been set in place, and all surrounded by a membrane, this is still only a simulacrum of life, as inanimate as a marionette without its strings. The spark that is needed is energy, a supply of fuel to drive the whole motor of self-reproduction. Without this animation the bacterial protocell would have blown away to obscurity on its distant Precambrian shore; or the small pond would have been both life's cradle and its grave.

Life is a thief. To feed its growth, an energy-yielding reaction is stolen by the living cell, which has learned how to divide molecules in order to appropriate electrons for itself. We all know from Einstein about the energy that is stored in atoms and the collections of atoms known as molecules. Primitive life reduces molecules—strips them down, as it were—and thereby manufactures new chemicals as a by-product. Many bacteria use the most commonplace of materials to derive their energy. Some use hydrogen and carbon dioxide to produce the inflammable gas methane, together with water—these are the methanogens. Their activity in fetid swamps and pools produces gas bubbles—you can occasionally see large methane bubbles arising from the bottom of ponds, as if some mysterious aquatic creature had suddenly exhaled within the deep mud. These bubbles sometimes ignite, when they burn with a pallid, bluish flame that is known as will-o'-the-wisp. Its luminous beauty has led lonely travellers across marshy places to horrible deaths in treacherous bogs. Other bacteria reduce sulphates; in consequence, they produce another poisonous gas, hydrogen sulphide (H_2S, which schoolchildren relish for its smell of rotten eggs). Different bacteria again use sulphur itself to produce the same, unpleasant gas. There are even bacteria which exploit the simplest chemical reaction of them all, combining hydrogen and oxygen to produce water. Many of these bacteria need only the raw materials of Earth itself to thrive—they feed themselves, rather than having to feed on other biochemicals produced by different bacteria—though there are plenty of those as well. These most primitive of bacteria are known by the most wondrous jargon, mastery of which is guaranteed to cause jaws to drop at social functions, for the correct designation of many of them is "chemolithoautotrophic hyperthermophiles." Since these bacteria are often only a thousandth of a millimetre or so long—minute rods, discs, or cocci (spheres)—this affords an example of a rule well-known to biologists: that the length of the description is inversely proportional to the size of the organism.

But the second part of the name reveals the most astonishing fact about these most primitive of living entities: they are heat lovers. This does not mean that they bask in the kind of temperatures that we might just tolerate on a warm day in the Sahara Desert. These organisms are *hyper*thermophiles—they need extreme heat, and die if they are deprived of it. Many species cannot reproduce if the temperature drops below 80 degrees Centigrade—and many of them thrive under virtually boiling conditions. They are found today around volcanic vents, mud holes and hot seeps on the ocean floor, and at depth in the very body of the Earth. Their names—*Thermoproteus*, *Thermofilium*, *Pyrobaculum*—accurately reveal their tastes. *Pyrodictyum* grows best at 105 degrees Centigrade. These are the creatures of Hades itself, happy in hot vats that are torture for all other life. By the mud holes in Yellowstone National Park, or the geysers in New Zealand, or the fuming vents at Mount Etna, where acid, volcanic waters break through from plutonic depths, they thrive in their billions. But they reveal themselves only as coloured smears—red, orange; even blue—on the surface of the rocks, or as a subtle mistiness in a smoking pool.

C. S. Lewis imagined thermophilic humanoids or gnomes in *The Silver Chair*: these Earthmen were exiled from the deep land of Bism to which they ultimately returned.

> The depth of the chasm was so bright that at first it dazzled their eyes and they could see nothing. When they got used to it they thought they could make out a river of fire, and, on the banks of that river, what seemed to be fields and groves of an unbearable, hot brilliance . . . There were blues, reds, greens, and whites all jumbled together: a very good stained glass window with the tropical sun staring straight through it at midday might have something of the same effect.

The *real* inhabitants of the netherworld make pasture of the implacable rock, and dine upon sulphur, which might be thought even more fantastical than the homunculi themselves. That some of them must be closely related to life born more than 3,800 million years ago makes Bism seem a modest fancy.

Professor Stetter of Regensburg collects these heat-loving bacteria, and described many of them for the first time. He takes samples from boiling springs and deep water seeps. He is enthused by "solfataric hydrothermal fields lying above hot magma chambers." When you sup with the devil you must use a long spoon, and you must be equally cautious with these semi-creatures surviving from infernal fires near the beginning of the world. You must sample them with a long-handled scoop, and, once sampled, you must keep them hot. Imagine the change of heart required to realize that in order to study your experimental organism you must keep it boiling! At first this might seem as curious a notion as keeping chimpanzees alive by regularly feeding them with arsenic. But the breakthrough in studying and identifying these ancient bacteria came with replicating and sustaining their apparently improbable life conditions in the safety of the laboratory.

The next, but probably not the final twist in this tale of microbes is that the heat-loving species, among them the sulphur-eating and methane-brewing bacteria,* proved to be near the root of all life. Their place in the story was revealed by the branching pattern of the tree of descent, a tree drawn out from gauging the relative similarities of ribosomal RNA molecules, and the genes which provide a code for certain enzymes common to all life. The basic truth of genealogy (see illustration on page 45) is that all life as we know it descended from living things that could only be content in extremely hot environments. Furthermore, most of them are anaerobes—which means that their biochemistry works only in the *absence* of oxygen. In fact, oxygen—the very element which has become known as one of the essentials of life and nourishment—is lethally poisonous to many of these bacteria.

Now it is time to reflect on an extraordinary picture which has been sketched for the origin of life. Far from Darwin's benign, almost cosy, "small pond," we have a torrid cauldron, acidic, emitting the sharp whiff of sulphur; and we have an atmosphere almost lacking oxygen. Almost everything in this biological Eden would have been damaging to most of the animals and plants alive today. In the beginning, there was dust and chaos and the relentless bombardment of meteorites. These also brought the seeds of life, no doubt, but then the important stuff of enzymes and energetics and nucleic acids and proteins and cell membranes was most likely cooked terrestrially; and all this between about 4,500 and 3,800 million years ago. The early cells were tacky with mucus: life was as much slime as soup. Where the memory of this distant world lingers on it is in the most inhospitable places on Earth, in hot springs and volcanic vents—in emanations from the Underworld, a sulphurous surrogate for Hades—and there, too, the descendants of the most primitive organisms still cook with hydrogen sulphide and methane, and many an arcane recipe besides. Truly, life began in something approximating to the

---

*Many of these will be termed Archaebacteria, or Archaea, in articles and scientific papers on this theme.

medieval idea of Hell. Some of Darwin's ingredients were correct, but how different the cookery! And what would he have thought about the fumes of brimstone swirling about the site of that first genesis? It is curious to recall the disappointed intentions of the alchemists concerning the elevation of matter: for surely what has happened to life since its inception—and, with it, the planet—is a more fantastical transmogrification than would have troubled the dreams of Paracelsus.

As to *where* the first cell was born—was there a cradle, a nursery, some sequestered spot where the near-miraculous assembly of all the components of life could happen without being destroyed as soon as accomplished? Mike Russell from Glasgow University insists with the fervour of a latter-day alchemist that the assembly site was a seep, a pyritiferous seep, a place where emanations from the interior of the Earth debouched into the primeval sea. The fundamental energy-giving reaction may have been the formation of the common mineral iron pyrites (the simple reaction can be portrayed as $FeS + H_2S = FeS_2$ [pyrite] $+ 2H = 2e$ [energy]). The energy released by this reaction can be used to break up carbon dioxide, to steal the carbon in order to make the carbon compounds which are essential to life. The beauty of this theory is that cell membrane–like structures also form in the same sites: they assemble themselves where the iron pyrites and the surrounding water meet. Was this the template upon which a true lipid cell wall was grown?

Hot springs at great depth in today's oceans are known as "black smokers," and that is just how they look: dark fumes belch out into the seawater along mid-ocean ridges where volcanic fluids are free to escape. They are picked out in the headlamps of bathyscapes as if one of the smoking chimneys of Hades itself had been discovered by torchlight. They are hot and sulphurous. Inconceivable numbers of sulphur bacteria thrive there, and form the base of a unique food chain. Animals (some as weird as any mannikin devised by C. S. Lewis) feed on the bacteria, or have them within their tissues as invited guests. It is like a dark, crazy garden, cultivated away from light, and bathed in a profligate torrent of sulphur. Clams which live there grow to an enormous size, like pampered dahlias. And chimneys and pipes of pyrites form in the ocean around the deep-sea vents and seeps. They cannot help it: the process is driven by the inexorable laws of chemistry. If life was cradled in some primordial hot spring rich in sulphur, then maybe these peculiar "smokers" provide us with a model for its earliest days.

Most of this tale of life's beginning has been inferred from its aftermath, which is still imprinted upon the life of the living planet. The rocks tell us comparatively little about the mysterious period of genesis itself. But the kinds of rocks found in the Archaean crust (dating from 2,500 million years ago or more) which have survived intact do serve to confirm that the early atmosphere lacked oxygen, or at least contained very, very little. For example, a unique kind of iron ore—densely striped and banded—is very commonly found in rocks of about 2,500 million years' antiquity. In these banded ores the iron is in an oxidized mineral form (haematite) which is likely to have formed in the presence of free oxygen in the atmosphere. The iron sucked up oxygen, and the ores were formed as a fine rain of minute, rusty particles accumulated on the ancient ocean floor. But these iron ores were distinctly rare in still older sites. This is most easily attributed to a paucity of oxygen in those early days—a time when reducing conditions prevailed almost everywhere. Red sandy and silty rocks which formed on ancient continents under oxygenating conditions became distinctly commoner after about 2,000

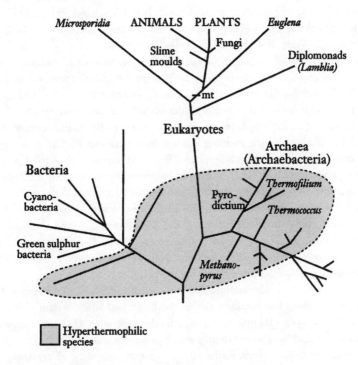

The fundamental tree of life showing how organisms at its base include a range of heat-loving (hyperthermophilic) bacteria

million years ago. The conclusion seems inescapable: that the early Earth lacked oxygen, and that oxygen slowly built up, over at least a billion years. The anaerobes had their day—or, to be more accurate, their thousands upon thousands of millennia. They thrived in this early world, perhaps almost everywhere; now they linger on only where conditions still resemble those of the primordial Earth. Unbelievably tough, they alone would survive any future holocaust or meteorite impact. They have seen it all, or worse, when the world was in its infancy.

This, though, was only the beginning.

IN THE EARLY 1980S I WAS SUMMONED by the Director of the Natural History Museum in London. He instructed me to go to Cardiff to attend a meeting on his behalf. At this meeting, Professor Hans Phlug was to announce direct evidence of life from meteorites—and from the oldest (if somewhat altered) sedimentary rocks from Isua, Greenland. Phlug's host was Professor Wickramasinghe, professor of mathematical astronomy at the University of Wales in Cardiff. "Wickram" is himself a protégé of Sir Fred Hoyle, who had a long and mostly distinguished career as an astronomer, and a second one as a science fiction writer. After the lecture there was to be a smart dinner for Sir Fred and other guests, at which I was permitted to be a guest. A large number of other academics were invited. I sat next to a jolly professor of Celtic languages who explained the bardic tradition to me. But the real business of the meeting was a lecture detailing alleged evidence of extraterrestrial life—including spheres from carbonaceous chondrites, which were, it was claimed, obviously organic. Bacteria came direct from outer space! Life, far from being a singular phenomenon and Earth-bound, the product of our position in the solar system and all the rest, was reported to be almost commonplace—distributed through the Universe and dusted through the stars. Life could be seeded here and there as chance ordained: comets which passed close to planets, including our own, were a favoured source for this cosmic interaction with the terrestrial world.

This notion of "Panspermia" has had a long tradition, and this was its latest incarnation. The subtext was that the Universe had a constant diameter, the "steady state" theory, as opposed to the expanding Universe proposed under the "big bang"—the theory which gained ascendancy in the latter half of the twentieth century. Rather than a dramatic moment of creation, the steady state theory held that matter appeared throughout the Universe. As a convinced steady stater, Sir Fred was playing an ambitious game—nothing less than the origin of everything. Life, in the steady Universe, also appeared widely, and was disseminated generally. Its appearance in comets or meteorites was nothing less than the confirmation of the steady state. Hoyle and Wickram also believed that the analysis of the spectrum obtained from interstellar space was like that obtained from bacteria. And so the theory built, step by step. A book was even published claiming that evolution itself was mediated by the extraterrestrial visitors—and extinctions could be attributed to the same cause. The periodicity of epidemics was tied to the sporadic, but predictable visits of comets: the great influenza attack which swept the world after the First World War was brought to us on the tail of a comet; evolution proceeded by a kind of viral infection of DNA . . . And so the great bran tub of natural history was rummaged for supportive data. Most charmingly, the origin of long nasal passages was claimed as a device to filter out those nasty viruses; the most spectacular schnozz of them all was that of the proboscis monkey—and this species lived at the *top* of trees, where the nose was constantly needed to filter out germs! (Shorter-nosed monkeys lived under the protective umbrellas of leaves.) It was a wonderful confection, cooked to a turn.

The Natural History Museum became involved because of *Archaeopteryx*. *Archaeopteryx* is arguably the best-known fossil anywhere, with the possible exception of *Tyrannosaurus rex*. For many years it has been known as "the first bird." Its exquisite skeletons are wonderfully well preserved on the surface of fine-grained limestone slabs from a formation of Jurassic age (about 170 million years old) known as the Lithographic Limestone. The specimen in the Natural History Museum was collected from a quarry at Solnhofen, in Bavaria, in 1861. It is a medium-sized bird about the size of a magpie, but it shows an extraordinary mixture of characteristics. If a knowledgeable observer looked at the skeleton alone the first comparison to come to mind would be with one of the smaller dinosaurs, rather than a bird: just at a glance, the characteristic breastbone of birds was hardly developed, and in the jaw there were quite unmistakable teeth, hardly comparable with the bill of a crow or a toucan. But *Archaeopteryx had* to be a bird, for surrounding the specimen like a delicate halo were the obvious impressions of feathers, and feathers are unique to birds. The chance of such a delicate, yet complicated structure as a feather evolving more than once was infinitesimal. So in its construction *Archaeopteryx* straddled dinosaurs and birds, one of those rare examples of an animal that bridged a gap in history. It became a *cause célèbre*, an example of the evolutionary process at work, and a chronometer for the origin of one of the great classes of living animals.

The problem with *Archaeopteryx*, according to the Hoyle theory, was that birds came into being *after* the great extinction of the dinosaurs (which, of course, was caused by massive, lethal infections initiated from space), whereas the Solnhofen limestone was deposited about halfway through the period when dinosaurs dominated the Earth. They had tens of millions of years of ascendancy still to come. This meant, so it was claimed, that the small dinosaur-like skeleton of *Archaeopteryx* really *was* a dinosaur—it was just the bird that was wrong. In a nutshell, the fossil simply *had* to be a fake. Possibly this idea was inspired by the famous Piltdown hoax (see Chapter 11)—a manufactured "missing link" that had been fabricated to provide what the fossil record had so far failed to yield. Whatever the inspiration for the claim, it was stated that the feathers had been "added" to the graceful little dinosaur by pressing feathers—that is, feathers of living birds—into a skin of limewash or cement which had been painted on to the surface of the slab. Scientists were either in on the fakery, or at any rate conspired to perpetuate it. This was heady stuff. The press likes nothing better than a "boffin baffled" or "experts fooled" story, and Fred Hoyle was a name that journalists knew would guarantee a good story. It was not long before headlines of the "FOSSIL GETS THE BIRD" variety were in all the newspapers. Denials were issued by the British Museum (Natural History), as it was then known, which only served to spice the story. In palaeontology, as in politics, there is nothing like a vigorous denial to convince the public of guilt. Amidst the brouhaha, the calmer voices of several rather quiet palaeontologists went unheeded. Few people grasped the connection with Hoyle's old battle for the steady state Universe. And even when Sir Fred was heard to say that he knew it was a fake even before he saw it, this was taken as an indication of omniscience rather than prejudice.

But it was all so much nonsense. Unlike Piltdown "man" *Archaeopteryx* was not a unique discovery. Another fine specimen, now in Berlin, was found in 1877. Since then, three more have turned up, the last in 1951. Forgeries do not operate over three or four generations! One would have to envisage a kind of coterie of fakers, each generation fanatically loyal, yet so self-disciplined that only now and then was another fraud released into the world. And all for no (or very little) monetary gain. It is rather like a bank robber taking endless pains to devise the perfect robbery and then demanding a bag of small change at the *dénouement*. My colleague Alan Charig spent far more time than he should have done performing the most minute physical and chemical examination of the rock surface to ensure that it was all of a piece—which it was. The feathers were real fossils.

In pursuit of the unfashionable steady state hypothesis a slur had been cast upon the reputation of several generations of palaeontologists of previously impeccable character. But the refutation of the forgery claims was beyond dispute. As a *coup de grâce*, different fossil feathers were discovered in the Santana Formation of Brazil, somewhat younger than *Archaeopteryx*, but still contemporary with dinosaurs. The red herring had been beached.

But there is a curious twist to this tale. It seems that Sir Fred was right about some important things. To be sure, the principle of evolution by "infection" of bits of DNA from comets or meteorites was improbable at best. Nor are viruses the primitive organisms they were claimed to be—rather they are specialized, reduced organisms, absolutely dependent on the prior existence of cells for their livelihood and propagation. But the involvement of extraterrestrial bodies with the origin of life was almost certainly correct. Some of the synthesis of carbon molecules vital to life may have happened in the crucible of creation at the inception of the Universe. There is indeed organic material dusted between the stars. We shall discover that the involvement of comets and meteorites with the fate of the Earth did not cease once living cells were established; from time to time they interceded dramatically with the narrative of our planet. As the astronauts observed, with all the force of a truism minted for the first time, we are just a small planet suspended in the vastness of indifferent space, but space is not empty—our fate is bound up with that of stars, comets, meteorites, all drenched in a bath of radiation, some of which originated at the dawn of time itself.

When researching this book I read a massive compendium on early life published as a Nobel Symposium in 1994. There is a reference list with more than 1,500 entries, and nowhere does any item of the Hoyle/Wickramasinghe opus get a mention. With the loss of respectability adhering to a discredited allegation, even the coincidental virtues get ignored. Maybe the alchemists were also tarred by their own mumbo-jumbo to the extent that their real achievements went unacknowledged. Science is only interested in winners, although the losers sometimes provide more entertainment.

In that first world, the winner was a bacterium that used light to generate energy. There may have been more than one way to do this, but the bacterium that was to influence the whole history of the planet used light upon the green pigment chlorophyll to break gaseous carbon dioxide into two parts: carbon for its own nourishment and growth, and oxygen. The oxygen was released into the atmosphere as a by-product, the most precious waste in the firmament. This is the process known as photosynthesis. It is the motor that powers all green plants. We can see the importance of light by the effects of its deprivation; see how spindly and pale plants deprived of light soon become.* They are starving to death. "The force that through the green fuse drives the flower," as Dylan Thomas perfectly described it, is a force fuelled by photosynthesis. The vital chlorophyll molecules are arranged in plates inside the bacterium to maximize their efficiency, like solar panels. And what if its colour had been other than green? We might have had brown glades, or celebrated the restfulness of violet fields.

But after the brilliant, even lurid thermophilic world the colour was green—wherever light penetrated and photosynthesizing life could survive. Cyanobacteria ("blue-greens") were the first of these miracle workers. The atmosphere was rich in CO_2 and there was an adequate supply of the other elements needed for life, such as phosphorus. As cells grew they divided— simply split into two, and into two again . . . and again. Population growth could be extremely rapid under favourable circumstances (and bacterial "blooms" still happen today). Some of these early cyanobacteria were no bigger than their sulphur-eating antecedents. Others formed quite long, but slender threads only a few thousandths of a millimetre wide. They still abound where the bath turns greenish beneath a dripping tap, or in ponds, in springs, in arctic climes, anywhere. These provide the first fossils—or rather, threads and spheres that look just *like* the living blue-greens have been recovered from rocks assuredly older than 3,000 million years, and possibly as old as 3,500 million. It seems reasonable to assume that these similarities do not mislead us, and that the first little spheres and threads recovered from the rocks can be interpreted as photosynthesizing cells.

I should reiterate that time—3,500 million years ago—in order to try to come to terms with its immensity. First, imagine each cell exhaling the merest puff of oxygen, such as would fill a balloon smaller than a pin head. Then imagine a world thick with such cells, billions of them, dividing and dividing again, and each time they divide another minute puff of oxygen is given to the air. Then this process continues through generations that can only be reckoned as numerous as the stars in the Universe. And for every generation a thousand billion tiny balloons of oxygen released . . .

And now the environment attempted to remove the oxygen as fast as it was formed, for natural materials love such a reactive element. Limestone rocks captured oxygen in the form of calcium carbonate. Minerals oxidized,

*The exception may be the aspidistra, which seems to thrive in the darkest corners of ill-lit rooms. George Orwell amusingly described attempts to kill this plant in his suburban tale *Keep the Aspidistra Flying*. All methods proved futile, no matter how cruel: even stubbing out cigarettes in its pot only stimulated it to put out more leaves. The truth is that this plant is an astonishingly efficient light gatherer (its natural habitat is beneath dense tree cover), but it depends on light just as much as any green plant.

in the process gobbling oxygen into iron ores or rust. Oxygen can combine with so many elements that there are a myriad ways to sequester it. And against this continuous haemorrhaging only the bacteria—the minuscule bacteria a few thousandths of a millimetre across—puffed and puffed their tiny draughts of breath into the world.

It is possible to estimate some very rough figures. A thousand million years is 10 raised to the power 9; that is equivalent to about 3.5×10 to the power 11 days. If the bacterium had a life of just one day (many life cycles are much shorter) a measure of the quantity of oxygen released over 10^9 years can be obtained if this "day's" figure is multiplied by the numbers of bacteria that flourished over the world at any one time. Because thousands of bacteria can exist in the most insignificant smear around a tidal pool, the true number is almost inestimably large, and no two authorities give the same figure. But whatever the figure, the effect is incontrovertible: oxygen was produced faster than it could be sequestered, and thus changed the world.

Remember that the early atmosphere was quite unlike that of today and hostile to most life that thrives at the present. By this slow, inexorable process of photosynthesis the early atmosphere was modified; gasp by gasp, oxygen was added, and carbon dioxide commensurately reduced. It was life processes that shaped the atmosphere, that paved the way for other, more advanced organisms. These early "blue-greens" were simple, single cells (prokaryotes), lacking both a nucleus and the ability to cohere into larger bodies. They are reminiscent of a line from that curious folksong: "One is one and all alone and ever more shall be so." But it was not to be so, or at least not for ever more: more complexity *was* possible once the world was ready for it, but not for a long, long time. If the length of the chapters in this book were to represent geological time elapsed, these early chapters would have to take up 85 per cent of my text or even more. It might prove rather dull, because the endless repetition of bacterial cell division would occupy much of it.

This was also the time when the face of the world was sculpted. The Earth's crust had long since solidified, but the surface of the globe would have carried no recognizable traces of the continents we know today. The earliest continents were probably small and temporary ("microcontinents") separated by ocean basins replete with their hydrothermal vents, where all the heat-loving bacteria could thrive. Volcanic islands rose above the primordial seas, belching sulphurous smoke and fumes and releasing more gases from the interior of the Earth. But the hot springs that bubbled from their flanks were living stews, where simple life used simple atoms for uncomplicated propagation. In cooler pools—where clear water transmitted life-giving light—

or even between grains of sand on shores, there cyanobacteria lived their momentous but repetitive lives, immortal in the sense that some dim chain of division links these early cells with those that still undergo their mindless fission in rock pools around Hawaii as you read this. Later, perhaps by the middle of the Archaean (3,000 million years ago), microcontinents came together to form the nuclei of the continents we know today. They were clots of relatively light, tough, annealed rocks that held together, moved about as single pieces as they were propelled by the motor of plate tectonics. The centre of southern Africa, the most ancient part of the Canadian Shield, pieces of Finland and central Asia, and, of course, timeless Australia were probably formed at this time, and endured the vicissitudes and turbulence of the world to become the ultimate rocky survivors. It was on these ancient pieces of continent that some fragments of life are found as fossils. They are ancient passengers stranded upon rafts of nascent continents that have been carried around the globe like lost mariners. Rafting has been the secret of their longevity.

The higher concentration of carbon dioxide may have produced a hot, "greenhouse" world, where the thick atmosphere slowed the loss of solar radiation. Imagine the silence, except for the hiss of fumaroles, and the slow crash of waves breaking upon endless naked and inhospitable shores.

The first obvious biological structure was a humble, slightly tacky or slimy skin, comprising a community of microbes, something that covered the sediments in a tenuous bandage. These mats were our ultimate cradle. They can still be found on saline flats in the tropics, in moist places where grazing animals are scarce. This is an astoundingly ancient ecology. The commonplace aspect of such mats should not blind us to the remarkable fact of their durability. If endurance were the criterion of success, then the most successful organisms on Earth are a few thousandths of a millimetre across and found in mats.

The thin top layer of the mat is where the photosynthesizing threads and spheres are interweaved, forming a skin. This is the part that breathes out oxygen. Below this surface layer there is little oxygen, and that is where primitive or specialized bacteria that thrive in its absence (anaerobes) take over, partly feeding off the remains of the cyanobacteria above them; this is a food chain in miniature. Today, cyanobacteria can live in a wider range of habitats than in mats, and maybe this was true even in the Archaean; they can even survive prolonged drying. In the Atacama Desert, near Calama in Chile, several species have been found in a site where there has never been any rainfall in recorded history. They can resist harmful radiation, as they would have had to do at the dawn of time. They are tough.

Mats grow. A new skin of bacteria forms above the old. Tiny particles of clay and dust are incorporated into the mats, then become a kind of composite; nature and silt combined. They grow layer by layer, a millimetre or two at a time. Some grow into domed heads the size and shape of the leather poufs that my grandmother used to rest her legs on. If you cut a section through the rounded heads they are finely, and characteristically banded—or *laminated*. They are called stromatolites, and they have been present on Earth for well over 3,000 million years.

At last we can begin the narrative with objects that can be held in the hand, or stumbled over in boots. Stromatolites are conspicuous. I saw them on the Arctic island of Spitsbergen, where I began the history in this book. There, in rocks dating from the later Precambrian, they form pale mounds and pleated columns. In places, you can walk over the mounds, which stick out from their enclosing limestone as if they were the bald pates of buried skulls. Where they are cut through by melt streams you can see the fine layers of which are composed, like so many sheets of filo pastry. Others were more delicate, cauliflower-like, with small, lumpy protuberances; others again were like stacked pipes or chimneys just a few centimetres across, but the striped laminations were curved into arcs that revealed the pattern of growth, like the grain in wood or the wrinkles on a shell. For all that they are cast in nothing more than everyday limestone they somehow *looked* organic, too regular for a mere sedimentary perturbation, yet not so regular as to reflect the atomic inexorability of crystals. Life produces patterns, but they are not the facets of a diamond, but curves and spirals, and curls of the kind that Art Nouveau artists appreciated. The closest inorganic structures to stromatolites might be the pillows formed by haematite, an iron ore, but those do not reveal the same internal laminations. Nor do banded agates show the regularity of form displayed by stromatolites, which have a distinctive design all their own.

But when they were discovered and described as *Cryptozoon* (from the Greek "hidden life") by James Hall, a great nineteenth-century North American palaeontologist, there were many who claimed that they simply *had* to be inorganic. They were too old, it was thought, to be organic. Everybody knew that there was no trace of life in the Precambrian; *ergo*, this was no trace of life. As Darwin wrote in *On the Origin of Species* (1859): "to the question why we do not find rich fossiliferous deposits belonging to these assumed earliest periods prior to the Cambrian System, I can give no satisfactory answer." For a long time after Darwin wrote this, no matter how thoroughly investigators searched there was nothing associated with stromatolites which could be definitely identified as of organic origin. Perhaps this is not surprising, because the delicate and minute cells that produce stromatolites now—and probably produced them in the Precambrian—are not readily preserved. It took a century to find remains that everyone could agree upon as being truly organic. Even so, it is easy to be misled by contaminants and chimeras. I once corresponded with an old gentleman who was convinced of the organic nature of clusters of tiny, wholly inorganic minerals embedded within larger crystals from metamorphic rocks that had been heated under pressure to several hundred degrees Celsius. He believed that they were radiolaria from an ancient Precambrian sea. Conviction provokes perceptual optimism, just as scepticism taken to extremes will disavow the obvious.

The fossils that were recovered were detected in thin sections made through the rocks. The best rocks for such studies are cherts, hard rocks made of impalpably fine-grained silica (silicon dioxide, best known as the mineral quartz). Early impregnation of the mats by silica served to enclose the filaments and cells of the Precambrian organisms in a time capsule. It prevented further decay. After the earliest discoveries, in the mid 1960s NASA, the space funding agency in the USA, supported research into the earliest history of life, which must be a testimony to the ingenuity of the research workers involved. Their ruse evidently worked. Now there are nearly 3,000 recorded occurrences of microscopic fossils in Precambrian rocks.

The oldest of these microfossils look like the modest organisms they are. The most ancient of all are 3,500-million-year-old objects recovered from the Fig Tree rocks of the Transvaal, South Africa, and from the Pilbara rocks of Western Australia. In the South African rocks there are beds covered with the fossil remains of ripples which can only have formed under water, looking much like the ripples you can find on a sandy beach after the retreat of the tide. There can be no more telling proof of the existence of water at this remote time. As for the fossils, little threads are typical, which reveal their organic nature by the regularity of the transverse walls that cross them; others look like strings of minute beads twisted into sinuous shapes. In spite of their simplicity, the regularity of their size and form are striking. Even more remarkably, they are virtually indistinguishable from living species of cyanobacteria. Sceptics say that this proves only that simple things came first, and that it tells us little about their metabolism or close relations. But it does seem possible that some of the earliest organisms on Earth are still with us, having bided their time in mats for thousands of millions of years. Like their living relatives, these bacteria would have secreted a protective coating of slime, which served to bind them together, lending coherence to consortia of the minute cells in their enduring mats.

Australia has a reputation for preserving ancient things, for being on nodding terms with the primeval, a haven where time has continued to pass by on the other side. Perhaps the stromatolites of the Pilbara rocks and Bitter Springs in the centre of Australia are just the first example of this unwillingness to keep up with the rest of the world; maybe the duck-billed platypus is just one of the latest to follow in the tradition. The time raft that carried the cones and pillows of these most ancient organic structures happened to come to rest in Western Australia.

The outback where these ancient rocks crop out is an extraordinary place. It comes as a shock to discover its vastness. Driving across semi-desert soon becomes mesmeric and disorienting, as hot hours dissolve into days. Tracks appear briefly, only to disappear into ambiguous washes, where the bewildered and anxious passenger is instructed to hang out of the window to look for broken twigs which might indicate where a vehicle has passed before. On some journeys, every twig seems to have been broken. Much of the land has little topography, so that landmarks which might become recognizable or familiar are rare. It is appallingly easy to get lost. Bruce Chatwin described how the aboriginal peoples (whom the ranchers call "black fellas") learned to sing their routes across the huge distances, buoyed on a kind of musical topography. The European visitor is tone deaf. In the outback the beautiful, white-barked ghost gums (one of the hundreds of *Eucalyptus* species) offer occasional shade; you swear you would know that one again if you saw it, but the next ghost gum confuses, and the next again looks like the first, and the one after that looks like the previous one, and so it goes on, until you wonder whether you have been driving around in circles.

The maps can be discouraging. They may mark occasional water boreholes, where wind-driven pumps tap into deeply subterranean water to fill huge, circular tubs of corrugated iron. They are marked with names like "Dribbling Bore." Even more dispiriting are ones that say "Bore (Dry)" or even "Bore (Arsenical)." But if you find one containing water, it is a marvellous thing to be able to swim in the artificial pool under the relentless sun—no millionaire in the spa at the Waldorf-Astoria could so relish the touch of cool water on skin. Occasionally there are creeks with temporary pools. These are the billabongs where the jolly swagman rested his tuckerbag. I can vouch for the shade of the coolibah tree (another *Eucalyptus*). Here you can swag out under the stars, and a blessed peace comes over you. The stars stretch from horizon to horizon for there is no pollution here, nor any mist, to obscure their visibility. I never realized that there were so many stars, and no doubt in the panoply there are more galaxies than can be reckoned. There is probably no better place to reflect upon the extraordinary circumstances that gave rise to life on our small planet, and the wonder of being a conscious soul able to speculate upon it. In the night you can hear the creak of time, feel how this place survived from the early days of the Earth, smell a whiff of the sulphur from its beginnings, and imagine the peneplained landscape flooded by shallow waters spilling over endless microbial mats. The vast scale of this empty place allows the imagination to appreciate the great exhalation of oxygen that took place over thousands of millions of years. Some of the light from those distant, twinkling stars started its journey to your retina at a time not far removed from the age of stromatolites. The starlight is so strong it can even make shadows. This is the place to try to grasp the processes that allowed the impalpably slow transformation of the atmosphere through the very business of life, molecule by molecule, and all the while the indifferent stars shone just as they do now. In the bush, night insects buzz incessantly in the trees. From time to time a dingo yelps strangely in the far distance, a disquieting cry to make a shiver run up your spine and prompt you to pull the sleeping bag up around your neck.

Dingos have to be tough to survive here, a toughness shared by those early cyanobacteria. Every organism in the outback is a survivor. Trees can endure drought or fire. Kangaroos can prosper on dry vegetation. The white cockatoo known as the galah is apocryphally tough. The first outback joke I learned was: "How do you cook a galah?" Well, you take the plucked galah and place it in a billy, along with a brick and pepper and salt, and cook and cook until the brick is soft when tested with a fork. Then you throw away the galah and eat the brick . . .

Even the people are a special tough breed—those that live way out from the main roads. The aboriginal tribes have had thousands of years to adapt, and now their problem is the reverse one of coping with the dubious privileges of "civilization." But the white fellows are a curious species. Many seem to have Scottish ancestry. They are tough in the way that is portrayed in traditional westerns starring John Wayne or Alan Ladd.

The ancient rocks come to the surface in ridges and hills that break the endless plains. Such is the remoteness of this place that the oldest fossil-bearing locality, discovered by Stan Awramik, was never found again (others have come to light since). The stromatolites are typical low mounds showing dense laminations. Some of the siliceous rocks associated with them have yielded the remains of delicate threads only a few thousandths of a millimetre wide which are very similar to those produced by living "blue-greens."

Of course it had to be in Western Australia that *living* stromatolites were discovered. Some 650 kilometres along the coast north-northwest of Perth, in Shark Bay, there are mounds, low laminated hillocks and columns about half a metre in height, composed of carbonate, which would not have looked out of place in the Precambrian. They grow slowly, especially between the tide-marks in this strange, hot site, and when the salty sea drains through them they stand proud like so many overstuffed cushions. They hardly look like the cradle of life; they more closely resemble some strange and tumid erosional remnants. In the early 1960s it was realized that these distinctive mounds were the living analogue of the odd structures which were by then being discovered widely in the Precambrian strata. These mounds allow us to see back through time to the start of life, rather like looking through some extraordinary and paradoxical periscope. They are firm to the touch, and warm under the heat of the sun—and slightly tacky. It is an extraordinary thought that our fingers could have touched a similar mound 3,000 million years earlier. Like the coloured stains around hot springs this is persistence personified, survival through a thousand plagues, through changes of climate beyond number, and through the devastations that removed far grander organisms from the Earth for ever.

It was, to say the least, an interesting question why stromatolites should have survived in Shark Bay, other than through Australian atavism. After all, there were thousands of other tropical bays which might harbour them, but there must be some special ingredient on this distant coastline. The answer seems to be the *absence* of other organisms. Stromatolites grow because there is nothing to remove them. Many kinds of living animals graze algae; some kinds of snails specialize in rasping at algae with their spiky jaws. You can see such small snails sliding over rocks and glass in fish tanks, polishing away the algae and bacteria they feed upon, incessantly clearing away their food, much as an obsessively houseproud individual might run over and over a small flat looking for specks or crumbs. These kinds of animals eat away the coherence of stromatolites. Only when some factor such as elevated salinity combined with excessive heat removes the grazers can these ancient communities re-establish themselves. The stromatolites in the Precambrian were so much more widespread because at that time there were no raspers and grazers at all: the growths could prosper unhindered. In that world of biological mats nothing disturbed bubbling photosynthesis, and only storms ripped away the fabric of the cushions and pillars.

It was not long before other examples of living stromatolites were found, especially in the ocean, where conditions were appropriate to permit their unhindered growth. They were found in the hot, protected inland gulf of Baja California, Mexico. They were found at greater depths in the seas around the Bahamas. These living examples permitted an understanding of the rather subtle ecology of mound life. It was proved that the living, photo-synthesizing "skin"—as thin as a sheet of paper—was underlain by other lay-ers inhabited by different kinds of bacteria, some of which lived by breaking down the waste products of the upper layer by fermentation. Trapped grains of sediment contributed minerals that helped their growth. It was a sustain-able system, an ecosystem in miniature. If this truly reflected the state of the nascent biological world it is clear that cooperation and coexistence were a part of life close to its inception. Existence at base can be thought of as recip-rocal rather than competitive. If the first living systems worked as teams, by mutual nourishment, it must have been the arrival of animals that hotted up the competitive aspect of evolution. To be sure, each new innovation that happened in the mats must have spread and prospered, but the mere fact that the living mats seem to resemble their ancient progenitors in so many ways—among them the organisms they include—argues for a profound conser-vatism, a kind of inertia to change. These humble structures are the birth of ecology.

For all the simplicity of their microbial community, stromatolites pro-duced a wondrous variety of shapes. After their modest start in the ear-lier part of the Archaean they had blossomed into many different forms by 2,700 million years ago—a panoply of columns, and every variation of crimp-ing. Different kinds of stromatolites inhabited different ancient environ-ments. Little, delicate, fingered ones lived along the shallow pools and tidal flats, bolder pillows and gentle mounds could face the open sea. In rocks about 2,000 million years old the acme of stromatolite construction was to be found in quiet habitats beyond the reach of destructive storms. These were forests, or reefs, of great cone-shaped stromatolites that stretched for hun-dreds of miles. Individual cones could be tens of metres high, even up to 100 metres. Malcolm Walter, who during the 1970s made known these extraordi-nary structures to a marvelling geological community, memorably described these *Conophytons* as "like rocket nose cones placed edge to edge." This was the apotheosis of the algal mat, the first demonstration (there were to be many more) of life's propensity to hanker after the gigantic. All this was ac-complished by the tiniest organisms embroiling mud and slime in a grandiose scheme of construction. Sadly, none of the living stromatolites approaches the dimensions of the Precambrian giants.

Stromatolites are to be found stranded upon the ancient core of North America: the Canadian Shield. This Precambrian centre of the continent covers a huge area between the Great Lakes and Hudson Bay. The very name conjures up thoughts of a vigorous defence against the inroads of time, a shield against the inevitable obliteration that faces most ancient things. The central core of the Shield accreted early in Earth's history, and stromatolites recall the first waters that lapped upon the early continent. On the shores of the Great Slave Lake early Proterozoic stromatolites are laid out in fields of fossil cushions. It is a re-creation of a primordial seascape, exhumed by erosion. Veil your eyes to cut out the modern vegetation along the shore, and you could be looking at the Precambrian scene, nearly 2,000 million years ago. This is truly time travelling, and if the world were attuned to its real wonders this sight would be as well-known as the pyramids of Giza, for all its modest dimensions.

The remoter part of the Canadian Shield is as interminable in its fashion as the outback of Australia. Endless coniferous forest is comprised of a handful of species of dark fir that can endure the cruel winters and the thin soil; but where there are clearings there may be delicate birches or aspens. Networks of bogs and ponds and rivers bewilder and baffle anyone who strays from the tracks, of which there are few. The tough fossils of stromatolites survived scouring by the great ice sheets of the last Ice Age, which retreated from this landscape only a few thousand years ago. The debris of its retreat litters the landscape with boulders and moraines. Drainage follows the whim of this history. But through this irregular patina of glacial waste the ancient rocks are exposed at the surface in many places, like the backs of some great school of rocky whales. Often there are banded rocks, gneisses or greenstones, which have been hardened, baked and squeezed in the great vice of the Earth's interior. These were the ancient basement upon which the stromatolites built their slimy pillars under the glare of the youthful Sun. Some of the most beautiful remains of early life have been recovered from the Gunflint Chert of the Lake Superior region. This hard and fine-grained rock had the property of furnishing good sparks when struck (as do many cherts and flints), and hence found employment in early firearms long before it found its way into the laboratory. Maybe by the time the Gunflint Chert was being formed the mats had donated enough oxygen to the atmosphere to initiate the start of the ozone layer: a high atmosphere layer of modified oxygen that serves to filter out harmful parts of solar light. Ultraviolet light can cause damage or harmful mutations in many cells. Bacteria not only provided the breath of life, but also its overarching shield. The recent hole in the layer over the Antarctic has been one of the few phenomena scary enough to provoke governments to worry further ahead than the next election. But tough bacteria are protected by their slimy coat from the worst effects of the Sun. It has been shown by experiment that some kinds of bacteria can withstand massive doses of ultraviolet radiation. Without this resilience there would have been no future.

IT WAS WITHIN SOME STICKY MAT on some long-obliterated shore that the first more complex cells were born. Such cells contain discrete bodies, the best known of which is the nucleus, which holds the genetic blueprint and is present, for example, in every cell in our own bodies. Other kinds of discrete intracellular bodies with different functions are known as organelles. It was shown that some of these organelles closely resembled bacteria that were capable of leading an independent existence outside the cell. So the idea arose that complex cells may have arisen by capture—one bacterium embracing another to produce a new level of complexity. Bacteria were tucked away, complete with their life functions, like swag in a burglar's sack. By conjoining different small and simple organisms a more complex and effective structure could be built. This extraordinary development is a kind of symbiosis*— Nature's mutual back-scratching. It affords the ultimate example of the co-operativeness of early life.

The bacteria that I have written about so far lacked nuclei: they were prokaryotes, reproducing by splitting, or fission, to produce (often exact) copies of themselves. Although they were capable of building large structures like stromatolites, most were only a few thousandths of a millimetre in diameter. Nucleated (eukaryotic) cells are more complex, and many times larger. The large size allows them to include organelles such as mitochondria and chloroplasts—tiny, membrane-bounded bodies which perform vital functions: photosynthesis, enzyme production, and the like. A pioneer in this theory, Lynn Margulis, identified such organelles with particular species of free-living bacteria. Thus it was that the photosynthesizing eukaryotic cell was able to function—by virtue of capturing the appropriate "blue green" bacterium. Once included in the cell, the bacterium became the organelle

*From the Greek for "living together."

known as a chloroplast, responsible for photosynthetic processes. Chloroplasts endure and function in every green plant cell of grasses, herbs and trees, so the capture was for ever. The cell can be compared with an engine run by chemical fuels, but powered by motors that were capable of running on their own. Life hoisted itself up by its own bootstraps. It can be questioned whether the process began by consumption—a gobbling of one bacterium by another species that stopped short of digestion—or whether the introduction of bacteria started for some other, coincidental purpose, and then acquired a useful permanence. Whatever the history, this was the next miracle. Without it, a slimy, undulating mat would have remained the acme of biological development on the Earth. Maybe in some galaxy that I could see from my swag in the Australian desert there is a gloomy and undeveloped world where all the exuberance of life stopped at the stromatolite. The oxygen produced in this monotonous milieu would be consumed and lost in the deposition of limestones; the continents would remain interminably fallow, no plants would prosper in the soils, nor the most insignificant invertebrate disturb the slow growth of the biological pillows. Pillows, millions of pillows, would truly be appropriate to a world so half-asleep.

Lynn Margulis maintains that the harnessing of bacteria happened not once, but many times—indeed, was almost a routine property of protoplasm. This may not be so bizarre as it might seem. The world is full of symbiosis—some of it obviously ancient. Lichens are the toughest of all living things, able to perch on baking, dead stones in deserts and feed on a moment's fortuitous dew. They are a collaboration between fungi and algae. Only when little fungoid cups appear at the time of reproduction is the nature of the deception revealed. Many corals have photosynthetic algae within their tissues that feed the simple animal with important nutrients. Fungi nurture the roots of orchids. The stomachs of ruminant mammals house a great, digestive soup of microorganisms that break down the cellulose of grass, and without which the animals would soon starve. The spectacular bovine fart is the living proof of fermentation. It seems that there is a propensity for life to adopt working relationships with bacteria. Disease is the dark reflection of the same process, whereby the microorganisms turn upon the host, much as the fledgling cuckoo does upon the chicks it once lay next to.

The biological world is a massive cooperative powered by the small organisms that began it. The primordial chemistry lives on, in life processes like photosynthesis, in enzymes, in membranes. But larger, nucleated cells had new potential. They could build big organisms: ultimately, they could join to form complexes of cells with different functions, *tissues* that Eliot's Dr.

Lydgate would have recognized. They became differentiated into sexes, with the possibility of exchange of genetic material—with the implication that natural variation could increase to allow greater adaptability. But all such cells were aerobes—that is, they needed oxygen to flourish. The world became oxygenated enough through the incessant efforts of photosynthesizing mats to be suitable for their development about 2,000 million years ago. Then the long overture played with bacteria could give way to the first full chorus.

There is a conception of primitive life which could be described as that of the "protoplasmic blob." I suppose that when the cooking metaphor was dominant, writers might have conjured up a picture of this "blob" crawling ready-made from the primeval soup. *Amoeba* is the organism that closely approximates this archetype. It is a cell that creeps by putting out evanescent arms—pseudopodia—that mutate and change in shape as the animal makes its slow passage; its protoplasm flows along with its metamorphosing form as it moves. It can sense food, which it engulfs in its fluidity. It feeds on other things; it does not manufacture its own food. It has a nucleus. There are old horror films that feature the protoplasmic blob. One had enormous blobs coming out of holes in the ground, wobbling like jellies, before going on to do much engulfing. In such films the scientist is always referred to as "doctor." (I can say that in all my years in science I have only once been referred to as "doctor"—and that was when I was mistaken for a cardiologist.) The "doctor's" job is to look bewildered and grim by turns for the first seventy minutes before having a blinding insight of astonishing banality in the last five, which dispatches all the blobs in short order. One does not have to be a theoretical post-modernist to appreciate that these amusing and not very horrific films reveal a commonly held concept of what is, at the same time, both primitive and threatening. Threatening, probably, because lack of form, sheer blobbiness, is a kind of sinister mutability—after all, sensible, higher forms of life have *structure*. It may have been some similar notion that inspired H. G. Wells to choose big octopuses as the aliens in *The War of the Worlds*, although Wells also had enough biologically literate friends to appreciate that these molluscs were also highly intelligent (as invertebrates go). However, there is in reality no question of single cells achieving the enormous dimensions of the blob—this is impossible without the benefit of organized tissues: muscles, nervous communication and the like. And how far in time, and in organization, from the first life-giving broth is the amoebic animalcule—it probably appeared halfway through Earth's long history. It must have been the slowest crawl from the "soup" imaginable. As for evidence, there is little enough in such protoplasmic organisms that is capable of being preserved as a fossil.

Instead, the eukaryotes that can be found preserved are mostly plants—algae of various kinds. Some formed threads not unlike those of the cyano-bacteria, but much larger and more complex. Some joined the stromatolite communities (which, however, continued to be dominated by prokaryotes). But spherical objects also appeared, often about a fifth of a millimetre across. Some are equipped with spines, making them look like tiny floating mines. These are likely to have lived in the plankton, as part of an open ocean community. They had tough, organic walls, which probably served to protect the cell during "resting" phases. But these walls fossilized readily: the tiny, durable fossils can be recovered when shales are digested in hydrofluoric acid (which removes almost everything, including the fingernails of the investigators if they fail to take the most stringent precautions). It is a simple technique, but it challenged the view that such ancient rocks were devoid of life, as these fossils were found in great numbers—especially in rocks laid down beneath the sea between 1,000 and 850 million years ago. Some giants appeared at this time, monsters of the single cell: *Chuaria* was a centimetre across. This may not compare with the blob, but for an algal sphere it is astonishing.

The importance of such humble spheres to the evolving Earth can scarcely be overstated. The colonization of the open oceans rapidly increased the area inhabited by photosynthesizing organisms. If the Earth then, as now, was two-thirds ocean, the dilute soup of oxygen-giving life spread over something like thrice its original area. The new opportunities of this strange habitat spawned new designs of cells. As geological time moved through the thousands of million years from the Archaean into the Proterozoic, and thence through another 1,500 million years, there was a slow, impalpably slow increase in variety. The oceans began to breathe—but they also acted as a "sink" into which nutrients could drop out of sight, as the bodies of the floating organisms drifted to their sedimentary graves on the sea floor. The world was growing into an interconnected web of life, chemistry, oceans and geology—even the most sceptical reader will be able to accept that this much of Professor Lovelock's *Gaia* must be true. Life made the surface of the Earth what it is, even while it was Earth's tenant.

Like bacteria, some of the water-dwelling algae seem to be amazingly persistent and tough. Distinctive fossils can be matched by distinctive members of the living flora. This is true of prasinophycean green alga, or certain red algae. In some cases the ancient fossils can be closely compared with species still living. The primitive plant world has sat out the dance of time.

Some of the scientists who work on ancient life show a different kind of toughness. It is beyond dispute that the hardest scratchiest place in the world to geologize is the Australian outback. Fieldwork there requires almost prokaryotic endurance. Every shrub is equipped with spines, and those that are not are equipped with burrs. There is a terrible weed called *Spinifex* which grows in glaucous hemispheres a metre across and is apparently composed of nothing *but* spines; its weapons are tipped with silica (the same material that makes the fossiliferous cherts). To tumble into a *Spinifex* bush is to experience an accident with a cartload of syringes tipped in vitriol; one comes to regret the evolutionary *chutzpah* of the plant kingdom. It is also alleged that many of the insects in the outback are poisonous. On my perambulations through the bush I did not see any—yet *Spinifex* was in my path everywhere I turned. The reverse is true in the conifer wilderness of Canada, but fieldwork there requires similar endurance. There are no particularly nasty plants; rather, the malevolent insect rules. There are mosquitoes for every occasion: ones with brown, fuzzy heads that come out in the day; huge, black, extra greedy ones that come out on still nights, with an appetite for placing the proboscis into armpits or behind the ears; there are nearly invisible "no-seeums," which must be first cousin to the Scottish midge. Worst of all is the implacable blackfly, which takes a small bite out of the living flesh, and at the same time paints on a drop of anti-coagulant so that the little nip bleeds. A small rivulet of blood trails down your face or arm. Pamphlets issued by the National Parks of Canada offer the reader the nugatory assurance that these creatures are vital to the ecosystem.

Another kind of persistence inspires the doggedness that drives scientists to seek fossils a few thousandths of a millimetre across in 10,000 square kilometres of Precambrian rocks. Professor Bill Schopf has done as much for this hunt as anyone, and his jocularity is not incompatible with a special kind of toughness. He is an ebullient man, but an indefatigable one. His office light burned long after mine had been switched off for the day during his sabbatical year at the Natural History Museum in London. Determination to find the paydirt leads to the examination of hundreds of microscope slides. This can be boring and dispiriting work. Gradually, though, expertise in recognizing the right kind of chert in the right geological setting shortens the search. But patience is more than a virtue: it is a necessity. A lot of palaeontology requires this kind of persistence. We shall meet it again in describing the search for fossils of our own ancestors. Sometimes the line that divides an untiring search from a wild goose chase is a subtle one. Disappointed hunches never make the news. But even a failed search tells the next searcher to look elsewhere. It is not often that a biographer celebrates an exceptional failure: the person who always hacked the rocks to the left when he should have been hacking the rocks to the right. Monomania probably helps ultimate success. If it were possible to recruit pagan gods into the Christian canon then the patron saint of Precambrian palaeobotanists should be Sisyphus, condemned for all time to push a boulder uphill, only to have it repeatedly roll back again to where it started. However, just once in a while these flesh and blood sisypheans reach the top of the hill.

THIS CHAPTER HAS LOOKED AT some 3,500 million years of the history of the Earth, and we have arrived at about 1,000 million years from the present day. This is the greater part of time. Life has now reached a single, large, but complex cell, the kind of small thing that might once have been imagined arising from the progenitive broth. Can this be all? The events that took place during this Brobdingnagian stretch of time might seem small taken one at a time. But if they are rushed by again in a desperately speeded up reprise it will look quite different. At first, organic chemistry dominated the story, the juggling of a thousand compounds based on the capacity of carbon to form chains, rings and polymers. Some of the ingredients of the fertile broth were donated from beyond our planet, secreted in the bodies of meteorites. Synthesis of carbon compounds was an inheritance from the early moments of the Universe itself. But if we were conceived among the stars the next stages were Earth-bound. There was the generation of cell membranes—the envelope of life. Maybe an RNA world preceded the first, true cell. There was the recruitment of phosphate compounds to channel the energy of life. The simple chemical reactions which might supply the energy by which life propagates itself were incorporated into the cell—perhaps the fundamental process was as simple as the production of the common mineral iron pyrites. Enzymes guaranteed the chemical transformations necessary for metabolism. DNA secured reproduction, and the synthesis of proteins. And all this happened in an atmosphere hostile—lethal, even—to almost everything that lives on Earth today. A hot world, even a nearly boiling world, spawned the first bacteria—the Archaea—which revelled in hot springs and dined on fumaroles. The world was coloured bright with films of bacteria for which oxygen was a poison. Vents and seeps in a restless jigsaw puzzle of small oceans nourished these tiny cells. But still they share genetic links with all other life—including ourselves—that prove that this most extraordinary event, this singularity, this genesis of life, happened only once.

And then, in some small corner of some stew of bacterial slime, there was a cell that utilized light to capture carbon from carbon dioxide—not much more than a simple chemical slicing. But the same cell produced oxygen as a waste product. Thus was the metamorphosis of the world secured. As the continents began to coagulate into masses that would endure until the present, then, too, did the community of bacteria collaborate in constructing mats—and most of life's history has been conducted inside these mats. Sealed in slime from the cruelty of ultraviolet radiation, mats formed mounds, and columns, and pillows, and fingers; mats formed great cones in the deeper sea, the like of which have not been seen on Earth for 1,500 million years. Ultimately, mats maketh man.

Oxygen released by photosynthesis slowly transformed the atmosphere, and some of the original atmospheric gases which might have prevented the further development of life were neutralized. It may not have happened with the dramatic flourish of a revolution, but the increase in size and complexity of cells was revolutionary beyond calculation. This was accomplished by capture, a rape of the cell, as bacteria were tucked *inside* the larger cell, complete with their special skills. Photosynthesizing bacteria became photosynthesizing organelles. The cell, which had been like a solo musical instrument, affecting but limited, like a pan pipe thinly trilling a melody, suddenly became an orchestra. With this new resource what symphonies might be composed, what oratorios, what infinitely varied pieces? The subsequent history of life was conducted in richer sonorities, in increasingly full harmonies. One can argue about the meaning of "advancement," but what is not at issue is that life became more complex and more interesting.

This enrichment was accelerated by sex. The crucial innovation was the combination of genetic information from separate parent cells, followed by the donation to the offspring of a new combination derived from both parents. This process of mutual exchange enhanced the chances of natural variation—new and useful genetic mutational changes could spread rapidly through the population. Bacteria reproduced by simple fission which engendered clones; when mutations occurred, these in turn were then propagated by fission. But combining DNA from two parents readily produced new and rich combinations of genetic information. Professor John Maynard Smith of Sussex University, the doyen of evolutionary biologists, regards the "invention" of sex as perhaps the crucial development in the history of life. I recently heard him give a lecture in Budapest, which was, or so he said, the first he had given for twenty years in which the word "sex" was not mentioned. To those who believe in the hegemony of DNA, sex was the step that sealed its global sway. Exactly *how* sexual differentiation originated is still hotly debated, and a book as long as this one could be devoted to this issue alone. Because much of this story relates to details of the internal engineering of the cell and leaves no record in the rocks, I shall not give it the attention here that it undoubtedly deserves. It must be stated that it was evidently an ancient, Precambrian innovation because so many plants and animals—even fungi—show evidence of sexual reproduction. The sexual imperative runs into deep time.

The peculiar fact remains that primitive forms still endure—those cells that might have been assumed to have disappeared with the changing world, outpaced by cells more complex and more adapted to advanced environments. But the simple design was effective, tough and durable. Where conditions like those on the early Earth still persist, so too do the tiny cells that feel comfortable there. Deep in boreholes, in rock buried away from the sight of man for millions of years, there are bacteria. Wherever grazers relax their attention on the edges of the sea, in salty bays, or tropical mud flats, there are mats growing that might have graced the shores of the Precambrian. The bald truth is that the old persists alongside the new, albeit tucked away into forgotten niches or hot stews where only the dedicated bacteriologist feels tempted to linger. The Earth did not shuffle off its anaerobic past; it secreted it in odd corners. I am told that an ancient bacterium was rediscovered against an inn wall where men had urinated for generations. The lingering ammonia associated with human waste excretion resurrected a memory of an ammonia (NH_3) metabolic pathway which originated deep in the archives of the Archaean. The Earth does not forget—or at least its amnesia is selective.

If the first great threshold was the origin of the first bacterium with its genetic package, and the second the dawn of photosynthesis, then the third was the appearance of a range of complex cells. We can still sample these thresholds today in simple living things that survived. What arrogance it is to assume that we are some kind of pinnacle of creation, when these biological Methuselahs still live on. If the pinnacle of athletic achievement is the marathon, then these little rods and threads are the marathon runners of existence itself. Their names do not trip easily off the tongue: *Methanococcus*, *Pyrobaculum*, *Lyngbya*, *Eoentophysalis*. Perhaps we should be as familiar with them as with *Archaeopteryx* or *Tyrannosaurus*. Such names may seem a burden dressed in all their classical polysyllables, but for those with an ear for language they soon become informative: surely *Pyrobaculum* must have something to do with heat and fire; *Methanococcus* must be associated with methane. The very names reveal something of their history.

In the beginning there was dust, and one day the great, improbable experiment of life will return to dust. We are not secure. Just as our ultimate genesis was entangled with the birth of suns, and the terrifying tumult of asteroids and meteorites, so we are still bound to the cosmos. Cyanobacteria grew the cocoon of an oxygenated atmosphere, a thin shield against the void. The unimaginable stretches of the indifferent Universe lie beyond this envelope. As we shall see, from time to time the immensity outside reasserted its power, and may yet do so again, to lethal effect. But whatever asteroid assaults the Earth in the future, surely there will be torrid vents and seeps that will once again allow the primeval cells to spread their coloured slime over the Earth, even as creatures of complexity and elegance know their last days.

Chapter 3

Cells, Tissues, Bodies

E DWARD LEAR PUBLISHED HIS *Nonsense Botany* in 1870 (see drawings on next page). He had great fun cocking a snook at the Latin names of plants, and at the serious business of naming and classifying the vegetable kingdom. *Cockatooca superba* is a lily with a parrot for a flower, while *Piggiwiggia pyramidalis* sports a splendid raceme of porcines. *Nasti-creechia krorluppia* is a plant nobody would wish to sit next to on a picnic. Botanizing had become a respectable pastime for ladies of a certain stratum in English society, and Dr. Joseph Hooker's *Flora* helped the novices with their identifications, and inculcated Latin labels upon a wider circle. "Goodness me!" the cry might go, "what a fine specimen of *Geranium robertianum*," thereby lending a certain significance to the discovery of herb robert. Although a number of Lear's spurious plants incorporated artefacts (it does not require much effort to imagine *Washtubbia circularis*), most of them mix animals and plants. Lear's humour may seem a little contrived to us, but it evidently struck hard upon the middle-class Victorian funny bone, for the *Botany* ran to many editions.

The gulf between animals and plants seems obvious to us, but it was not always so. The ancient Greek *Weltanschauung* included animate spirits that routinely merged with the vegetable kingdom, and the nymphs known as hamadryads were as much tree as sprite. The gods visited biological transformations

Piggiwiggia Pyramidalis.

Nasticreechia Krorluppia.

out of revenge or compassion: the daffodil was the metamorphosed outcome
of human narcissism; even the gnarled olive tree had a human biography.
Categories were not fixed, and animism was probably a general belief. J. R. R.
Tolkien revived the notion of conscious, animate tree spirits as the beneficent
but ponderous "ents" in *The Lord of the Rings*; it was an appealing conceit that
these plant spirits could take days to have a conversation because their lifespan was so drawn out that hurry was never necessary. Hamadryads reputedly
lived ten "palm-tree" lives, which classicists state to be 9,720 years; the ents
had lost their partners—the entwives—to a distant and devastated land, and
were doomed to extinction. This is reminiscent of the saddest tree in the
Royal Botanic Gardens, Kew, a primitive cycad collected from Zululand,
Encephalartos woodii, which is extinct in the wild, one of a handful of specimens of this species. It resembles a spiky-leaved palm, but it is not a close
relative of this advanced family: it is a dioecious species and, like the lost
entwives, the female of the species has apparently not been preserved. Surely
this is the most solitary organism in the world, growing older, alone, and
fated to have no successors. Nobody knows how long it will live. Perhaps it is
a mercy that, unlike the ents, it cannot speak.

But the recognition of natural order in organisms served to separate plants from animals, as surely as it separated conscious mankind from instinctive beast. Lear, of course, created his chimeras a decade after Darwin's *Origin of Species* had been published, but the acknowledgement of evolution was not an essential prerequisite to render his botany nonsensical. The notion of the proper order of things was rooted in many previous classifications of animals and plants, the higher distinguished from the lower: what the great eighteenth-century naturalist of Selborne, Gilbert White, knew as the System (a term which survives in the name of the science of classification: systematics). The System divided the natural world into kingdoms, the most major division of life, and the two kingdoms were plant and animal (Plantae and Animalia). Across this great divide only the humorist or the fantasist dare step.

In this narrative several more kingdoms or even greater divisions of life have already been encountered than Gilbert White would have been able to recognize with the equipment available to him. These organisms include the ancient bacteria—the Archaea, the bacteria (Eubacteria) themselves, crucially including the "blue-greens" (Cyanobacteria); and the Eucarya—nucleated cells carrying the basic apparatus of inheritance, which typifies all higher and complex life. Animals, plants and fungi are all *sub*divisions of Eucarya. The only reference to the great divide between animals and plants has been the description of archetypical, amoeba-like organisms—animals, surely—sliding furtively between the threads binding some slimy mat.

The division between plant and animal is in truth much less clear-cut than *Nasticreechia krorluppia* would portend. However, in the bookshop in the Natural History Museum in South Kensington the visitor soon learns that the simple bipartite world lingers on in the minds of shelf-stackers. If you wish to find a book on fungi then you have to go to the shelves marked "Botany." Yet fungi are probably closer to animals than they are to the bindweed or crocus. The most obvious feature they share with plants is that they are fixed in their growing site—they do not move around to feed. However, like animals, they do not manufacture their own food, but rather feast on the efforts of others, devouring wood and vegetable debris; they have enzymes capable of breaking down cellulose—the carbohydrate substance which provides the structural strength for much of the vegetable kingdom. Fungi feast on decay. Perhaps their vegetable reputation owes as much to where they appear on the menu as to any innate characteristic; after all, mushrooms are stacked between the aubergines and chives in the supermarket. Fungi obviously lack the vital pigment that is the hallmark of all "greens"*—chlorophyll—but then they scarcely look like animals, and so vegetables, presumably, is what they must be. They do, after all, reproduce by means of impalpably tiny spores—and primitive plants such as ferns also bear spores. But "spore" is no more than a name for a small propagule, and air or water-borne dispersal virtually compels such small size; the presence of a spore tells us nothing—by itself—about biological relationships. For every large mushroom-like fungus there are many more species that are microscopic; and some of these have mobile stages in their development which invite comparison with single-celled animals; thus, the distinction between fungus and the smaller animals is less insurmountable than might have been supposed. Now that the signature of the sequences of amino-acids in nucleic acids themselves has been investigated it is even more likely that fungi are closer to certain animals than they are to plants. At the very least, they deserve to be reckoned a kingdom of their own. Maybe the twee Chinese mushrooms that danced to Tchaikovsky in Walt Disney's *Fantasia* were more biologically apposite than *Nasticreechia*.

Thus it is that easy distinctions between kingdoms are blurred. When we go back to the Precambrian the differences were probably even less clear. The question of what makes an animal is neither so easy nor so obvious as comparing a lion with a leek. There are many kinds of living, single-celled eukaryotes which have left absolutely no fossil record, being both too small and too lacking in substance. As a whole, these simple organisms are known as protists. The amoeba is only one of them; many more live in bacterial mats, in pools, or in damp litter. The old name protozoa ("pre-animals") is redundant, since it is unclear whether they are any kind of animal (zoa) at all, and it is becoming increasingly obvious that they comprise two or more groups of organisms, which may be only rather distantly related. Using the term protozoa is rather like referring to the inhabitants of all former civilizations as ancients—it does little good apart from conferring a generalized antiquity. It is clear from their simple protistan organization that they belong low in the tree of life. But exactly how the branches on that early tree were disposed is still unclear.

Margulis's theory of the capture of organelles allows for progressively

---

*Ironically, one of the few green mushrooms is also the most poisonous. This is the Death Cap (*Amanita phalloides*)—but the sickly green of this species is a pigment which bears no relation to chlorophyll. Modern medicine has not discovered a cure for the deadly alkaloids that this species contains; death follows after protracted suffering interrupted by occasional false dawns. Unlike poisonous properties in the animal kingdom, it is hard to see that the fungus derives any benefit from its deadliness; if we were still imbued with animism it might seem perfectly reasonable to attribute its properties to a malevolent spirit.

complex cells to be created. Increasing numbers of simple cells, which probably once lived independently as prokaryotes, were stowed away inside the
host cell. In this way, mitochondria were permanently installed, and these
remain essential for the energy metabolism of all advanced cells. A single
liver cell may include a thousand mitochondria. There are other organelles:
Golgi bodies, lysosomes, glyoxysomes, all performing vital cellular functions.
The tools were assembled, and the cells were ready to steal upon the Earth.

Many of the simple cells propel themselves by means of a whip-like organ,
a flagella. Some scientists lay great stress upon the structure of this tiny
whiplash in the organization of primitive life: for example, some flagellae are
covered with minute hairs; others have a collar. The giant, freshwater amoeba
Pelomyxa palustris lacks mitochondria, and has an appealing resemblance to
the protoplasmic Blob, but it is already too specialized to be a candidate for
the common ancestor of all protists. There is quite an array of these, bearing
names that are familiar only to students of recondite organisms: *Euglena, Tri-
chomonads, Diplomonads Apicomplexans, Ciliates.* None the less, they are important to an understanding of the various ways in which simple cells may have
developed. From a consideration of the evolutionary history of *all* the eukaryotes Margulis claimed that there were as many as five kingdoms wrapped up
in various cellular packages: only two of these would correspond with conventional animals and plants. How far we seem to have come from a clean-
cut, two-branched biology! The evolutionary events leading from one to
another of the protists must have happened in the Precambrian, probably
down among the mats: indeed, bacterial mats are still a good place to find a
variety of protists. It is sad that we will probably never have direct evidence of
the steps between them in the shape of fossils.

The important principle for our perception of the order of things is this:
that plants, animals and fungi—the many-celled organisms—have been
derived from simpler protists. Put another way, we share with the palm tree
and the puffball a common ancestor. The amoeba lies below the plants (meta-
phytes) in a kind of evolutionary basement. And if the amoeba is an animal,
then, logically, the higher plants were derived from an animal! There are
those scientists who think of a plant as some kind of amoeba that gobbled a
chloroplast.

Slime moulds have a special place in this argument. On forest floors in
late, wet summers they glide through the soil and leaf litter, or over damp
wood, disguised as a trail of amoebae, shapeless, or at least continually mutable, growing and dividing as they absorb nutrients from the leafy detritus.
They can grow as large as a soup plate. But then, as the leaves take colour and

fall from the trees, a strange transformation happens. The fluid protoplasm slows down, thickens, congeals. It hardens into a definite shape. In some species there is a series of inflated balloons, in others a line of little "spinning tops" borne on short stems, in others again there are delicate spindles supported by a tracery of rods. The slime mould is fruiting, and, as it does so, it ceases its animal-like wandering to become something like a fungus. The protoplasm re-forms as a mass of brown spores which, like all spores, can disperse to produce new individuals on new forest floors. Slime moulds have been called Myxomycetes; the latter part of the name (mycete) reveals the fungoid relationships they have been considered to possess—an organism, it might be thought, in transit between two kingdoms. Some develop lurid colours as they mature: pinks and reds and yellows. My daughter acquired a mild phobia about them when she encountered *Lycogala epidendron* in the process of ripening: it made bright pink cushions sitting on rotting pine stumps, looking as if some forgetful child had dropped a handful of sweets on the soft wood. When she prodded one cushion out of curiosity it exploded, and spat pink slime over her fingers: a *Nasticreechia* if ever there was one. Other slime moulds are larger: they can form white, soft mounds as large as a bun, which nestle on the undersides of damp branches. Professor Bonner of Harvard University has proposed that slime moulds can provide a model of life at the stage where it began to distinguish direction, to tell fore from aft, at a junction close to the inception of specialized tissues. He cultivates slime moulds with the attention that pigeon fanciers give to their flamboyant charges. Their molecular sequences confirm that they occupy a place close to the branching point where higher plants split off from fungi and multicellular animals; they serve to remind us of that distant join between the mobile and the rooted, and their amoeboid phase speaks of a time when all animal life streamed in protoplasmic amorphousness, a world before shape.

So what are animals? In the first place, they feed: they feed upon plants, either directly, or by feeding upon some other animal that feeds upon plants. Animals are spongers on the hard work of photosynthesizers: they graze or hunt, or else they absorb nutrients from an organic soup which was ultimately brewed by plants. If plants (autotrophs) manufacture their own growth and substance, animals exploit this self-sufficiency in their role as heterotrophs. Many animals are mobile, and primitively were certainly so; thus they move about like slime moulds to find their nourishment. They rasp, they entrap, they exploit, they absorb. And they respire—they use oxygen to fuel their metabolism, the oxygen that cyanobacteria and algae manufactured with such patience over the billions of years since the early Archaean. Eating organic food is a kind of burning, mediated by oxygen. This is why food values are measured in terms of calories, the unit of burning. Without endless photosynthetic labour producing oxygen, animals would simply have been impossible. Living protists vary dramatically in the amount of oxygen they need to thrive, but it is obvious that photosynthesizers would have had to precede the first animal in order to modify the original atmosphere and enrich it in sustaining oxygen. Cyanobacteria and algae made the world a place fit for animals, and the grazing has continued ever since. Size was important, because single-celled organisms preceded multicellular ones, in both plants and animals. Where plants led, animals followed. We shall discover that this precedence applied more than once in the history of life. Animals rely on plants as much today as when they began in the Precambrian. Just as we human beings have to take in nutrients from plants for the good of our enzymes and our resistance to infection, so animals have always been in thrall to their diet; a poor diet produces a malnourished and unfit animal, and when the food runs out the animal dies. If plants compete mostly for light, the precious fuel of photosynthesis, many kinds of animals can run or crawl from one place to another to assert their dominance as a species or as an individual: they hotted up the competition. The price of failure to compete successfully might be extinction.

There *are* fossils of animals in the later Precambrian rocks which are now becoming widely known from localities scattered around the world. Yet not many decades ago all such evidence of past life was unknown. Perhaps people merely followed Charles Darwin's dismay at the lack of evidence of life at this crucial phase. Like Lewis Carroll's Snark, the reward would not repay the effort of the search, because whatever was found would only turn out to be a Boojum in the end. Why search for what you know in your bones isn't there? Maybe it took the naive vision of a schoolboy to spot what was obvious. The first Precambrian animal fossil in England—and one of the first in the world—was discovered by such a young man, John Mason, in 1957, in Charnwood Forest, Leicestershire.

The forest juts through the English Midlands, an island of wilderness, scattered with gorse, birch trees and undergrowth. Around it there are the farms, comfortable suburbs and light industries of the Midlands. It seems like an atavistic relic, for its rocks are ancient indeed, breaking through the general cover of younger, softer Mesozoic rocks that stretch for miles on every side. In one quarry, John Mason saw a large, leaf-like impression, the size of a small frond. No snark or boojum this, but an organic remain impressed upon the hard, sandy sediment. Time and earth movements had upended the later

Precambrian sea floor, so that the frond was hanging nearly vertically, like a picture in a gallery. Mason was fortunate in having a scientist to hand in the University of Leicester, Trevor Ford, who was open-minded enough not to pooh-pooh the discovery out of hand. A rubber cast was made, which showed more clearly that the structure had a stem, continued forwards as a kind of midrib, which served to divide the frond into two halves: these, too, were subdivided crosswise into a number of lobes, upon which striations could be seen. It was all too elaborate and structured to be a "sport of nature"—a chance effect of wind and rain. A return to the quarry revealed further impressions. There was a circular, disc-like one to which, it was suggested later, the frond had once been attached, like a seaweed to a holdfast. The frond somewhat resembled the living "sea pens." Thus it was that in 1958 *Charnia* and *Charniodiscus* were first described; young Mr. Mason had his reward when the species was named after him, *Charnia masoni*. Not many schoolboys are so immortalized, and there is a law, approved by the International Commission of Zoological Nomenclature, that once a scientific name has been given it cannot be changed, so Mr. Mason's perspicacity carries a guaranteed celebrity for ever.

One of the minor pleasures arising from describing fossil animals is giving them a scientific name. There are rules about this which have to be observed, and are guaranteed by the same body that certified Mr. Mason's nomenclatural celebration. For example, a Latin or Greek form of words is obligatory. The names have to be formally published in an accredited scientific journal; one cannot just wake up on a Monday morning and coin a whole new battery of names for species. No offence is permitted, say the rules. In the past, it was possible to exercise some ingenuity in seeking revenge on those who had done you down by naming horrid organisms after them, or making unflattering comparisons. *Blenkinsopia foetens* might imply that your old foe Blenkinsop was a stinker (*foetens* being Latin for malodorous). No more. It might still be possible to name a new leech after someone notorious for free-loading—but then it might turn out that the recipient was rather flattered by the attention. There is a rather insignificant trilobite called *Forteyops*. Jokes are still possible. A mollusc worker waited for years for a new species of the genus *Abra* to turn up—so that he could propose the species name *cadabra*. However, in many cases the name justly records a discoverer, or a contributor to the science.

The same is true of Mr. Sprigg. In the Ediacara Hills of the Flinders Mountains in South Australia there is a fine-grained, hard, sandy rock known as the Pound Quartzite. It was here that a varied array of late Precambrian animals was discovered, some of them remarkably like the ones discovered in Charnwood Forest on the other side of the globe. This ancient assemblage of fossil animals has come to be known as the Ediacara fauna, and, in consequence, the late Precambrian age is sometimes called the Ediacaran. Mr. Sprigg was a diligent collector who recognized the organic nature of these remains, which, like the Charnwood ones, were preserved as impressions cast by the sandstones of the Pound Quartzite. They were soft-bodied animals, preserved in a most unusual way. R. C. Sprigg had already described some of the fossils as early as 1947, but regarded them as being early Cambrian in age—because at that time the existence of still older large fossils was considered virtually impossible. He published several papers in the *Transactions of the Royal Society of South Australia*, a journal received by few libraries. As a result, these animals did not attract the attention they deserved, and were shelved away with other jellyfish-like curiosities in one of the obscure and unvisited corners of palaeontological consciousness. Publicity, sad to say, is often as important as truth in science. It was really only after the discovery of other Ediacaran fossils in the 1960s, and their advocacy by a forceful professor, Martin Glaessner, combined with a general realization that they underlay rocks of Cambrian age and had to predate them, that the Ediacara fauna started to approach its present fame. One of the animals was eventually named *Spriggina*.

Spriggina was smaller than *Charnia*, with a crescentic "head" and a slightly flexuous body divided into numerous segments. Clearly, this was an animal with bilateral symmetry, like many multicellular organisms including ourselves, and unlike, for example, jellyfish, which have *radial* symmetry, like a bicycle wheel or a pork pie. There were also several kinds of these in the Ediacara fauna, some of them as large as a plate. Many of these jellyfish-like animals have radial "spokes," or concentric ridges. Their circular impressions look almost splashed over on the rock surface, as if they had been created by pouring liquid plaster from a height of several feet. Some show evidence of a stem-like attachment centrally, and *Cyclomedusa* and *Ediacara* are thought to have been attached, sedentary, rather than free-floating in the fashion of the planktonic, bell-like jellyfish that I saw cruising off Spitsbergen. Jellyfish were perhaps the least surprising animals to find in the Ediacara fauna. After all, most evolutionary biologists had already placed them on a low rung on the ladder of life. This was not merely due to their lack of substance and their radial symmetry. Their tissue organization is also minimal: they lack blood, they have a primitive nervous system. Their body wall is composed of only two layers of cells separated by "jelly"; the inner layer lines a stomach cavity, often little more than a pouch-like bag, which also absorbs the food; the mouth is a simple opening with oral arms surrounded by tentacles—but these can be fantastically varied, from short stubby fingers to elegant and plumose fringes. The tentacles of ocean-going medusae grow to yards in length. They carry the stinging cells (nematocysts) which are typical of this whole group of organisms. The venom they can deliver varies from one kind of jellyfish to another, but swimmers off the Great Barrier Reef know to avoid the "blue bottle"—the sting of which is a reliable killer. One of Sherlock Holmes's last cases was "The Adventure of the Lion's Mane" concerning the lethal properties of "the fearful stinger, *Cyanea capillata*"* on the Sussex coast. "It did indeed look like a tangled mass torn from the mane of a lion. It lay upon a rocky shelf . . . a curious waving, vibrating hairy creature with streaks of silver among its yellow tresses. It pulsated with a slow, heavy dilation and contraction." Such dilation and contraction serves to buoy the medusa in its slow oceanic passage. I saw such pulsating monsters leisurely cruising the Arctic seas as I leaned over the side of the boat on my first trip to Spitsbergen; I could have been looking back through 700 million years. The stings serve to paralyse the jellyfish's planktonic food, which can then be passed into the stomach cavity for digestion. Such a simple exploitation of the plankton would have been possible from any time in the Proterozoic after the oceans were colonized by simple algal cells that took to free-floating life. The medusae were probably among the first harvesters of photosynthetic bounty. Another group of jellyfish took to life on the sea floor, electing instead to harvest the plankton from the security of a permanent base. Polyps like these included sea anemones, and animals that stabilized themselves on sandy sea floors. They are commoner in the Ediacaran fauna than their floating relatives. The late Precambrian must have been a trial by nematocyst, a stinging forest. Relatives of these animals still survive in profusion, like the bacteria and algae before them, proving again that simple organisms can be durable.

But others did not survive. *Spriggina* was one. Another was the largest creature in the Ediacara fauna, a bizarre object called *Dickinsonia*, first named by Mr. Sprigg. It is a flattish disc with an elliptical outline like a pitta loaf, divided crosswise into numerous segments or compartments; at the narrow ends of the ellipse the divisions swing round to become parallel to the long axis of the ellipse. *Dickinsonia* grew by progressively adding divisions, chang-

---

*Arthur Conan Doyle got his rendering of biological scientific names correct; the name should be italicized, and the genus name is always capitalized, the species name not! Because it is a name, *Cyanea* does not require "the" before it, any more than does George Ford or Bill Blenkinsop.

ing by degrees from the size of a coin to the size of a large platter. Like that of *Spriggina*, the body is symmetrical, the left-hand side being almost a mirror image of the right. Along its midrib there is a narrow line. But you would be hard pushed to recognize either a head end or a tail, or an up or a down. Unlike *Charnia*, there is no stalk. In the yellow sandstones, *Dickinsonias* look almost like giant handprints. In life, they probably simply lay upon the surface of the sandy sediment, like soft, structureless and inanimate flatfish. There is nothing closely similar alive today.

There are stranger things still. *Tribrachidium* comes as a medallion the size of a small coin, but its surface is sculpted by three radiating ridges which curve as they approach the edge of the disc; they bear a remarkable resemblance to the flag of the Isle of Man. Well-preserved specimens show a network of finely branched "veins" between the main "arms." *Anfesta* is a similar creature discovered by Misha Fedonkin on the White Sea, in northern Russia, which has straight rather than curved ridges. In a sense these are radially symmetrical animals, although the symmetry has become twisted. *Arkarua* is another disc with a five-fold array of "arms." But there are other bilateral animals with suggestions of a "head" and body: *Parvancorina* and *Vendia* have a kind of crescentic "head" and a midrib behind; in the latter there are half a dozen lobes or segments on either side of the axis.

These animals have been worth describing in detail to reveal the important fact that they display an extraordinary range of shapes and sizes—there is not just one multicellular soft-bodied animal, but a whole gallery of them; with the exception of some of the jellyfish, they are all mysterious. We have upright pens, prostrate radial discs, bilateral animals with no front and back, and others with head ends and bodies behind . . . some are large, and many probably did not move. Any notion that some kind of gigantic slime mould of infinite plasticity produced a single ancestor for all higher animals (metazoa) is clearly absurd. There was much variety, and new discoveries are still being made every year. The Ediacara fauna has now been found in many sites: particularly rich faunas have been discovered in Russia, especially Siberia, some of them better preserved than the Australian fossils. They have turned up in Newfoundland, in China, in California and in West Africa. Where these kinds of fossil faunas occur they are found, in most localities, not far below the earliest rocks that yield familiar skeletal fossils which have been known from Cambrian age rocks for more than a century. Their proper position in the geological column is assured.

There are some still earlier multicellular organisms, most of which have been interpreted as plants rather than animals, not least because they are often preserved as carbon impressions in the fashion of later plant remains. They are rare and sporadic in their occurrence. The oldest known are curled-up spirals about a centimetre across named as *Grypania*, which extend back about 2,000 million years. This provides a minimum date for the inception of multicellular life. There is a strange balloon-like organism, tethered by a string-like stem called *Longfengshania* (it is, of course, from China). Other organic shapes, variously curled or cylindrical, have been discovered and named; nobody really knows if they represent true "species." They seem to have had unusually large cells. Why it should take something like 1,400 million years from the appearance of many-celled life until the varied animals of the Ediacara fauna is still disputed. It is beyond question that there was a massive lag between the appearance of animals and plants. The simplest explanation is that the "blue-greens" and algae had not yet finished their work of oxygenating the atmosphere to the level where larger animals could respire. Animals could not summon up the energy to work.

But when the Ediacara animals appeared it is certain that some kind of threshold had been crossed, and animals would never go away again, although they were to suffer reverses that contribute much to the drama in this narrative. It was still true that bacterial mats were widespread, because there is no evidence that grazing animals had yet appeared to destroy them. It has been claimed that the preservation of the Ediacara fauna itself *is* evidence for the existence of such mats, which retained sufficient rigidity to take on the fugitive impressions of soft-bodied animals which would otherwise have been destined to leave no trace. Why, after all, are there no "Ediacara faunas" of jellyfish and their ilk in sandstones of younger age? We know these animals were there, for they thrive still. There must have been something special about late Precambrian times, and that was probably the widespread endurance of the same mats that had been crusting the shores of arid lands for more than 2,000 million years previously.

I HAVE BEEN AVOIDING the most difficult question, which is what the Ediacaran animals actually *are*. Jellyfish we know, and there are those who believe that *Charnia* is related to the living sea pens; but many of the others are conundrums. Are they monsters, hopeful ones that were seminal to what followed, or doomed experiments committed to oblivion? Samuel Taylor Coleridge's Ancient Mariner had a vision of the weirdest and most monstrous excrescences of nature:

1. En route to Spitsbergen—a rare sight of Bear Island

2. Ny Alesund in the late 1960s—a northern outpost of semi-civilization in Spitsbergen

3. Pitching tent in the far north—guy ropes attached to driftwood spars

4. Rock section along the bleak shore—the rocks get progressively younger to the right following the succession of strata.

5. Sanitary arrangements—our makeshift commode with a sea view

6. Frozen underclothes: in the icy breeze they billowed out before freezing. Note the long shadows produced by the low Arctic sun.

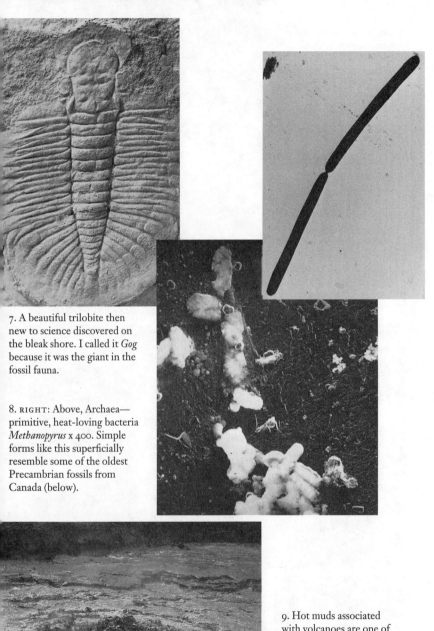

7. A beautiful trilobite then new to science discovered on the bleak shore. I called it *Gog* because it was the giant in the fossil fauna.

8. RIGHT: Above, Archaea—primitive, heat-loving bacteria *Methanopyrus* x 400. Simple forms like this superficially resemble some of the oldest Precambrian fossils from Canada (below).

9. Hot muds associated with volcanoes are one of the natural habitats of the thermophilic bacteria. This is the edge of a sulphurous spring near Mount St. Helens; a sulphurous crust lines the shore.

10. A "black smoker," a submarine vent belching forth hot volcanic fluids, developed along a mid-ocean ridge—one contender for life's cradle.

11. Stromatolites—the structures generated by bacterial mats, the most ancient community. Main picture shows recent stromatolites forming in Shark Bay, Western Australia. Fossil stromatolites in limestones of the Precambrian Bulawayo Group are shown in pictures at bottom.

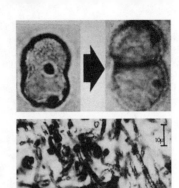

12. These simple cells are Precambrian fossils preserved in cherts. These Bitter Springs Chert (850 million years old) examples actually show cells of what are identified with a living group of algae in the process of division (right, top). Thread-like forms from the Gunflint Chert, Canada (right), are less than one hundredth of a millimetre across. Below, living "blue green" *Schizothrix*

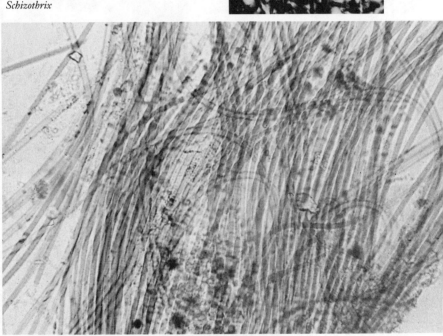

13. Ancient Precambrian stromatolites, now part of the Shores of the Great Slave Lake, Canada

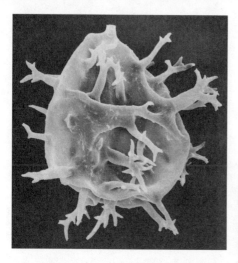

14. Acritarchs—probably the remains of single-celled oceanic planktonic algae—appear in the later Precambrian. These examples, magnified approximately 2,000 times, are from the Devonian.

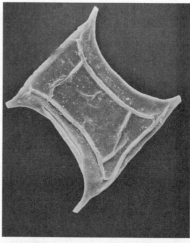

15. Slime mould *Fuligo septica* in its habitat on the forest floor

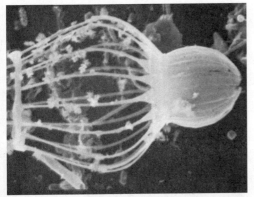

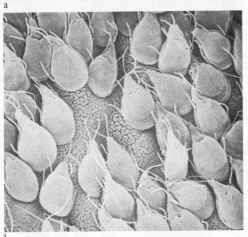

16. A cavalcade of protists—such fundamental variety in single cells makes any simple division of life into animals and plants implausible. The choanoflagellate *Diplotheca* (a) is now considered to lie at the base of the whole "tree" of animals; *Giardia* (b) is a controversial protist here shown living in the gut of a mouse; (c) dinoflagellate; (d) an amoeba with one of its tentacle-like pseudopodia extended; (e) paramecium showing its cilia

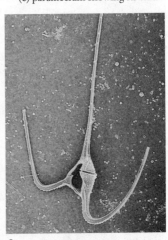

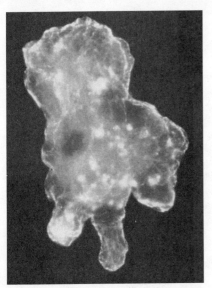

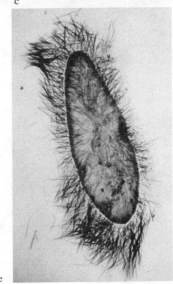

17. Living primitive jellyfish (*Mastigias*)

18. *Dickinsonia*—one of the larger soft-bodied animals from the Ediacara fauna

19. *Spriggina*—an enigmatic soft-bodied animal from the Ediacara fauna of the Flinders Ranges, South Australia

> Yea, slimy things did crawl with legs
> Upon the slimy sea.

Somehow it is the legs that make them so horrific, and there are no limbs preserved in Precambrian animals (most of them certainly did not have any). Slime there was in plenty, for it was both armour and adhesive for bacteria. But Ediacaran animals are neither bacteria nor romantic monsters. *Spriggina* and *Parvancorina* have been claimed as ancestral arthropods—those jointed-legged animals that include everything from crabs to beetles. *Tribrachidium* has been thought of as a kind of ancestral sea urchin. But if they are such seminal organisms there is something awry, because they do not resemble the ancestors one might have predicted; they are too odd and specialized. This has led to claims that the Ediacara fauna is entirely a sideline to the history of life—*none* of these animals was our ancestor, but all of them were closely related one to another, comprising a kind of dominant life form before all the animals we know today had appeared in any recognizable shape.

Professor Dolf Seilacher of the University of Tübingen is the chief proponent of this theory. Seilacher is a marvellous observer of things. He has a habit of peering minutely at the most commonplace items in rocks and perceiving things in a way that no naturalist has done before. I once sat with him on a seashore in Oman as he explained how molluscs catch food only as the waves drain over them. In the Australian desert he described how trilobites scratched their grooves in sandstones, or how aboriginal people made stone tools. The patterns of galaxies was clarified for me, and the names of stars recounted. Seilacher is tall, with short grey hair that just stops short of a crew cut, and enormously active for his seventy years. He has only two weaknesses: one is for cigars, which are enormous and pungent; the other is a didacticism which comes from a conviction that he is right about every issue. One's most perspicacious observation is countered by: "No, you see . . ." and thence to a lucid explanation of why you have to be wrong. I am not sure whether this is a mannerism of great German professors in general; in all other respects he is a man of the utmost charm and generosity. The best place to be, he maintains, is somewhere nobody has been before—intellectually speaking, of course. This is where his explanation of the Ediacara fauna comes in: according to him, they all—*Dickinsonia*, *Spriggina* and the rest—reveal a different structure from that of all other organisms, a kind of internal quilting. The segment-like partitions that Mr. Mason noticed on *Charnia*, or the bands on *Dickinsonia*, are the traces of internal walls that maintained turgor when the animal was alive. This gave substance that could be impressed on the surface of mats. In this way he sought neatly to explain both the preservation of the Ediacara fauna and its peculiarities. These curious organisms were passive; their large surface areas were ideal for harbouring symbiotic bacteria—maybe photosynthetic, or perhaps sulphur bacteria in other situations. Such passive creatures did not need much oxygen as much of their metabolic work was done for them by their bacterial inmates—in a sense this was a reprise of Lynn Margulis's theory of engulfment of small bacterial organisms by larger entities that lay close to the origin of the eukaryote cell. The Ediacaran world was a different one, a world in which large, but inactive animals untroubled by grazers or predators lay out and basked in the shallow seas under the late Precambrian sun (or deeper in sulphurous seas)—animals, yes, but barely; organisms that lay almost androgynously between the kingdoms.* This world has been called the Garden of Ediacara, which carries with it a sense of a time before the fall, when true animals—predators, exploiters, fuelled by sex and aggression—came to disturb the ancient symbiotic peace with their loutish ways.

It is, like all Seilacher's theories, both highly original and plausible. It is also, however, unproductive, since it merely serves to discourage any attempt to "fit" the Ediacara animals into the world of the Cambrian and beyond. In a paradoxical way it serves to re-erect that wall between the Precambrian and everything that came later that Charles Darwin had wondered at more than a century ago. Much of the story I have told so far has been about the demolition of that wall in the wake of new discoveries.

When I was in Australia, and in an outback not so different from the locality where *Dickinsonia* was collected, I was shown carvings made in the rocks by aboriginal nomads. Spirals and simple shapes, ellipses among them, that were almost reminiscent of Ediacaran organisms. I wondered idly whether those observant artists had seen the puzzling fossil marks on the rocks and accorded them magical significance. But there are in fact still people living in the desert who can explain the meaning of the carved symbols in the context of their Dream Time myth. However, imagine if the last aboriginal had vanished from that dry land; then the meaning of these drawings would truly be lost for ever. It would be a time beyond recall where imagination alone could pencil in meaning for lost symbols, and my story would have been a plausible one. Doubtless there would be other, more coherent stories,

---

*As this page was being written another heterodox opinion about Ediacara was published—explaining that the "animals" were actually lichens—thus invoking a second group of organisms that is neither one thing nor t'other.

but without a link to other times and other cultures there would be no way of testing their truth. In an analogous way, if the Ediacara organisms are designated as wholly "other," they can be shuffled into an obscure, if fascinating, sideline of history, where, on principle, nothing in our own world would be relevant to their interpretation. It would be difficult to deny that many of the late Precambrian animals *are* hard to interpret. Seilacher was right, as so often, in noticing that the left-hand and right-hand sides of the segments in these animals often seemed out of kilter—as if they had been shifted half a step. But equally, it seems perverse not to recognize the "head" and body division of *Spriggina*—features which place this creature firmly in the realm of bilateral animals. Maybe the answer is that some, but not all, of the Ediacara animals were truly sidelined curiosities, while others were on the road to something we might find more familiar. In historical controversies the compromise answer is often the one which proves nearest to the truth in the end, although without hypotheses originating at the edge of knowledge fewer new perceptions would be born. It is at this edge that Dolf Seilacher has always worked.

Some of the most peculiar and distinctive of these animals failed to survive the Precambrian. Certainly there are no *Dickinsonias* or *Tribrachidiums* from younger rocks. There was an episode of extinction between the ancient world—the world of the previous 2,000 million years—and the new world order that began in the Cambrian period. Our knowledge of just how great this extinction was will obviously depend upon how the Ediacara animals are eventually understood. But if it is recalled that a surprising continuity with the living planet was typical of the lowly bacteria and algae that dominated the early history of life, then this is an important development: for the first time, something has been taken away, never to reappear; previously, change was wrought by the successive *addition* of more and more kinds of organisms. It is beyond dispute that one of the greatest transformations in the history of the planet took place at the end of the Precambrian and close to the base of the Cambrian. This is when Earth moved from childhood to adolescence, and when the lineaments of adulthood became inscribed upon its character. The changes happened quickly. So far, I have brandished great swathes of time with an insouciance that comes from leisure, from knowing that biological turnover was ineffably slow, and that profound changes were wrought over great measures of time. This is true even if the dating and measurement of significant events (the appearance of the first eukaryotic cell, for example) are often approximate, because events so remote are inevitably poorly pinned down in absolute time. There was nothing urgent about the slow cleansing of the atmosphere and the cumulative, photosynthetic build-up of oxygen. But at the end of the Precambrian there was a revolution.

Within a few million years animals with shells appeared, and, what is more, a considerable variety of them. Included among them were forms which we are familiar with in the seas today: if not their direct ancestors, then at least they are the long-lost cousins of much of the marine life that still flourishes in reef and rock pool. But a host of other animals were there, too, some of which are not to be found in any rock pool, or picked out in the headlamps of a bathyscape. They are as mysterious as anything from Ediacara.

THE STORY HAS BEEN READ from the rocks in many parts of the world besides Australia. In China, in the province of Yunan, there are wonderful rock sections which run through from the Precambrian to the Cambrian; in places along both the east and west coasts of North America there are sections nearly as good. There are small quarries in Sweden, and whole riverbanks in the Siberian Arctic, stretching for eighty kilometres or more. In England there are scruffy pits and stream beds in Shropshire, close to Church Stretton in the Welsh Borders. But it is on the island of Newfoundland that the base of the Cambrian has been officially defined, after much tetchy academic bickering. I am delighted by this decision (made in 1990), not least because it adds another reason to visit this extraordinary island. On the map it seems to be tethered off North America, but this hardly reflects its true origins. In spirit, Newfoundland is a piece of Ireland towed to the other side of the Atlantic Ocean, with a leavening of Cornwall, and perhaps a pinch of France in one or two places. The "Newfie" accent is an arcane hibernian one, peppered with "by jeez" and references to dories and cods' tongues. The Rock (as they call it) was one of the first places in North America to be colonized; it then remained culturally frozen almost until the 1940s, when it joined the Canadian Federation. It seems curiously appropriate that this island should also furnish ancient fossils, which have survived unaltered in its cliffs.

Newfoundland place-names reflect its history: there are towns like Trepassey, which could easily be Cornish, and St. Shotts—for here, too, as in that other rather odd part of the world, there are saints never heard of anywhere else. It is said that the natives of St. Shotts, on the south coast of Newfoundland, formerly indulged in the notorious Cornish habit of wrecking—luring passing boats on to the rocks by means of false lights.

Perhaps the inhabitants needed something to amuse them during the extraordinarily long winter, which still seems to be in slushy session as late as May.
Newfoundlanders will tell you that St. John's is on the same latitude as Paris,
and this is undoubtedly a fact. Were it not for the global weather system it
might be a *boulevardier's* paradise. In fall the cliffs are heavy with blueberries,
and windless days are occasionally perfect: tranquil and warm. One can
understand why some early colonists were tempted to stay—but only long
enough to learn the dismal climatic truth (they went on to found Maryland).
Inland, the plants that grow best are stunted firs, and birch and aspen; the
wilderness can be quite impenetrable. But the coastal scenery is everywhere a
wonder, a matchless alternation of sheer precipices with gentle coves, rich
with inlets and spits, which the Newfoundlanders call "tickles." I have visited
Goose Tickle and Dildo Run (the origin of the latter name is obscure). On
the sides of the coves there are small fishing villages with square, clapper-
board wooden houses painted white or blue or pink, and often as simply
designed as a child's drawing: door in the middle and four windows on the
flat, unsophisticated frontage. By the sea there are the harbour buildings in
plainer wood, often built out over the water on crazy stilts. Cod was the staple
crop, so much so that when I was there in the late 1970s and early 1980s it
was simply synonymous with "fish"—ask for a fish and cod was what you got.
The only other species recognized was salmon. Now overfishing—not by the
Newfoundlanders but by the factory ships equipped with sonar—has meant
that the boats are grounded, and these bustling and distinctive villages are in
suspended animation once again. But they are durable people, Newfoundlanders, and now at least they have the dole to sustain them where their fore-
fathers would have starved. They will not abandon their lonely shores.

On the Burin Peninsula, in the south of the island near Mistaken Point,
the cliffs that dip into the sea are covered with prints of fossils of the Ediacara
fauna. It is a good place to work, for the stiff breeze keeps the mosquitoes and
blackflies at bay. These fossils are a little different from the Australian ones,
since they evidently lived in deeper waters and are distinct species. None the
less, there are animals that are clearly like *Charnia*. The rock is desperately
hard, and there is no possibility of collecting the creatures to take them back
to the laboratory. Instead, pots of latex solution are lugged over the cliffs, and
painted on to solidify in the wind. After several coats have been applied the
casts can be peeled off to give a perfect replica of what has been weathering
out slowly for hundreds of years. Then, around the coast, you can follow the
rock successively upwards, looking at progressively younger slices of history,
exactly the procedure Geoff and I had followed in Spitsbergen along another

bleak and lonely shore. There is no substitute for sharp eyes and the tap, tap of the geological hammer. The first thing to notice is that all traces of the Ediacara animals disappear. Next, something extraordinary happens. The palaeontologist carefully crawling or plodding with his nose pressed to the rocks begins to notice marks on the sandy or silty surfaces—not shells, but regular swirls or sinuous wiggles a few centimetres long. They are formed from the sediment itself, an impression on the surface maybe, or occasionally tube-like structures. These are, he begins to understand, the marks made by wandering or possibly burrowing organisms, and definitely the kind of thing only produced by real, bona fide animals. You can go to any sandy strand today at low tide and see shapes which are not dissimilar; Dolf Seilacher would know the names of all the animals that made them. These are trace fossils: the petrified evidence of the activities of animals squirming in the sand or mud 600 million years ago. Whether or not we can find the bodies of the animals that produced them (and we can't), we still know that the appropriate kinds of animals must have been around to make them. Only things that walk, slide, crawl or burrow are capable of making traces. Like the footprints left behind at the scene of a crime they speak of deeds that might one day be called to account.

Chapter 4

My Animals and Other Families

I N THE ROCKS ABOVE THE TRACES and tracks left by soft-bodied animals, there are tiny tubes. In hand specimen they often show as little more than mottling on the bedding surface of the rock. But where limestone formations are developed at precisely this geological interval, as they are in many parts of the world—in China, Russia and Australia—the rocks can be dissolved in dilute acid and the tubes are recovered from the residue by sieving. Many of them are composed of calcium phosphate, which does not dissolve in the acid. A glance quickly reveals that there are several more shapes than simple tubes—there are coils, and little asymmetrical scales, and twisty shapes looking much like the favourite fried snack the Spanish call *churros*. There are a few things that are obviously tiny shells. Then there are some little spines, often branched, which are recognizable as the skeletal components of sponges, known as spicules. Tube, shell and spicule are all hard—animals had evidently learned to secrete skeletons. This means that we have passed into the Cambrian period, which is marked for the first time by shells. And most extraordinary of all is that in *all* the appropriate rock sections around the world, whether they are exposed in the bleak cliffs of Newfoundland, or the hot, dry creeks of Australia, the story is the same. There is a sudden appearance of what palaeontologists call "small shelly fossils" above the Precambrian rocks. Since the earliest fossils include closely similar species throughout widely separated and scattered localities, the conclusion is inescapable: this momentous development of shells happened all over the world within a short time, just a few million years. From now on, for the next 540 million years, the fossil record would become prolific, because hard minerals are routinely preserved. Animals with skeletons appeared for the first time,* and with them the first hints of all the achievements of biological engineering which were to come. In the biosphere size really matters, and large size requires mechanical support. The possibility of the blue whale was implicit in these first humble remains.

The very earliest tube of all is now known to just predate the great diversification at the base of the Cambrian. It is called *Cloudina*, and its name celebrates Preston E. Cloud, one of the pioneer investigators of this great watershed in the history of life. It is an odd little thing, just a millimetre or two in length. Its mineralized skeleton was probably composed of calcium carbonate (calcite), like most sea shells today. Sheets of such calcite were arranged in stacks rather like those piles of paper cups that are found beside soda fountains. Nobody knows to what living animal *Cloudina* might be related.

In the rapid diversification of the "small shellies" at the base of the Cambrian there are many other unsolved identities. For example, many different things could live in tubes, and the shapes of the tubes provide no clue as to their identity. As at one of those ghastly parties where everyone is supposed to turn up as Count Dracula, the more effective the disguise the harder it is to recognize who anybody is. Ellis Yochelson, a grizzled and often sagacious mollusc expert at the Smithsonian Institution, in Washington, DC, was fond of saying, "What the hell, call them all worms," which has some pragmatical advantages, although it does rather beg the biological questions. In some cases the disguises have, in part, been penetrated. It used to be thought that the phosphatic skeleton of many of the early shells was an original feature—and hence that this material (which is related to the material of our own bones) was at a distinct advantage in the genesis of mineralized tissue. Now, after a careful study of the microscopic preservation of the shells, tubes and plates, investigators are convinced that in some cases the phosphate actually *replaces* an earlier shell, which was likely to have been composed of calcium carbonate, much the commonest shell material among living animals. There were rather more phosphatic shells at the start of the Cambrian than there are now, to be sure, but calcium carbonate was found with similar frequency

---

*The technical term for this phenomenon is "biomineralization."

at the time of the inception of shells. Sponges and certain single-celled animals, like radiolarians, used silica (SiO_2) for their skeletons, just as they do today. Curiouser and curiouser, for it seems that all common skeletal materials got going together, synchronously, even though common sense might tell us that they should require different conditions for their secretion by living tissues.

Now there are some more puzzles. It seems possible that the early shelled animals were not as small as the tiny fragments recovered in sieves might suggest. In 1990 a remarkable creature several centimetres long was discovered in the Lower Cambrian rocks of Greenland. This strange *Halkieria* was composed of literally hundreds of tiny plates arranged in ranks, like chain mail, over its back. Even more extraordinary, there were two additional, very much larger, nearly circular plates positioned at either end of this tessellated oddity. The small plates were totally different in shape from the larger, and there was no reason to associate them until the amazing find was made. The isolated bits—or sclerites—had been discovered many years previously, and had already been given different names. Whatever this odd animal was, it evidently fell to pieces when it died, and in this fashion one individual spawned numerous fossils. The new discovery thus banished several mystery organisms at a stroke, only to generate a single, more obdurate enigma. It also highlighted the possibility that other small shelly fossils might be pieces of something larger. The distinguished Swedish palaeontologist Stefan Bengtson, for example, believes that *Halkieria* was a mollusc, and records an important stage in the evolution of shells: he thinks they started out as little elements which became fused to form a continuous mineralized covering. One of the most primitive living molluscs, the chiton, is still composed of a small number of rows of plates: *Halkieria* allegedly recorded an even earlier stage when the rows were more plentiful. It is not an entirely satisfactory answer, because it still does not explain the big plates that *Halkieria* also carried.

It is now becoming clear that other species of the earliest skeletized animals were merely sclerites belonging to something larger. There are the curious tommotiids, genuinely phosphatic animals with very irregular little plates, but so far nobody has found more than one or two plates hanging together. Sponges are not so arcane because living examples still put their skeletons together from fine struts and rods (spicules). They interlink in much the same way as I described organic molecules bonding to form larger and more complex chemical constructions. When the beams of a bathyscape pick out the inhabitants of the ocean floor in the abyss, a remote world far beyond the reach of any daylight, one of the organisms that does not run away is the glass sponge. They are perforated bags, tethered to the sea floor, and are so called because their spicular skeleton is siliceous, like glass, but spun into a tracery so delicate and subtle that no glass blower could ever imitate it. Many of the spicules are six-rayed. Sponges have chambers within their bodies through which seawater passes, and where any edible particles are removed. The walls of the chambers are lined with cells carrying whip-like flagella, which, as the name implies, beat to encourage currents into the chambers. Individually, these cells look very like free-living protist species, and the sponge is rather a loose association of such cells—forming a kind of proto-tissue. Some sponges will even re-assemble themselves if they are mangled up into tiny pieces. It is scarcely a matter of wonder to find these lowly patchworks of cells among the earliest fossils—but their endurance to the present day *is* remarkable, not least because you can scarcely distinguish a spicule plucked from a *Hyalonema* caught in a bathyscape's lights from a specimen dissolved from a Cambrian rock. Like stromatolites, glass sponges have lived on in corners of the world which have endured with little change. The familiar bath sponge does not have siliceous spicules—it would be an agony to wash with if it did. It is composed instead of a tough organic material, spongin. The prodigious capacity of the internal chambers to hold water has bred the metaphor which links sponges with freeloading, which is rather unfair: they are certainly not parasites.* Worms, toads and dogs have been employed in a similarly derogatory fashion, and so far as I know the first two lead exemplary lives (with the possible exception of the bloated cane toad, although even that creature has its enthusiasts); dogs are no more ignoble than their masters. I notice that in current genre movies the bad guys are cursed with the ultimate insult: "You . . . *slime!*" It will be recalled that slime had a crucial part to play in the stabilization of bacterial mats, without which this book would have been very brief. I can only conclude that insults are proportional to the position of the organism on the evolutionary ladder, and have nothing to do with its virtues or failings. The game is given away by the great division between animals with backbones and those without—for who wants to be described as "spineless"?

Among the earliest Cambrian small shells there were a number that were certainly not a mere scrap of something larger. There are several kinds of diminutive molluscs, distant relatives of living snails and clams and slugs.

---

*A pedantic reader might insist that the sponge *Cliona* effectively *is* a parasite, since it bores into and weakens the shells of the clams it colonizes—often fatally.

Some of these almost certainly grazed the algae which formed mats, and are implicated in the decline of that quintessentially Precambrian community. There were the earliest brachiopods, another kind of animal enclosed in two shells, which we shall discover again among the rich marine life in ensuing periods of geological time. Brachiopods filtered out small organic particles from the seawater. In Australia there may be the earliest corals, which did likewise.

Just a little bit higher again in the rock sections in Newfoundland, China, Siberia, Morocco, Australia—and many other places—the small shells are succeeded by an even greater variety of larger, skeletized organisms. Most prominent among these are the trilobites. These are arthropods, animals with jointed legs. Spiders, flies, fleas, crabs (both kinds), mites, beetles without end—indeed, almost everything that crawls, stings or scares is an arthropod. They are not exactly primitive organisms in comparison with most of those mentioned previously. They have a fully developed nervous system, a brain of sorts, eyes, limbs, gills, antennae, and some of them hunted. Trilobites are *my* family. They probably adopted me long before I went to Spitsbergen, when, as a lad armed with optimism and a coal hammer, I tapped along Welsh cliffs and riverbanks in their pursuit.

The three divisions that give these animals their name divide their hard, calcite carapace lengthways into a prominent axis which housed guts and brain, flanked on both sides by flatter lobes. There is a semicircular head region equipped with eyes, which often have a crescentic outline, and backwardly directed spines at its lateral corners. Behind the head a thorax includes a number of articulated segments, thus lending the body a certain flexibility. It is this flexible thorax that gives the trilobites their superficial resemblance to woodlice (also known as slaters, or pillbugs, according to where you were raised). At the hind end there is a tail, often a shield somewhat like the head, which in some species is large, but in primitive species is usually very small. The carapace is composed of calcium carbonate, like those of clams and brachiopods. It is this shell which guarantees the trilobites such an excellent fossil record. For many years they were thought to be the *only* Cambrian arthropods. In any case they are far and away the most common as fossils, and the shallow Cambrian seas must have been fairly swarming with them. The sad fact is that some of the most important parts of the anatomy of trilobites are hardly ever preserved. These are the limbs, which the overlying calcite carapace served to screen from harm. The limbs themselves were not covered with calcium carbonate, and hence left little that could be routinely preserved in the rocks. Rarely, though, they *are* found as fossil impressions, and these show that trilobites had many pairs of legs, a pair to each segment, as well as antennae at the front. The limbs are themselves further divided into many small segments, often with bristles between them. If you pick up a small leg of a lobster you can see just the same arrangement of bristles close to joints, which serve to detect movement, and sometimes also to "smell," because some of the hairs are tipped by sensitive little organs. So if you had scooped up a trilobite in your hand—and some of them would have been sluggish enough to catch easily—you could have turned it on to its back and watched the limbs scrabbling desperately to gain a purchase and scuttle away back into the sea. The eyes were unique, as well as being the oldest visual systems known. They resembled those of flies or lobsters in being compound—that is, composed of many lenses (I have counted up to 3,000 in a single eye). But they were unique in that each lens was a calcite rod. The stuff of common or garden limestone was used to see with—its hardness and durability permitted the preservation of these early eyes. The optical mechanism was not complicated. Each lens was carefully orientated in such a way that one particular crystallographic axis ran along its length. Calcite is a rather complex mineral, but its transparent crystals have this one preferred axis along which light passes in straight lines, unrefracted. Thus, calcite lenses "see" in the direction of their long axes—hence it is possible to reconstruct the fields of view of animals dead for 500 million years or more, the most literal way of seeing into the past. The image produced by the eye was a kind of mosaic.

In his book *Supernature* Lyall Watson regarded the trilobite eye as too sophisticated for its age, a precocious structure in advance of its appropriate use. On the contrary, it is as clear as calcite that the eyes were needed: acuteness of vision was necessary either for seeing the approach of hunters or to pursue prey. The tracks of Cambrian trilobites have been seen converging upon a worm trail—and only the trilobite emerges after the engagement. It was certainly not the worm that was doing the hunting. It seems that even one geological second after the appearance of Cambrian skeletons there also arose something recognizably like a modern marine ecology, with hunters and hunted, grazers and filterers. Stefan Bengtson has even recognized what look like the borings of predators into the very first shells belonging to *Cloudina*. Far from being a precocious luxury, the trilobite eye was a piece of equipment necessary to cope with a hostile world. Thus it was that the world of Ediacara and the Precambrian passed away, and another great step across a threshold in the history of life was taken, never to be retraced. Cellularity had become a food chain, gobbling began, and voracity has never gone away. If there were a point in history at which Tennyson's famous phrase "Nature, red in tooth and claw" could be said first to apply, this was it, not the age of the dinosaurs—still less that of mammals. Perhaps the first predation happened as the result of an erstwhile cooperation that went awry: the biographical details are not recorded. The era of photosynthetic passivity and peaceful coexistence among bacteria and algae had passed from the Earth, and the hierarchy of power has never subsequently been forgotten.

It would be misleading to claim that the appearance of mineralized skeletons was the crucial event which reconstructed the living world; for this is only part of the story, although an important part. After all, most marine animals do *not* have hard skeletons, even today, and these creatures are just as important in the ecology as those that do. In a few places there are special geological sites where we get a fuller insight into what was living in the Cambrian. These are places where shells are accompanied by impressions of soft-bodied animals, often preserved in miraculous detail. Dolf Seilacher called these special rock beds *Konservat Lagerstätten*, a pompous-sounding term which has none the less passed into general scientific usage. And the most famous of these is without doubt the Burgess Shale.

High in the Canadian Rockies of British Columbia, in the Yoho National Park on the flanks of Mount Stephen, there is a small excavation on a steep hillside. A modest apron of spoil reveals the quarrying activities of several generations of palaeontologists. It is a hard march uphill all the way from the town of Field, and it is cold until well into summer, and frequently inhospitable even then. The site was first made famous in the late nineteenth and early twentieth century by Charles Doolittle Walcott, a man with as inappropriate a middle name as could be devised. Walcott did almost everything. He described Cambrian fossils from all around the world, including China (if placed side by side his books must stretch to a yard of shelf space); he was a considerable administrator; and he discovered the Burgess Shale. The apocryphal tale has it that his mule lost a shoe at the critical spot, but he was already prospecting these mountains for their ancient, middle Cambrian spoils, so any intervention by serendipity was mediated by design. Walcott noticed something glistening on the surface of a black slab of shale; only when the light struck the bedding surface from a particular direction did the remarkable fossils display themselves. They are revealed as flattened, silvery films, which glisten momentarily when the light is exactly right—otherwise they are dark black on slightly lighter black, and very hard to see. Walcott was quick to appreciate the importance of what he had found, which was nothing less than a completely preserved biota of larger organisms from the Cambrian. The animals had been buried so fast in their entombing sediment that their soft tissues had, for once, escaped decay. He traced the mother lode up the hillside, which is how the quarry came to be opened. Hundreds and hundreds of specimens were later collected, and stored in the US National Museum in Washington, DC. Walcott thought the world might be sceptical of the reality of his astounding discovery, and as a precaution he sent batches of specimens to other museums around the world so that the curators there, too, could be dumbfounded. He could spare a few specimens, because some of the species were not at all rare. When I made a visit to the site (which is now rigorously protected) I was able to find specimens of Walcott's delicate "lace crab" *Marrella** after a few minutes' searching among the slabs lying on the slopes. Clearly, they were common enough to throw away.

The black shales preserved all the limbs and the delicate antennae of a trilobite, *Olenoides*. It preserved sponges and seaweeds and brachiopods and jellyfish. It preserved one *Charnia*-like survivor from the Ediacaran days. It preserved an array of worm-like animals, some of them readily recognized as belonging to living groups like annelids and priapulids, but also others which were much more puzzling, and possibly allied to the early Cambrian *Halkieria*. It preserved at least one animal much like the laboratory *Amphioxus*, a creature which used to be dissected by biology students seeking to learn the rudiments of organization in animals with a dorsal nerve cord like our own; from an anthropocentric viewpoint this little *Pikaia* might be the most significant animal of them all. But most spectacularly it preserved arthropods— the jointed-legged animals of which trilobites had once been regarded as almost the sole Cambrian exemplars. How wrong that idea was. For on the dark shales there was a fishmonger's slabful of arthropods: big ones and small ones, blind ones and sighted ones, some with grasping claws, others with spiny legs, twenty-six or more different kinds. Of these, *Marrella*, a dainty, feathery creature with no real carapace, was much the commonest—but it was known from nowhere else. The sea must have teemed with these little swimming creatures of which there was no hint from other rocks. Finally, there were bizarre enigmas, animals which seemed to resist classification in any of the biological filing cabinets into which zoologists were accustomed to tuck away the animals they recognized. The very names of these animals spoke of arcane mysteries unsolved: *Hallucigenia* and *Anomalocaris*. How would these perverse and paradoxical Cambrian curiosities fit into the story

---

*Walcott was clearly taking no chances. *Marrella* was named for Johnny Marr, the Woodwardian Professor at the University of Cambridge, the senior position in palaeontology in Britain. However, he named various other species from the Burgess Shale after his family, of which *Sidneya inexpectans* is the most charming.

of life? *Anomalocaris* was a large predator the size of a lobster with two great, spiny leading arms and a mouth surrounded by an apparently unique circlet of plates. As for *Hallucigenia*, it seemed to be a gut topped by wiggly tubes propped up on spikes. Had Edward Lear seen it he might have considered that it needed no embellishment; it was quite nonsensical enough as it was. A true glimpse into the Cambrian seas permitted an appreciation of how partial was the view from skeletons alone; how much richer the world had been, and how much less predictable.

The modern study of the Burgess Shale was led by Professor Whittington, who held Marr's position as Woodwardian Professor at Cambridge University during the 1970s. It was Harry Whittington who made the first identifications of the Ordovician trilobites from Spitsbergen, observations which were instrumental in determining my own future. He described a number of the Burgess animals, in the most painstaking way, and guided the research students who studied *Anomalocaris* and *Hallucigenia* and the varied coterie of Burgess Shale arthropods. He and his students prepared an inventory of Cambrian anatomy which was like having a dredge plunge down into the hidden seas of 520 million years ago. Teasing details out with a pin by gently flaking off little pieces of the covering rock, preparing drawings of every limb under a *camera lucida*, testing reconstructions until they made sense in three dimensions—this is slow and laborious work. But it meant that Burgess animals are now known nearly as well as animals dredged from the deeps only yesterday.

What did these animals contribute to our understanding of evolution? What was their place in the story of life? Lying so close to the beginnings of the record of complex animals they almost certainly hold some special message, if only we could read it. Harry Whittington had noticed how curious the Burgess arthropods were. They did not seem to him to display the kind of anatomy appropriate to ancestors of living arthropods—they were altogether too peculiar. His natural caution made him reluctant to interpret more than one or two as close relatives of anything more familiar. Gradually, an idea took root in the 1980s that perhaps the Burgess Shale revealed not ancestors (as Walcott may have believed), but designs long vanished from the world. The reconstructions of the remarkable animals *Hallucigenia* and *Anomalocaris* seemed to set the seal on this speculation. These animals did not look like *anything* figured in zoology textbooks.

The greatest division of animal life is into different phyla (singular phylum): Mollusca are one phylum, arthropods another, jellyfish and allies another, brachiopods another, and so on. The record of nearly all these phyla begins with dramatic suddenness at or near the base of the Cambrian, like a curtain suddenly being pulled aside to reveal a drama in the middle of the first act. Perhaps, so the speculation went, in the Cambrian there were dramas undreamed of, phyla uncharted, a richness beyond compare. *Hallucigenia* must be one of these vanished designs, *Anomalocaris* another; several other Burgess creatures were lined up for the honour of being experiments in life that failed. There is something romantic in the notion of a "lost world" of strange creatures shuffling over the sea floor. Some, possibly favoured by no more than good luck, would survive and prosper to populate the hundreds of millions of years that followed. Others, the unlucky ones, survive only as misty shadows on the surface of shale slabs prised from a Canadian mountain-side, and, but for the chance stumble of Walcott's mule, might have remained unknown for ever, unrecorded might-have-beens. One thinks of Thomas Gray's rumination on obscurity in *Elegy in a Country Churchyard*: "Some mute inglorious Milton here may rest." Mute and obscure indeed were these mysterious fossils. How could they be made to speak of their identity?

At this point Stephen Jay Gould entered the Cambrian charivari. In 1989 he published a book explaining the Burgess Shale to the world. *Wonderful Life* became a bestseller and did much to promulgate the excitement of investigating the past. It also broadcast the notion of an unparalleled variety of body designs in the Cambrian: his narrative took the former students of Harry Whittington, Derek Briggs and Simon Conway Morris as the heroes in a detective story, the *dénouement* being nothing less than a new picture of the tree of life. The Burgess fossils, as he deciphered them, meant that this most familiar metaphor of descent should be turned upside down: the tree of life had been viewed in the past as a kind of bush that branched ever outwards and upwards. Now that tree could be inverted. In order to retain the comfortable arboreal comparison the image of a Christmas tree was substituted— a tree wider at the base and thinning upwards. More designs had been present in the early days near the base of the tree. The sudden bushing-out above the "trunk" of common descent portrayed the evolutionary explosion at the base of the Cambrian into a panoply of unrivalled variety. Subsequent history has involved a thinning of this early wealth of designs, so that only a handful continued to evolve to populate the world as it is today. If it were possible to replay history, chance alone would have "weeded out" a different set of animals, and the living world would be utterly different today. The course of life was not inevitable. If one of the silvery streaks on a Burgess slab had not failed, its history would have been neither mute, nor inglorious. It is an idea that is easily grasped and readily explained, and was championed in virtuoso style by Gould.

The idea of replaying life stories was rehearsed in Frank Capra's 1946 movie *It's a Wonderful Life*, which gave Gould his title. A short story by Ray Bradbury—not mentioned by Gould—provides an exact illustration of his case. The story described how marksmen of the future return to the geological past to hunt dinosaurs. These trophy hunters are strictly forbidden to disturb history and so rules are fixed such that only moribund dinosaurs may be killed; in this way nothing will interfere with the subsequent unravelling of history. But one member of the shooting party steps off the protective walkway and treads on an insect. When they return to their present time the hunting party is met by a race of giant ants. . . .

If palaeontology has a priesthood, then Steve Gould is the pontiff. The Burgess Shale, however, was one case where he has, I think, been fallible. The excitement of the ideas being promulgated was so seductive that he simply passed over the real evidence presented by the Burgess fossils. This in no way diminishes their importance. Part of his misinterpretation was the result of following those researchers who took Whittington's doubts about the relationships of the more curious animals and inflated them into claims about hitherto unknown animal phyla, complete with appropriately mysterious names. More thorough examination showed that they were less radically different in organization than an initial appraisal had suggested. *Hallucigenia*— the typical example of a "weird wonder"—turned out to have been interpreted upside down! The wiggly organs on its back proved to be, somewhat more prosaically, mere legs; a second set of legs was discovered when the fossil was appropriately excavated, and the animal was rapidly reinterpreted as a stumpy-legged creature with spikes on its back. Some other discoveries from China made in the late 1980s showed that animals of this kind were widespread in the Cambrian, and were probably related to a curious group of living stumpy-legged animals, the velvet worms (Onychophora), which now survive under rotting logs in many places in the southern hemisphere. These primitive animals had long been regarded as occupying a lowly place in the hierarchy of life; indeed, many writers had indicated that they might represent a kind of proto-arthropodan organization—just the kind of bug, in fact, that you might have *expected* to discover crawling around in the Cambrian seas. It is now becoming clearer that the velvet worms were much more varied and diverse in the Cambrian—they had more designs and were more disparate then than they are today. Spiky forms no longer survive, for example.

This was no different from a discovery that Simon Conway Morris had made in 1980 about another obscure group of living "worms," the priapulids, which were much more varied in the Burgess Shale than in our seas now. This is an important and exciting finding: we could no more predict what kinds of velvet worms were living in the Cambrian than we could infer the shapes of dinosaurs from thought alone. Only the fossils could chart the story. *Hallucigenia* was no hallucination, but rather a vision. So, too, with *Anomalocaris*. To those like myself who had studied arthropods its grasping arms always spoke of that great, versatile ragbag of jointed-legged designs. It now seems that they, too, were widespread. Anomalocarids have been found on Kangaroo Island in South Australia, in early Cambrian rocks—even older than the Burgess Shale. Their compound eyes and limb forms make more sense as an arthropod than anything else, but one that was already highly specialized as a hunter (6 FOOT SHRIMP RULED ANCIENT WORLD, as one of the tabloids put it). One could not have predicted *Anomalocaris'* existence, of course, without the wonderful insights provided by special fossil preservations, but no purpose is served by assigning it to a phylum unknown. One may still marvel at the fecundity of nature without making wild assertions about every fossil belonging to a different world.

Many of the other arthropods, too, seemed less strange when they were interpreted through what they *shared* with other, known arthropods, rather than being celebrated only for their peculiarities. None was as strange as a present-day barnacle, nor as grotesque as a queen termite. To be sure, there remain many problematic animals, but their problems seem more to do with how they are assembled than with doubts about common ancestry. The jointed limbs of arthropods are arranged in different ways, often having different functions, rather like an articulated tool kit. Some have grasping claws, others fine hairs for grooming, others again spines for filtering. It is no coincidence that some of our attempts at mechanical robots clank around on spindly legs, with different arms for different jobs. Exactly how the tools are kitted out and in what order they were assembled is what understanding these Cambrian animals is truly about. In the process of early animal evolution, many viable, but to our eyes odd-looking creatures came into being and thrived. We need to know about these animals in order to gain some idea of the richness of the history of life. It remains true that if—but what a big *if* (to re-use Darwin's phrase about life's genesis)—one rather than another of these animals had survived, then the subsequent course of history would have been different: the consequences of an early branch would have been a different tree, ultimately in a different forest. If there was ever a claim that evolutionary history would slavishly replay itself if it could somehow be restarted, I have never come across it. No doubt crucial crossroads were decided on a turn of chance. And no doubt a different turn would have had quite different consequences.

These somewhat arcane arguments are of crucial importance. Any history of life is torn between portraying the narrative of successive species as orderly, almost a logical progression, and as something trawled from mighty disorder and upheaval from which chance alone plucks survivors. There is no question of the rapidity of the change we see at the base of the Cambrian. It was a time when animals could be constructed in ways unknown before or since, perhaps because their genetics were more flexible than at any other time in history. But many Cambrian animals actually do make more sense in the light both of what came after them and of what is still alive today, particularly when compared with the puzzling Ediacara animals. Despite the claims of the "new phylum" enthusiasts (one book claims at least 100 Cambrian phyla which did not survive) there are relatively few Cambrian designs which are wholly unfamiliar to us. Many, like *Hallucigenia*, elaborate in unexpected and inventive ways upon a ground plan which we already know, rather like improvisations upon an underlying musical theme that we can only recognize if we listen very carefully.

The "English Mozart," George Frederick Pinto, was born in Lambeth, London, in 1786. Samuel Wesley said of him: "Greater musical Genius has not been known . . . England would have had the honour of producing a second Mozart." Like Wolfgang Amadeus he had astonishing natural ability. As was the fashion among prodigies, he had already performed brilliantly in public by the time he was eleven years old, and over the next few years he composed piano and violin sonatas of remarkable maturity. His songs have been likened to those of the young Schubert. He was celebrated by his contemporaries, who had every expectation of his future greatness. Sadly, he was to disappoint those expectations, for he died before his twenty-second birthday. He has become a peripheral figure in musical history, neither mute nor inglorious, to be sure, but none the less a might-have-been. The historian can only speculate about the course musical development might have taken had Pinto lived longer, but one cannot say that because his potential was that of a Mozart, his place in history *deserves* to be the same. Mozart's stature is the result of what he did rather than of what he was. Our judgement is inevitably coloured by historical perspective—how can it be otherwise? But, equally, we should recognize that those who listened in awe to the "English Mozart" as the eighteenth century drew to a close were not mistaken when they surmised a glorious future for the handsome prodigy playing the piano so brilliantly. But for the attentions of a cruel bacterium it could well have been so. Heard again today, Pinto's brief contribution fits into the context of the music of its time, just as Mozart's followed from Joseph Haydn's example. Poor Pinto might be compared with a Cambrian animal that was destined to produce just a handful of descendants: it could well have been many more, but for a brief stroke of fate. While chance—mere random luck—may have picked off this animal rather than that one, there was also a kind of orderliness, because each animal was adapted after its own lights to life in the Cambrian sea, just as music belongs recognizably to its age regardless of the individuality of the genius who wrote it.

With predators like *Anomalocaris* swimming around (to say nothing of old-fashioned jellyfish) there was no room for limping, maladapted and vulnerable morsels even at this early stage in metazoan history. But crawling over the Cambrian muds there were also creatures which fill in for us some of the gaps between groups of animals that later seem widely separated. These are hybrids between trilobites and crustaceans, between clams and snails, not so much a gallery of missing links as a pot-pourri testifying to common ancestry, survivors of a mutable age. The later weeding-out of early forms served incidentally to separate the surviving animals more distinctly, but, in the Cambrian, categories were still blurred. There was no question of Lear-like chimeras spanning plant and animal, but there were mixed-up molluscs and ambiguous arthropods. The velvet worms alive now, like *Peripatus*, are all terrestrial animals, but *Aysheaia*, which included their Cambrian relatives, were marine animals equipped with gills. You would *expect* a marine relative of a *Peripatus* to have gills. Whittington's uncertainty about how to classify these animals came about not merely because of their strangeness, but because of their ambiguity, their previously unknown assembly of characteristics. Gould's mistake was to accord them a status they did not deserve because of these same novel combinations of features. The importance of the Cambrian fossils is not as a pot-pourri of zoological strangeness but rather as a key to understanding the state of the animal world close to its birth. There could hardly be a more important insight.

I have dwelt upon this question of the Cambrian evolutionary "explosion" because it is probably the most significant threshold to be crossed since the first organic molecules joined together to polymerize the first molecular chains. This is where leisureliness disappeared from our story. The animals that evolved in the Cambrian would have crawled over one another to be first to mate, evading the attention of predators on the way. Competition was introduced into ecology: these animals led exciting lives, vying with one another, sculpted for fitness. By contrast, the mat-formers which populated the endless stretches of Precambrian time, and even the Ediacara fauna, may have led lives almost devoid of incident. When this changed, it changed for ever. Indeed, competitiveness still governs all biological life: "Getting and spending, we lay waste our powers." And all getting is at some other creature's expense.

As to the different interpretations of the meaning of the Cambrian fossils, they furnish an object lesson in the relativity of truth. Flann O'Brien, one of the greatest Irish writers of the twentieth century, created a natural philosopher, De Selby, who elaborated alternative theories of an atomic nature. He proposed, for example, that darkness was not, as convention has it, an absence of light, but in reality a progressive accumulation of coal-black particles, so that night fell by a kind of blizzard of accreted blackness. As almost everything can be described as particulate in modern physics, De Selby's conceit in turning a mirror on to light and dark has a certain attractiveness. Such philosophical ambiguity is appropriate to the question of the meaning of Cambrian animal designs, which can be regarded as strangely unfamiliar or strangely familiar, according to the way in which they are considered. It all depends what you mean by familiarity and oddity.

There remain several Cambrian animals which *have* defied all attempts to relate them to something more familiar. For example, *Opabinia* from the Burgess Shale had a number of eyes, and peculiar appendages; some of the animals which carried plates all over were neither sponge nor worm, and may, possibly, represent a large group of organisms of which there is no known close relative. But there is no question that the uniqueness of the Cambrian designs has been exaggerated by those with eyes for novelty rather than commonality of descent. To be sure, the Cambrian was a peculiar world, but a recognizable one for all that. There were as many links as enigmas.

There are a few palaeontologists who are as eccentric as *Opabinia*, and one wishes there were more. Usually, eccentricity is within the bounds of academic variation (quite a generous allowance, admittedly)—absent-mindedness, odd expressions or manners of speech, tramp-like garments worn for decades, etc. etc. There are a few who transcend such mundane bounds to become legendary figures. Dr. Rousseau H. Flower was an eccentric in the grand manner. He was an authority on nautiloids—the living pearly nautilus is the last representative of a formerly diverse group of marine animals distantly related to the octopus, whose straight or coiled shells are often found in rocks of Ordovician and younger age. From the 1960s until the 1980s he lived in New

Mexico, although he came originally from New York. Possibly as a consequence of his removal he became more western than the westerners. He always wore hand-tooled cowboy boots with elaborate curlicues in the stitching, and a hat and jacket to match. He was very short-sighted, and tended to stumble along in the purposeful way adopted by the cartoon character Mr. Magoo, while mumbling vigorously to himself. When some particularly appealing idea occurred to him he would stop and cry out "aha!" with a kind of surprised enthusiasm, as if he had just recognized an old and valued friend. In the field he carried a bull-whip, and was known to lash outcrops that refused to yield up their share of nautiloids. He was even said to try the same trick with a six-shooter, but his approximate aim reduced bystanders to gibbering terror. He was an inveterate smoker, which was eventually to lead to the emphysema that killed him. When I shared a hotel room with him, I recall him emerging from the shower with a bent cigarette still clenched between his teeth, and dripping sadly on to the floor. Yet he was also a masterly cello player—indeed, he could play the Dvořák concerto, the Mount Everest of cello concert works. For all his eccentricity, Rousseau Flower led a productive and satisfactory life. Perhaps his example should be remembered as we contemplate the alleged eccentricities of animals long extinct, for they, too, would have been capable of living successfully on the sea floors of their own vanished worlds. Our own perceptions are schooled in what we know, and are not lightly converted to imagine life in the Cambrian or before.

There remains the question of timing. It is true that the biological world changed, and changed utterly, between Ediacara and the Cambrian. This is only about 15 million years, at the most, according to recent estimates from radiometric dates. The appearance of skeletons certainly happened over a shorter time-period again. There can be little doubt that one change followed upon another in a chain reaction, as the sequence of rocks in Newfoundland or Siberia relate with the simple clarity of a folk story. The first predators appeared, and with them protective skeletons in their victims; which prompted better predators, which in turn gave an advantage to animals with thicker shells, and so on and so on, in a kind of arms race. But we also know that soft-bodied animals proliferated as fossils or traces at exactly the same time: hence the Cambrian threshold was marked not just by the acquisition of shells, although that surely was a part of it. Now a whole array of animals like those of the Burgess Shale have been found among the *earliest* rocks of the Cambrian at Chengjiang, in China, and at Sirius Passet, in Greenland, so it seems that the differentiation of the designs of animals must have been even earlier, and its progeny more widespread. I recently examined a PhD thesis written about the Greenland arthropods. The author was convinced that there were a variety of *Peripatus* relatives which were variously in the process of transforming themselves into more familiar arthropods. And yet how can it be that all this variety arose apparently *instantly* at the base of the Cambrian? We know that some of these animals are more primitive than others, just as surely as we know that living *Peripatus* is more primitive than a living butterfly. In short, there must inevitably have been a history of descent which is not recorded along the windy shores of the Rock, or in the tundra along the banks of the Lena river in Siberia. *Marrella* and trilobites and *Anomalocaris* and all the rest must have had a prior common ancestry to account for what they share, as inevitably as we and the chimpanzee share a previous history because we can use our thumbs in a similar way. Where, then, were these ancestors? Why were they apparently invisible?

There seems only one way out of this paradox: the late Precambrian ancestors of the arthropods, molluscs and all the rest were very small—a few millimetres long at most. It was only when the "small shellies" acquired shells that they became visible. At the same time they became large enough to leave their footprints or their tracks and trails. Maybe some of the small animals lived as part of the plankton, where larvae of many groups of animals live today. Most of these are virtually invisible in the fossil record. Lest this seem like special pleading, it can be shown that even among living groups of animals the most primitive species are often very diminutive. The most primitive insects are tiny, wingless springtails, for example; there are curious, primitive little worm-like molluscs called aplacophorans; even shrews are both the smallest and possibly the most primitive of mammals. Small is not only beautiful, but often durable.

The explosion in Cambrian diversity of life may have been the result of an increase in size from small ancestors. Much of the hard, evolutionary groundwork may have happened in the late Precambrian. The apparently magical curtain that was drawn aside in the Cambrian to reveal a wealth of life displayed the point in the drama when the actors suddenly donned their costumes to become visible to the audience. The work backstage in the Precambrian, the rehearsals abandoned, the writing and rewriting, are forgotten in the sudden drama of illumination as the curtain rises. Maybe the costumes—shells, for example—appeared as a result of increase in size: after all, stature requires support. A tiny organism can be supported by its own turgor, and requires neither struts nor enclosure by hard shells. Shells do many things besides offer protection; they enclose a chamber for feeding; they allow for life inside sediments on the sea floor without choking vital parts; they provide privacy for housing the organs of sexual reproduction. All this may have come about as a by-product, a chance gift, after the appearance of shells for entirely different, protective reasons. As with the discovery of fermentation, with its intoxicating primary products, it was a bonus to find that wine also tasted good, and provided a drink immune from those poisonous bacteria that infect unsafe water. The latter properties could scarcely have been the prime mover for the original innovation, but the connoisseur would doubtless claim that taste was its most durable consequence, especially when he is well into the second bottle. These originally incidental advantages became the main reason for further development and sophistication as wines competed to tickle the palate, until the thousands of châteaux, vineyards, and haciendas we know today had come to fruition. It is always tempting to describe such changes in a thoroughly purposive way; in the Cambrian instance, it would be as if a shell was an advantage waiting for its moment in time, and the animal schemed to exploit its various properties. This is to confuse purpose with effect. But there is no question that novel interactions between animals, including competition and predation, prompted the explosive proliferation of life in the Cambrian, even if the generation of fundamentally different designs had happened in animals of small size over tens of millions of years previously. There was a chain reaction, unstoppable once started, a bacchanalia of zoological inventiveness, which has never been matched again. This is the point in my biography when those forces which are still supposed to energize human economies first made themselves felt in the economy of Nature itself.

The explanation that the level of oxygen had reached a critical level for the respiration of large animals has a pleasing continuity with the role of humble, inconspicuous photosynthesizers, which had transformed the atmosphere through the previous vast stretches of geological time—the slime time. Oxygen may have increased to a point where it was sufficient to supply a protective ozone layer high in the atmosphere, which served to shield vulnerable animal tissues from harmful solar radiation; shells may have offered additional protection, another unpredictable benefit of this innovation. But geologists have discovered evidence for several further interesting events that happened close to the base of the Cambrian. This was when seas advanced over the continents to leave a rich sedimentary record—in some parts of the world (as in many areas in Scandinavia) this sea flooded a landscape which was already ancient and eroded. In such areas there was no opportunity for any geological record of the critical interval of time *until* the sea had made its inroads. The diversification of life may have accompanied the flooding as continental shelves, rich in nutrients, became available to the new and larger fauna, literally a feast of opportunities waiting to be exploited.

Then there were changes in the quantity of that vital nutrient, phosphate, in the ancient oceans. My companion in Australian fieldwork, John Shergold, had become interested in the abundance of latest Precambrian and Cambrian phosphate deposits, which are commercially important as a source of fertilizer and industrial chemicals. He took me to the mine at Mount Isa, almost exactly in the middle of Australia. Like all open-cast mines it has a blatant obtrusiveness which sits ill in the surrounding landscape, a hole where none should be. There is always a temporary feel about mines, which is often justified by their busting soon after booming. But it became clear to me at once how the Cambrian sea had flooded over Precambrian Australia, bringing with it dark, shaly rocks black with little nodules of phosphate; the same seas teemed with tiny, swimming trilobites, which must have clouded the water in their abundance. This was no struggling ecology, poised on the brink of things; rather, this was a cornucopia of biological activity. Comparable Cambrian deposits are known around the world. Clearly something had stimulated the production of these phosphates, and John Shergold would like to link this curious phenomenon with the appearance of many phosphate-shelled animals at more or less the same time. Today, an abundant source of phosphate still produces bumper harvests in the sea; in places where seawater charged with dissolved phosphate wells up to the surface, as it does off the western coast of South America, for example, there are limpets as big as soup plates. The whole guano industry is founded upon phosphate excreted by millions of seabirds, and derived ultimately from the recycled deeps. Perhaps the Cambrian enrichment was powered by phosphate, catalysed and accelerated by a rejuvenated ocean circulation system. Others have sought to turn the same argument completely on its head, claiming that it was the breakthrough in deposition of minerals made by animals and plants at the end of the Proterozoic which served to *trap* more phosphate, which could then be readily preserved in the rock record. They point to the fact that most geologically younger phosphate deposits are closely associated with biological agents, as in the case of guano.

Momentous events in the history of life were evidently inextricably bound to events in the world, and vice versa. Life and environment comprise one linked system, they have an umbilical connection. It is impossible to talk about a breakthrough in the story of organisms without looking at what was happening in the seas around, or at how the continents were distributed and what this might have done to the circulation of wind and water. Life itself created the conditions under which further life could thrive, and thus began the complex interdependencies that are the stuff of ecology.

I can imagine standing upon a Cambrian shore in the evening, much as I stood on the shore at Spitsbergen and wondered about the biography of life for the first time. The sea lapping at my feet would look and feel much the same. Where the sea meets the land there is a patch of slightly sticky, rounded stromatolite pillows, survivors from the vast groves of the Precambrian. The wind is whistling across the red plains behind me, where nothing visible lives, and I can feel the sharp sting of wind-blown sand on the back of my legs. But in the muddy sand at my feet I can see worm casts, little curled wiggles that look familiar. I can see trails of dimpled impressions left by the scuttling of crustacean-like animals. On the strand line a whole range of shells glistens— washed up by the last storm, I suppose—some of them mother-of-pearl, others darkly shining, made of calcium phosphate. At the edge of the sea a dead sponge washes back and forth in the waves, tumbling over and over in the foam. There are heaps of seaweed, red and brown, and several stranded jelly- fish, one, partly submerged, still feebly pulsing. Apart from the whistle of the breeze and the crash and suck of the breakers, it is completely silent, and nothing cries in the wind. I wade out into a rock pool. In the clear water I can see several creatures which could fit into the palm of my hand crawling or gliding very slowly along the bottom. Some of them carry an armour of plates on their backs. I can recognize a chiton, but others are unfamiliar. In the sand there are shy tube-worms. A trilobite the size of a crab has caught one of them and is shredding it with its limbs. Another one crawls across my foot, and I can feel the tickle of its numerous legs on my bare flesh—but wait, it is not a trilobite, but a different kind of arthropod with eyes on stalks at the front and delicate grasping "hands." Now that I look out to sea I can see a swarm of similar arthropods sculling together in the bright surface water— and can that dark shape with glistening eyes be *Anomalocaris* in pursuit? Yes, for the top of its body briefly breaks the surface, and I can glimpse its fierce arms for an instant. Where the water breaks it shines luminously for a while in the dying light—the seawater must be full of light-producing plankton— and I have to imagine millions more microscopic organisms in the shimmer- ing sea.

Chapter 5

Marine Riches

T HERE ARE, SO IT IS ALLEGED, only four different jokes: the
mother-in-law joke, the sexual innuendo, jokes about foreigners,
and jokes about politicians. The stand-up comic's routines are permutations and combinations of these fundamental wisecracks, peopled by
different characters, transferred to different situations—but, at root, only
four different jokes. So it is with ecology. In the world's oceans there are only
a small, finite number of habitats, but there are many variations in the way an
animal can earn a living in any one of these. Different organisms can play different parts in a given ecological niche—more, by far, than the variations of
the mother-in-law quip. But the situations in which the organisms find themselves have a recurring similarity after the Cambrian. What drives the whole
ecological engine is energy—energy derived from the sun and fixed in photosynthesis. Everything else depends upon it. For much of the Precambrian,
photosynthesis is where the story stopped. But after the profound changes at
the beginning of the Cambrian there were new roles for animals, and new
possibilities for their mutual relationships. The narrative became richer; it
was a proper show.

First, there were those creatures that fed directly upon plants. There was
a rich algal plankton, and it is certain that many animals in the Cambrian had
planktonic larvae that harvested this rich soup, just as their descendants do

now. The plankton today looks no more dense than a Chinese clear broth, but within this insipid bouillon there are microscopic dramas of capture and submission—the greed for food provokes much gobbling of one small creature by another, and slashes the final fecundity of animals that have laid a million eggs on the chance that one alone will survive to adulthood. Grazing was soon an option for bottom-dwelling animals, too, especially among the molluscs, all of which share a rasp-like radula within the mouth, a kind of microscopic chainsaw, with which to scrape off nutritious scum. Certain of the arthropods learned to perform the same trick with their sophisticated limbs, which could act as scrapers, files or sieves as the need arose. Other animals, like brachiopods and many echinoderms, chose to filter out edible materials directly from the seawater, just like comic uncles who were always supposed to get soup trapped on their moustaches. This is a simple process; all that is needed is a way of directing currents to some sort of feeding organ—so that food is wafted to where it can be consumed. Since these animals had planktonic lunch brought to them on currents, they were usually immobile, fixed to the sea floor. This feeding habit was already ancient by the Cambrian; it will be recalled that this is how sponges and attached jellyfish fed. The beating of tiny cilia in harmony was one way of creating feeding currents. Once captured, minute organisms and other organic debris could be fed into the stomach for processing. Some echinoderms put their arms up into currents, using the sea's own ventilation system as a kind of conveyor belt to bring edible morsels.

These primary consumers could then themselves be eaten by other animals. Weight for weight, animal food is more nutritious than plant food, and it follows that those creatures that ate other animals could grow larger. Large size alone means that there are spare corners to store excess food in the form of fat—thus providing a way of surviving hard times. Almost as an unexpected corollary of the growth of different feeding habits there arose *strategies* for survival, whether the organism concerned happened to be a filter feeder or a grazer, hunter or hunted. It is a matter of simple arithmetic to discover that the number of ways there are to interact increases geometrically with the complexity of the ecological system: introducing another tier does not merely double the possibilities but magnifies them mightily, as in some kind of chain reaction. It is like the wealth of a capitalist tycoon: every deal breeds a dozen further possibilities until there is no reckoning the movement of any single pound, escudo or dollar. By contrast, the pauper knows the whereabouts of his every penny.

When others grow rich there is a living to be made as a parasite. It is unknown exactly when this opportunity was first exploited. As any biographer knows, of all analogies to humankind drawn from the natural world "parasite" is the most derogatory. Decent animals (or plants) work for a living, it is implied, and anything else is morally reprehensible. It is likely that organisms discovered a short cut through the virtuous circle of production and consumption quite early in the history of life: there are examples of deformed Cambrian trilobites that probably display a response to being attacked by an ancient parasite (the bugs themselves do not preserve). It seems that as soon as food chains became a respectable length there were smart operators willing to jump a few links. Then a new chain of parasites developed in turn. As Dean Swift recorded (acerbically, with human analogues in mind):

> So, naturalists observe, a flea
> Hath smaller fleas that on him prey;
> And these have smaller fleas to bite 'em,
> And so proceed *ad infinitum*.*

The bottom end of this great chain of scrounging is probably a virus, since it is subcellular in size. There is no still smaller flea conceivable, and this one was certainly not known to the Dean. Although in some respects autonomous, viruses depend upon other cells to complete their life cycle—they hijack the cellular reproductive cycle to their own ends. Although they are simple, they are probably *simplified*, rather than representing a protocellular stage in evolution. If infection by parasites followed hard upon the Cambrian burst of evolution it has not gone away since. Disease and pathological deformations have constantly accompanied life's long story, and will doubtless dog us all to the stars.

So, feeding roles are comparatively few. The marine habitats in which organisms can act out their roles are finite, too, and they have been persistent through time. The sea, unlike dinosaurs and trilobites, has never gone away. The most fundamental division in the oceans is into an upper part through which light penetrates, and a deeper part which light cannot reach, a realm of

---

*By a nice irony this sentiment is probably rather better known from a couplet by Augustus de Morgan:

> Great fleas have little fleas upon their backs to bite 'em,
> And little fleas have lesser fleas, and so *ad infinitum*.

Since this was written in the century after Jonathan Swift's death it does itself represent a kind of literary parasitism. And like all parasites, Mr. de Morgan was diminutive in comparison with his host.

perpetual dark; even here, brief flashes of biologically manufactured illumination from special organs on the sides of crustaceans glimmer fitfully. Photosynthesis can only take place in the presence of light, so primary production is confined to the upper part of the sea where algae live. The open ocean is the realm of plankton. Most plankton is small (a millimetre is typical), and some of it comprises single-celled survivors from the Precambrian, including photosynthetic algae. Larvae of many marine animals that are seabottom dwellers as adults spend their babyhood in the plankton. The surface metres of the sea are both a nursery and a base for dispersal of species, and it has been so for at least 540 million years. Larvae routinely look different from the adults into which they will metamorphose. They are specialized for planktonic life, just as caterpillars are specialized for eating leaves. Some of the organisms in living plankton have been plying the same trade since the Cambrian. Delicate, spherical radiolarians, for example, have filigree skeletons made of silica, which support a protoplasmic body which can "fish" for prey among other planktonic animals. They did the same job in the Cambrian and Ordovician that they do now. It was once thought that, rather like those bacteria forming stromatolitic mats, nothing had happened to them during the course of their long history; that they were simply static, suspended in the amnesiac broth of the open sea. Now it is clear that this is not so, and that early radiolarians are different in the details of their skeleton construction from their living relatives. Although their role in the ocean has remained much the same, the species acting in that role have changed and changed again. This appears to be true of much of the plankton—which provides a constant scenario, but whose performers, acting out the characters, change repeatedly. One thinks of Agatha Christie's perennial play, *The Mousetrap*, which has outlasted all other productions, until its longevity has become reason enough to guarantee its perpetuation. A few actors survive in their roles long enough to collect their pensions; others take advantage of a change as soon as it presents itself. No matter. The play remains the same. Plankton is always with us, but it is not the *same* plankton. This is not an original metaphor, but it is a curiously appropriate one.

Inshore, the illuminated zone is the richest part of the marine habitat. Algae of all kinds sprout gardens; in productive seas, like those off California or Cornwall, brown ribbons of algae make growth to shame a paddy field or a bamboo grove. Animals can feed off these bounteous growths in countless ways, or use them as mooring pads to launch their own careers. Which child naturalist has not waded among rock pools, poking under the seaweeds to discover what lurks there, what scuttling or crawling creatures reside in fronds or crevices, only to elude young, grasping fingers? There are livings to be made here which have endured from the very beginning of animal life—adhering limpet-like to rocks to graze algae at high tide, or growing as a fan to extract nourishment from wafts of nutritious waters in the churning tides. Cambrian trilobites may have parodied today's shrimps, for all their separation by half the life of a star. Sandy and muddy sea bottoms have their own faunas, and the physical demands of these habitats impose constraints upon the designs of the creatures that can live there; neither have these constraints changed over geological time. Hence the designs of extinct animals can be understood if enough is known about their ancient habitats. If you lie down close to a sandy seashore as the waves wash across, you can see the shifting clouds of sand as each wave washes back and forth: how can anything live in such a mutable and restless world without being buried as it lived? Yet the sandy shore is littered with shells, so evidently living things can thrive there. The secret is to stabilize burrows; feeding happens only as the seawater washes over. Then, the water might be sucked into a feeding tube, or filtered by an organic net, to be relieved in one moment of its rich soup of plankton. It is a method that had already been learned by the early Cambrian, because there are inshore sands of that age—now preserved as hard sandstones—which are littered with vertical pipes that testify to the presence of batteries of specialized animals which had already mastered the sand trick. What were they? Possibly polychaete worms, but possibly not—the nature of the life habit is clear, but not necessarily the animal that did it, which leaves no trace of its body in the sands. What is clear is that the ecology was the same, as was the strategy of mastering the problems of living in shifting sand. These "pipe rocks" are found scattered through the record of the rocks. I saw my first examples along the shores of Hinlopenstretet in Spitsbergen. They were recognized in the last century on the beaches around Durness in the far northwest of Scotland, where they underlie rocks containing early Cambrian trilobites. At low tide, near Balnakeil, you can wander along the sands on a coast of unsullied perfection until you see a tough, yellow sandy rock which looks as if it has been scored vertically with a blunt knife. On the bedding plane there are little round holes, and you can soon see that each hole is connected with a tube. Surely, they were vertical holes, more or less regularly spaced over the sandy surface, and occupied in life by sand strategists. This Pipe Rock was the type example, to which the passing geologist has to pay his or her respects.

Muddy sea floors present a different set of problems: soft gooeyness does not provide good moorings for fixed organisms; but there are many creatures

that can crawl over mud, or trawl its black substance for nourishment—
molluscs, arthropods, polychaete worms, and delicate relatives of the starfish
called brittle stars. Muds often floor the sea offshore, where fine clay material
is carried in suspension, derived from rivers. But muds come ultimately from
the deep erosion of rocks. Thus is the substance of the world recycled, for
these muds will themselves again become rocks with the next geological
cycle, and then become eroded again in their turn.

We move further out across the continental shelves, deeper and deeper,
until the light is too feeble to support photosynthesis. But the fall-out from
the plants and animals living in the photosynthetic surface waters ensures a
rain of food, which can be exploited by what my childhood encyclopaedia
always called "the denizens of the deep." (As far as I know, denizens are confined to the deeps and to the criminal underworld.) The real deeps, however,
are off the edge of the continental shelves, deeper down than can be easily
imagined, down the continental slopes into the abyss. This is still no desert,
in spite of the darkness and the colossal hydrostatic pressure, for there are
many creatures specially adapted to life under these conditions: grisly fish
which are all mouth, always ready to engorge whatever scarce meal should
come their way; luminous shrimps and peculiar species of squid, too, and
many more species still undreamed of doubtless remain to be discovered.
Even less is known of this dark realm in the geological past. Because sea floors
are continually renewed at the ocean ridges by plate movements, their
ancient history has simply been subducted away. This renders invisible their
history more than 120 million years ago. We can populate the Ordovician
deep sea with any fancies of our choosing; our imaginations can never be
undermined by anything so dull and obdurate as a fact. But the deep sea was
as continuous as all the other marine environments, and it is likely that some
of the animals that live there today, like brittle stars, moved there quite early,
as to a kind of haven. Deep beneath the perturbations of the surface world
this lightless realm was a coolbox that preserved things. Some kinds of animals—like stalked sea-lilies—may have remained there long after their close
relatives had died out in shallower seas. It has always been my hope that, one
day, a bathyscape would bring up a trilobite, the last of its kind perhaps,
which scuttled to the depths hundreds of millions of years ago, and lived on
there while all its shallow water kin met their end. This is not such a wild
hope as might be thought. When the "black smokers" were discovered—
those deep-sea sulphurous vents which have become one of the models for
the genesis of early life—primitive animals were revealed clinging to their
rims, including lobsters which were known previously from Triassic fossils.

Maybe life forms clung to the edge here long after their natural span had expired at the surface. I would greet a living trilobite with the kind of incredulous gratitude that a relative feels when he discovers that some beloved cousin had, after all, survived a war.

IN SHALLOW SEAS, where light floods through clear waters and warmth bathes all life, reefs grow. In the present oceans, reefs are composed of corals and hard, limy algae. Reefs stand up to strong seas because they have a frame of tough limestone secreted by the organisms that construct them, ramparts that can withstand typhoons, and regenerate after hurricanes. Through the span of geological time, ever since the Cambrian, there have been reefs growing in tropical, clear seas. They have been made by different organisms at different times, but they share the same geometry; they make the same habitat: the structure endures. Like one of those jokes involving an Englishman, a Scotsman and an Irishman, it is possible to swap the protagonists around according to circumstance, but the punch-line shares a family resemblance. Because they were adapted to similar conditions, many of the animals that inhabited reefs came to resemble one another. Since they tended to grow towards the light, and many of them had symbiotic algae in their tissues which needed light to survive, a broad resemblance is not to be wondered at. There were cushions and fans and pipes, growing upwards and outwards, by turns robust or delicate according to how closely they had to confront crashing waves. The first reefs in the Cambrian were made of sponges and sponge-like animals, together with ubiquitous calcareous algae. I examined some of these in the wilderness of Labrador, where they still form mounds a few metres high.

Corals were added to reefs in the Ordovician. Corals are the limy skeletons of colonial relatives of the jellyfish, props which support the fleshy, insubstantial polyps so that they face towards the light. The fortunes of one kind of reef organism waxed as that of another one waned. There were times when reefs abounded. For example, coral reefs were abundant during the early part of the Carboniferous System that North Americans call Mississippian (because rocks of this age are well-developed along the Mississippi Valley); at other times they were comparatively rare. Before 230 million years ago the corals were of a different kind from those that live today—but even the most casual observer would recognize the broad biological *category* of the branching "organ pipes" that are seen in so many grey Carboniferous limestone cliffs. Abundance of reefs depended on the composition of the ancient atmosphere, and reefs were abundant at times when the climate was warm and there was carbon dioxide available (not least, CO_2 is needed to precipitate calcium carbonate for limy skeletons). The seaward edge of Carboniferous reefs was often reinforced with the crusty remains of lime-secreting algae, and massive sheets of odd, sponge-like organisms called stromatoporoids. From the Ordovician onwards reefs recruited many additional kinds of animals into the habitat. There were colonial bryozoans forming nets or low cushions, delicate individually ("sea mats," their common name, rather neglects their refinement), but often numerous enough to set up "baffles" in the surging water. There was a good living to be made for any animal that could extract nourishment by filtering out organic particles at the moment when waves broke over the reef in a turbulent rush. In reefs, the boundaries between individual organisms often become blurred, because they grow over and through one another, plant and animal cheek by jowl, in a search for light and nourishment. Corals, bryozoans and sponges are colonial animals, collections of individuals in pursuit of a common cause; the death of one polyp does not provoke the death of a colony. Like the Musketeers, colonial animals flourish on the principles of "all for one and one for all." Reefs are a cornucopia of life, with many species living together, mutually dependent, adding to the richness of the sea in the same way that tropical rain forests enhance the flora and fauna of the terrestrial biosphere. Their scale varies from the massive dimensions of a Great Barrier Reef hundreds of kilometres long, to reefs fringing atolls the size of a small town, down to structures no bigger than a house.

I have seen Silurian reefs in the Welsh Borderland, marking the hilltops at Wenlock Edge. Where A. E. Housman's Shropshire Lad might have lingered to look westwards into Wales, cars now rush along a busy road, but even in the public picnic spots you can see white patches of reef rock. If you kneel down and look more closely you can find honeycomb-like mottling which marks the coral colonies now embedded deep in limestone. Some are like miniature spiders' webs, others little chains or necklaces of tubes. Pinkish, fuzzy-looking patches may be stromatoporoids. These organisms together made the reefs that still stand up proudly along Wenlock Edge. Between the frame-builders lived dozens of kinds of trilobites and brachiopods, bryozoans and snails, and occasional sea lilies—the sea must have been sparkling with life. And remember these are only the animals with shells—there must have been hundreds more species again without readily fossilizable parts. None of these animals is alive today, and yet it is comparatively simple to recognize the community of extinct life for what it was—a reef, but one populated by vanished organisms acting out the familiar ecological scenario of the reef. Onwards up the geological column, I have seen similar reefs in western Canada, of Devonian age, with as rich a profusion of corals, trilobites, brachiopods and sea lilies, all of different species from those in the Silurian—but reefs none the less. The Carboniferous massive limestones that make up the "backbone" of England, in Derbyshire and Yorkshire, are partly fossil reefs. All the earlier species have themselves become extinct in turn, but new corals again replaced them; there are great clumps of pipes, and rounded pillows composed of dozens of hexagonal corallites packed together. There are solitary corals like horns, and compound ones making little hexagonal masses, and beds of smashed coral debris derived from the erosion of reef fronts as they were battered by waves stirred by hurricanes that swept over vanished seas. The Mississippian reefs in the central and southern states were the apotheosis of the sea lilies—no botanical lilies, these, but tethered relatives of the sea urchins and adapted to filtering the sea for microscopic food with their arms. During the Permian there were reefs in the Glass Mountains of Texas in which bryozoans and sponges were more prominent than corals. None of these peculiar reefs survived beyond the Permian. When Jurassic reefs appeared it was with a different cast of corals—distant relatives of those living today. Through France they skirt the wine-growing country of the Midi, and Jurassic reefs can be found eastwards in the Jura Mountains; they often have a warm, ochre colouring. Similar reefs are found in the Cretaceous of the Mediterranean, where you might encounter heads of corals in hard, white cliffs as you pad around the shores of a Greek island. So it continues. There were not always reefs, but when conditions were right, they constructed themselves by recruiting whatever suitable organisms were available at the time. Exactly where they grew in the world in any geological period depended on where the continents were in relation to the tropics, since warmth has always been necessary for reef growth.

This "change-in-continuity" is far from an obvious concept. It is quite different from "no-change-at-all" in the fashion of bacterial mats, and different again from the kind of breakthroughs—between single and multicellular animals for example—which tend to dominate any history, simply through the weight of their drama. The riches of the sea owe much to the constancy of its various habitats.

Some animals that occupied these several habitats have been enigmas—their role in the ecology has been mysterious because nobody has been quite sure what kind of animals they were, rather like the profounder puzzles of the Ediacara fauna. Prime among these were the conodonts. If you toss a lump of marine limestone of any age from late Cambrian to Triassic into acetic acid, it will fizz slowly as the acid attacks the calcium carbonate—but it will leave alone anything that happens to be made of calcium phosphate, which is insoluble in this kind of acid. If you rinse the residue that remains through a sieve, there may be some tiny, shiny, tooth-like objects a few millimetres long at most left resting on the grid. These are conodonts. When they are examined under an electron microscope they show a wonderful variety in the number, size and arrangement of their "teeth," and it was appreciated fifty years ago that they could provide an accurate way of dating rocks, because conodonts from one rock sample differed from those from any other slightly younger or older ones; they provided a natural chronometer. But they were as mysterious as they were practically useful. It was obvious that they were part of some larger animal—but what kind of animal? Just as it would be impossible to infer the shape of a human being from knowledge of the toenails or teeth alone, so the identity of the conodont animal tantalized biologists interested in ancient life. A few facts could be suggested. The conodont animal was presumably a good swimmer, because very similar conodonts were recovered from rocks of the same age collected from all over the world. They were capable of living happily in a range of environments from inshore shallow seas to considerable depths, judging this time from the kinds of rocks in which they were found; it was likely that at least some of them formed part of the plankton. They never invaded fresh water. They must have been abundant, too, because in some localities hundreds of specimens could be obtained from a sample which could be comfortably held in the palm of the hand. If we had thrown a large trawl into Silurian or Carboniferous seas it would have brought up lots of conodont animals; indeed, we would probably have thrown them back with cries of: "Why can I *never* catch anything but useless conodonts?"

It was discovered in the 1930s that different kinds of conodonts were consistently found together. Although this was apparent as a statistical fact, rare examples were discovered in which the conodonts, instead of being dissolved from limestone rocks, were found lying on a bedding surface together in a consistent arrangement, just as they must have been left after the mysterious conodont animal had died and decayed. The conodonts were arranged in pairs with their "teeth" opposed; furthermore, there were several different kinds of conodonts together in one natural complex—some simple, sharp, pointed ones, others with comb-like arrangements of finer "teeth." The conodont animals were beginning to seem both more enigmatic and more complex. It was as if some artisan from a vanished civilization had set down and abandoned a complicated set of tools, to be unearthed later by an archaeologist who finds no hint as to how they were to be used. The conodont specialists called these assemblages "apparatuses"—a delightful piece of nomenclature, with its intimation of precision. But an apparatus for what? The bafflement about their function continued. From time to time, attempts were made to reconstruct the conodont animal from a consideration of the apparatus. I sat through an entertaining lecture more than twenty years ago from the senior figure in conodonts at the time, Professor Mauritz Lindström, when he tried to infer the form of the conodont animal by rigorous application of reasoning applied to the apparatus. Professor Lindström is the very model of the Swedish professor—tall and cadaverous, with a grave and ponderous demeanor. It was a little incongruous, therefore, when the fruit of Professor Lindström's deductions was revealed to be a kind of small, fat toilet roll studded with paired conodonts on the outside; a floating, filter-feeding animal with a roly-poly look, yet spiky withall. Somehow, it did not seem to cut the mustard. There had been reports of the discovery of the conodont animal itself, one of which—a curious animal described by Drs. Melton and Scott from the Bear Gulch Formation of Montana—had turned out to be an animal with its stomach full of conodonts; something which had eaten, rather than solved, the enigma. Then, just as one could have been forgiven for thinking that the conodont animal would for ever elude both luck and reason—it was discovered!

Two of my arthropodous colleagues, Euan Clarkson and Derek Briggs, had been describing fossil shrimps in the early 1980s and their relatives from the Scottish Carboniferous rocks. The quality of preservation was remarkable, with details like limbs and antennae clearly visible on slabs of soft shales, large collections of which fill drawers in the Royal Scottish Museum. Euan noticed a strange smear, a passing shadow. Looking again he saw that there was the fossil of a worm-like object, long and slender, smaller than a little finger, a mere impression, but real enough. Many previous investigators may have passed it by without even spotting it. A closer inspection under the microscope showed that it contained, at one end, a collection of conodonts. Was it possible that this was another chimera, a chance association perhaps? Since neither Briggs nor Clarkson is a conodont expert, they called in Dr. Dick Aldridge, who is. I have to imagine the expletives that would have accompanied Dick's first glance down the microscope: the initial feelings of scepticism, the flash of recognition, the title of the scientific paper springing ready-made into the mind. It is a matter of regret that men (scientists particularly) no longer wear hats, because it must be good to throw them into the air at such moments, for not only did Dick recognize the conodonts as a known genus, but it was also obvious that what he was looking at was what had been termed an "apparatus"—that natural assemblage of conodonts which for half a century had been part of the riddle of the conodont animal. This elegant animal the size of a sand eel could scarcely have been more different from Professor Lindström's lugubrious construction. Not least, the space taken up by the conodont apparatus was only a fraction of the length of the animal, near the head end. The hind end had chevron-shaped muscle bars, and there may well have been fins on either side of the narrow body. This was evidently a flexible animal and, like a sand eel, a swimmer; maybe it, too, lived in shoals. The early guesses about what the conodont animal might have resembled were accurate in their way, although nobody could have imagined that this is how the animal would have looked in the details of its anatomy.

The corollary of this discovery was even more astonishing. The appearance of the entire conodont animal now revealed that it was likely to be related to the vertebrates: the great group of animals that embrace fish, fowl and mankind. It had more than a passing resemblance in its muscle structure to one of the staples of laboratory dissection in the schoolroom—the lancelet or Amphioxus (Branchiostoma). One of the Burgess Shale animals, Pikaia, had already been compared with this creature, but its status had been controversial. Amphioxus is well-known for providing an evolutionary link in the marine world between the lowliest vertebrates and their closest relations among sea squirts and other, even more obscure organisms. Thus the conodonts were likely to extend the history of the phylum of which we are a part back into the Cambrian, along with all the others. Oddly enough, one of the early suggestions about the identity of the conodont animal had been as a vertebrate relative, on the basis of its tooth-like apparatus, and because the elements were composed of calcium phosphate, the stuff of bones and teeth. Speculations are sometimes hardened into facts by circuitous routes.

My own footnote to this discovery is provided by the Giant Conodont Animal of South Africa. In the early 1980s I was sent some trilobites that had been discovered in a soft shale that crops out in the beautiful and rugged Table Mountain area of South Africa. The rocks from which they had come were alleged to be Silurian, but a detailed examination of the new finds proved instead that they were likely to be late Ordovician—a worthy small grace note on the theme of the history of the world, but scarcely worth more than a passing remark in a textbook. However, the new search through these shales turned up some other fossils, too, and one of these was described as a very early plant of terrestrial type, which was dubbed Promissum pulchrum.

It was, with hindsight, curious that such a plant should turn up alongside wholly marine trilobites. And indeed a few years later Dick Aldridge realized that the "plants" were in fact a collection of giant conodonts; more, that they were an "apparatus." Much breaking of Soom Shale eventually led to the discovery of another conodont animal, complete with muscle bars and fins (and there were a number of other soft-bodied animals besides). But this was a giant among conodont animals, the size of a small fish. At a stroke, it became plausible that conodont animals could have been more than mere plankton-eaters. And these new specimens showed something that had not been seen on the Scottish ones: at the front end there were a pair of globular bulbs—eyes! This was starting to look more and more fish-like: a reconstruction prepared from the Soom specimens (figure 26) has more than a passing resemblance to the fry of a salmon. The conodont animal was no blind groveller, but a visually sensitive swimmer that must have been as important in the Ordovician seas as small shoaling fish are in the Atlantic today. Thus it was that a plant became an animal, and an obscure enigma swam gradually towards the light.

I HAVE SPENT MUCH of the last twenty years trying to imagine what it would be like to wallow in Ordovician oceans; on my own journey through geological time this is where I have lingered most in my own halting voyage to enlightenment. It is impossible to catch more than glimpses of such long-vanished worlds. The excitement of discovery comes, as with the conodont story, when clarity breaks through the dimming confusion of time to vouchsafe a sight of something hitherto obscure, much as the summer sun dispels a morning mist to disclose an unexpected and pleasing landscape. In truth, the past is never completely revealed, for imperfections in the record of the rocks guarantee that parts of the view will remain indistinct. Nor can a few inquisitive souls reveal more than a handful of scenes from an Earth that was already complex and varied 450 million years ago. The constancy of ecology and marine habitats provides no more than crude shafts of light into the murky past. None the less, the endurance of these habitats and their varied ecological opportunities has provided at least a few familiar reference points, while so many other aspects of history have been more fickle and uncertain. Mutability has ruled: the face of the world has transformed many times, sea-changed again and again into something rich and strange.

The geography of the continents has changed repeatedly. The notion of what was once called continental drift, and is now called plate tectonics, has become almost familiar, for all that once upon a time it reduced serious scientists to apoplexy. It is now not difficult to piece together in the mind the distinctively crooked outlines of Africa and South America; nor yet to swing Australia back across the vast Pacific Ocean into its original place near to Africa; nor to close the Atlantic Ocean to anneal Europe and America into a vast whole. The idea of a single supercontinent in the Triassic, Pangaea, some 220 million years ago, is almost common knowledge. The continents split apart as the oceans slowly grew, a growth driven by the inexorable engines of convection deep in the interior of the Earth, which carried the continental plates on their backs, just as the Titan Atlas was once supposed to have supported the globe in its entirety. Earthquakes are the shudders of the reluctant crust as plates plunge under one another, or jerkily rub shoulders, buckling freeways in California or crushing hypermarkets in Japan in the process. Volcanoes are the fiery legacy of melting crust in sites where plates die on the edges of continents, or of virgin crust where plates are born at the centre of oceans. The restless Earth moves at a pace that is indifferent to human timescales and, once in a while, catastrophically indifferent to human life. Maps of the past have been drawn by plate tectonics, and life has had to respond to different maps as the geological history of the continents evolved.

But Pangaea was not the beginning, no stable Eden fixed before earthly plates began their ineluctable grind. Pangaea *itself* assembled from earlier fragments fringed by vanished oceans: change summed up upon change. The movement of continental plates is a continuous process extending backwards in time even into the obscurity of the Precambrian. The world before the Permian is progressively less certain, the maps more tentative. You have to imagine the vanished supercontinent of Pangaea first; then cut it further into new pieces. Sometimes the cuts will follow the lines of later oceans, being ancient lineaments of weakness set upon the face of the Earth for 1,000 million years or more. But there will be new lines, too, marked on the present globe by the remnants of ancient mountain ranges. In some cases these are rather obvious: the line of the Ural Mountains snaking across the middle of Asia looks like the crude stitching of an old scar. And so it is, the annealed join of two halves of Asia once separated by an ocean and driven together to form part of the core of Pangaea. In other cases, the joins are more subtle and controversial: across the middle of China, running through Shanxi Province, the Tsinling Mountains are now thought to mark the course of the edges of one of these vanished seas. I can remember an old Chinese professor politely telling me that any such notion was patent nonsense, but then it is scarcely a routine matter to reshuffle the Earth.

In the last twenty years I have devoted much thought trying to imagine the geography of the Ordovician, piecing together the shadows of vanished worlds. This is like completing a jigsaw puzzle fabricated of scraps and dreams. The fossils have been my guide. I have chased them through jungles and across deserts. I have tracked them along Arctic shores and behind abandoned barns in Wales. This fieldwork has kept my feet on the ground and my hands on the rocks. Theories come out of pounding hammer on shale and limestone, conjuring visions of the ancestral faces of continents.

In such a fashion I have followed the Ordovician equator. Hot climates produce specific types of sediments today as they always have, and these in turn become converted in time to particular rocks—massive limestones, or those which evaporated within isolated, inland seas under the tropical sun. Certain species of animals are confined to these same warm climates. So it was in the distant past. Hence it is possible to follow the ancient equator by identifying the right kinds of rocks, and finding fossils of the animals that relished tropical heat. In this way I have traced out a line running from the United States through Canada and icy Greenland, over the Arctic Ocean to eastern Siberia, always collecting and comparing the trilobites and other animals entombed in the rocks to certify the legacy of a vanished shallow, warm sea, charged with lime, where the sun periodically dried out pools and cracked the mud into crispy polygons. Now, these are limestones, stacked in beds full of algal pebbles and the spiral shells of heat-loving snails. This pursuit of the Ordovician equator led me further, into central Australia, where the heat still beat with little pause upon rocks originally warmed under the same sun nearly 500 million years before. The rocks looked the same here, in central Queensland, as those beneath the ice cap in Greenland or Spitsbergen: the world had changed about them, propelling them to their present sites by the vagaries of plate tectonics. How different the world looked in the Ordovician, with its continents all shuffled about like cards on a fortune-teller's table.

Eventually, my researches on the ancient equator led me to Thailand. Fieldwork can be comfortable or spartan. Ever since my initiation on the Arctic island of Spitsbergen it seems to have been my lot to endure the tough variety. My coral colleague Brian Rosen is always off to Italy to sample ancient reefs that are tucked beneath the ramparts of some stunning *palazzo*. Corals crop out in small bluffs among the olive trees and bougainvillaea; of an evening, he strolls down to the little *ristorante* for *antipasto* and *risotto al funghi*. His Italian friends ensure that every theory hatched there concerning the distant history of coral reefs is matched by an appropriate liqueur, and the science is accompanied by the churr of cicadas, and convivial chatter beneath the pergola. When he needs to study a living reef in order to gain a better understanding of the fossils, he has to take off to some tropical island—or, perhaps, the Great Barrier Reef. He returns from such a trip tanned and splendid in Hawaiian shirt hung around with artefacts of Oceania. When you ask how it all went he shakes his head and purses his lips: "Up at the crack of dawn . . . tropical sun . . . poisonous fish . . . poisonous colleagues . . . no picnic, I can assure you . . ." Further probing reveals that he was forced to stay at the Hotel Magnifico, and that the Symposium field trip happened to visit Mustique.

My own fieldwork is quite the contrary. For some reason, trilobites seem irresistibly drawn either to cold and soggy climates, or else to mercilessly hot ones. My anatomy is fated to be boiled or blistered or chapped, and sometimes all three at the same time. Cordon bleu cooks eschew the Arctic, or the centre of Australia, or the deserts of Nevada, or even the damper parts of North Wales. Tents are always the standard accommodation in cooler localities. In deserts, the stars are our blankets. Ancient rocks outcrop most prolifically precisely where those other stars, awarded by Michelin, are nowhere to be found. The older rocks which are my special passion are unknown on tropical islands where corals thrive, for oceanic islands are geological youngsters, thrown up by volcanoes that still smoulder and bubble and, from time to time, erupt. Ancient rocks carrying trilobites are found along the edges of Precambrian continents, tucked away, buried in inhospitable hinterlands, or along forgotten seashores. And it has been proven many times that the best specimens are always to be found furthest from camp.

I spent years scouring the world for a site where I could emulate Brian Rosen; some place as yet unvisited where I, too, could amble along coral strands in the name of work. Eventually I located such a paradise. Tarutao Island lies in the Andaman Sea at the southern edge of Thailand, and close to the border with Malaysia. As a consequence of some geological quirk, the rocks of the mainland kink seawards here, carrying with them a cargo of Ordovician fossils. Uniquely, this island is where living coral reefs and ancient sediments rub shoulders, and here at last I could collect another piece in the jigsaw puzzle of the Ordovician world while my toes wriggled in tropical sands. Enquiries made of some Australian friends who had visited Tarutao disclosed that here was what a brochure would doubtless describe as a tropical Eden fringed with palms and white sandy beaches, and unsullied by the tourist herd. I could scarcely wait to get there.

I was not disappointed. A small fishing smack carried us from the mainland port of Pak Bara. Late in the evening the Thai fishing fleet takes to the shallow and fertile seas off the coast. Each small boat—and there is a flotilla of them—is decorated with a string of lights, which reflect in the dark mirror of the sea, so that the seascape is a twinkling regatta, a floating parade. I idly dragged my fingers in the sea, which was still pleasantly warm. From time to time a flying fish would launch itself upwards, startled into flight by the lights, and gliding on its mighty pectoral fins. On occasion, one even landed in the boat, a self-served supper. There were small huts to sleep in, and only the scuttling of crabs in the dark to disturb us. It was the perfect Ordovician hunting ground.

It turned out that the critical rocks were not upon the idyllic island at all, but inland, some miles from the sea, in amongst the rubber trees. The locality proved to be hotter than the antechambers of Hell itself. Sharp rattan palms scored my legs, and from time to time a malaria-carrying mosquito would linger on my thigh. There were rumoured to be venomous snakes in the undergrowth. There were certainly venomous ants which stitched together nests of leaves, and which when disturbed ran up my arms biting furiously. This, naturally, was where the trilobites wanted to be. How could I have been so foolish as to think that they would prefer the seaside?

The sociology of fieldwork in Thailand is as different from that in the high Arctic as can be imagined. On an expedition the roles within the group are paramount. There *is* no society outside the tent. But in a strange land, working within a different culture, etiquette accounts for everything. I soon discovered that my Thai hosts were the most polite people I had ever met. A few rules had already been explained to me: always remove your shoes indoors; avoid crossing your legs and pointing your toes at your friends; do not raise your voice (the quieter you speak the higher your status); do not touch your male colleagues; above all, always be respectful of the King of Siam. But this was only the beginning. It seemed that I was groping my way unguided through a thicket of behavioural rules. My hosts' tolerance was endless, yet it was clear that I was offending them in some respect. During our evening meals, which lasted for hours, course after course was brought from a bottomless menu. I had the feeling I was doing something wrong, but what?

Thai food proved to be relentlessly "hot." What chilli does to the metabolism is interesting. In my case it provoked the lachrymal glands, so that I frequently wept with emotion while complimenting my hosts upon some particularly delicious steamed fish. It also stimulated the same gland in my nose, with similar effects. Thus I would spend much of the meal weeping and sniffing, while at the same time smiling inoffensively and conveying with appreciative gestures of the chopsticks the quality of the food (which was no lie), always being careful not to touch my colleague in the Thai Geological Survey or to point my foot in his direction. From time to time I blew my nose and wiped my eyes. When I did so, the driver giggled. I realized that not blowing the nose was another rule I should have been told about. This was what I had been doing wrong.

Blowing was obviously out, but this left the problem of how to cope with the chilli. I developed a compromise between politeness and necessity. I would sniff surreptitiously for as long as I dared, until I felt I was at the limits of my capacity for inhalation, usually signified by a moist drop gathering towards the tip of my nose. Then I could lay down my chopsticks in what I hoped was a high status and nonchalant way and walk towards the lavatories as fast as could be managed while balancing a growing drip. Once inside, all dignity was abandoned. But worse still, there was often no tissue paper. This compelled me to secrete about my person little scraps of the *Bangkok Times*, usually used for wrapping specimens, with which I could relieve my tumultuous nose. I had to plan my evenings carefully in advance, and could be found during the day by my puzzled Thai field assistant tearing up the advertisement section of the *Times* into small squares. These I rolled into cylinders and concealed in my pocket.

Matters became worse when the girls arrived.

Almost every restaurant in the provinces boasts girl singers, who render current hits from the Thai top thirty to the accompaniment of a Hammond organ. The singers are accustomed to sit with the guests after they have performed their numbers, and because foreigners are a rarity in those parts the singers usually arrived unbidden at our table. They were invariably beautiful, fragile, elegant. Some were of Chinese extraction, others Malay, but all were exquisite. They had no English whatsoever. The rules of politeness governing behaviour towards these singers were a matter of great puzzlement to me. I was evidently supposed to make conversation with them through my host as intermediary and translator. I surmised that ritual compliments were required—and they were not difficult to pay. I rehearsed such trite sentiments as, "You must be the most beautiful girl in Thailand," and the fact that, duly translated, this provoked dimpling and smiling seemed to indicate I was on the right lines. The next song was already in progress and my host spoke very quietly (on account of his high status), so I had only a hazy idea of his replies to me as relayed from the girl, but I concentrated on looking amiable without ogling and on keeping the drip at bay. I then discovered that the desire to blow my nose was stronger than my desire to flatter strange girls, no matter how beautiful. It was a shock when the girl rather delicately placed her hand on my thigh. This, I reasoned, must be another custom. What she thought when her hand encountered my rolled up squares of the *Bangkok Times* stuffed in my trouser pocket I cannot say, but I did notice that she immediately sprang up on stage to render another of the curiously plangent songs which seem to get to number one in Thailand. For some reason she never came back.

THROUGH SUCH PRIVATIONS scientific discoveries are made. I came away from Thailand with boxes full of Ordovician fossils which nobody had seen before. I found proof that this remote area had been close to the equator in Ordovician times; in fact, it had stayed more or less where it was, latitudinally speaking, in complete contrast to the Arctic localities I had studied previously. Then I was able to date the rocks. Similar fossil trilobite species from China had been described and quite precisely dated as late Ordovician. This use of fossils as chronometers is a simple and practical thing. It requires only experience; no sophisticated apparatus costing hundreds of thousands of dollars is needed; even a hand lens is sufficient to identify some fossils in the field. It is a demonstration of the overused dictum that *knowledge is power*, although one could scarcely have a more innocent species of power than the employment of fossils as a stopwatch for history.

The very best chronometers for the Ordovician and Silurian periods are not trilobites, or even conodonts, but graptolites. At first glance they could be mistaken for scribbles on the surface of a slab of shale, a drawn series of lines (hence the name, from the Greek—literally, "writing on the rocks"). Look closer, and a graptolite resembles nothing so much as a hacksaw blade, a thin blade a millimetre or so wide and a few centimetres long, carrying a series of saw-tooth indentations along one side. In many soft, shaly rocks graptolites abound. They are also widespread. Similar species can be found around the world. Hence it is possible to facilitate the division of geological time using these strange rock markings. More rock than organism, they resemble hieroglyphs written in the language of time.

Look closer still. Each tooth on the "saw" is actually a tube. So the graptolite was a colony of tiny tubes a millimetre or so long. There might be a hundred or more such tubes in the colony. It could be compared with a coral, were it not for its delicacy and regularity. Of the tenant that once occupied the tube nothing remains. It must have been some sort of soft-bodied "polyp," but there are no living graptolites, and so their soft parts are accessible only to surmise. Graptolites are to be found in almost all those sedimentary rocks that accumulated well away from turbulent seashores in the Ordovician and Silurian; their skeletons were evidently delicate, easily destroyed by surf. Like conodonts, they were for a long time mysterious. At first it was assumed that they must compare with one of the commoner colonial animals in the Recent seas. Comparisons were made with various relatives of the "jellyfish" which lived in colonies and inhabited tubes—animals such as hydroids. Other authorities argued that they resembled rather the "moss animals," bryozoans, whose branching colonies are often mistaken for "white seaweed" when washed up on beaches after storms. It was quite quickly recognized that graptolites must have been capable of life in the open oceans, because their remains accumulated in dark shales which were deposited deep on the ancient sea floor, far from shore, in places where there was little oxygen at depth. Because of this, these shales offered no evidence of any other sea-bottom life: there were no trilobites or snails, or any of the fossil animals I had collected from Thailand. The muddy sea bottom must have been devoid of life. How, then, could there be such a multitude of graptolites? There seemed to be only one possible explanation: the graptolites were part of the plankton, drifting passively through the Palaeozoic seas. They lived their lives thus in the oxygenated surface waters, and when they died they plummeted or sank slowly down towards the bottom of the sea, where they were entombed in dark mud in the lifeless deeps. Because nothing else lived there, graptolites were buried alone.

The graptolites were a great addition to the ecology of the open ocean, probably the most important modification since the appearance of the simple planktonic community that was typical of the Cambrian. The animals that occupied every tube along the saw-tooth edge must have harvested still smaller planktonic organisms, including such simple, single-celled plants as had survived 100 million years or more from the Precambrian. Such precisely engineered colonial animals were a unique adaptation to the problems of leading a planktonic life: there is nothing like them in the Atlantic or Pacific Ocean today. I imagine that the Ordovician sea was sometimes clogged with tangled masses of graptolites; maybe conodont animals nibbled at their vulnerable, softer tissues. Even the deep water boasted different species of these extraordinary colonial drifters. Whatever the tiny animals that lived in the graptolite's tubes were really like, they must have been efficient harvesters of microscopic plankton.

They were not, after all, likely to be bryozoans. Their tubes were made of an organic, protein-rich material, and there was no sign of the typical special-izations of individual "cells" in the bryozoan colony. Nor was the hydroid theory very satisfactory, because graptolite tubes had a curious zigzag pat-tern of growth unknown in hydroids, and the great regularity of graptolite colonies was quite unlike that of these irregular, encrusting animals. It seemed that they were altogether more highly organized: enigmatic, free-floating, colonial and ubiquitous, another geological conundrum, a paradox drawn in the rocks.

Whatever their biological relationships might prove to be, their practical use in dating rocks was sealed by Charles Lapworth.

The nineteenth century produced several scientists who revolutionized knowledge, but Lapworth is not on most people's list of them. This is an oversight. He contributed several extraordinary insights, and it is only be-cause he is not associated with a readily identifiable achievement that can be summed up as a single phrase—as Darwin can be identified with "evolu-tion" or Mendeleev with the "Periodic Table"—that he is not a name to be learned by schoolchildren. He *did* name the Ordovician System in 1879, and perhaps it is no bad memorial to have christened nearly 100 million years of geological time. Charles Lapworth started his professional life as a school-master, a local man applying his intelligence to geological enigmas. He fin-ished his career as Professor at Birmingham University, in a chair that still carries his name. He was a practical geologist, marvellously observant in the field. His notebooks are a joy to read, full of legible sketches that allow you to see the rocks as he saw them, and unravel their truth as he untangled it. He started always from first principles, and the first of them all was the evidence of his eyes.

Graptolites were a key to unravelling the structure of an intractable swathe of land which had defied previous attempts at geological interpreta-tion. The Southern Uplands are a wild tract of hills that run east to west across the southern part of Scotland, a land of sheep and tumbling burns; the rocks there are a monotonous series of tough grits and soft shales that alter-nate in endless stacks through mile after mile of remote moorland. There is no sound save the groan of wind and the mewing of buzzards, keening sounds that seem to mock any attempt by an investigator stumbling over the hillsides to understand the rocks in three dimensions. Lapworth realized that the pale hieroglyphs on the dark shales—the graptolites—held the key to understand-ing the mysteries of the ground. Species of graptolite appear in a particular order: "each species and variety of graptolite has a definite range in the verti-cal succession of strata," as Lapworth wrote in 1878. They could be used to

trace strata across country, their signature reliably impressed upon the shales
as proof of the identity of the rock formations. The rocks had been twisted
and distorted in the Caledonian Mountains, of which the Southern Uplands
were the merest rump. Once horizontally disposed upon the Ordovician and
Silurian sea floor, the rock beds now were vertical, folded, even turned bodily
upside down. But with their character affirmed by graptolites the endless
tracts of shales and grits suddenly acquired personality. Distinctive species
could be welcomed like old friends. Clusters of species found consistently
together would now form the basis of "zones," a kind of minimum unit of
geological time. When the usual order was inverted, why then so were the
rocks. Thus it was that Lapworth recognized that the apparently endless
strata were actually repeated packages, folded again and again. Without these
drifting chronometers the Southern Uplands of Scotland would have remained an enigma. The gist of Lapworth's interpretation remains sound, for
all that plate tectonics have changed the setting and the detail.

It is sad that Lapworth did not live to see the solution to the conundrum
of graptolite nature. His interest lay in their utility rather than their biology.
The solution to the problem of the latter was analogous to the conodont
story, in that discovery of perfectly preserved graptolites depended on their
being extracted from the rocks in which they were entombed. They are occasionally preserved in limestones. I have collected thin limestones from Lag-
gan Burn, in Ayr, grubbing for them with my hammer under luxuriant ferns
and mosses. They turn out to be unprepossessing stuff, dull, dark rocks, but
when thrown into a vat of acetic acid back in the laboratory they dissolve, and
floating to the surface of the container, freed from the rocks after more than
400 million years, there are little tubes, unmistakably belonging to graptolites. The organic walls of the colonies do not dissolve in the acid. Released
thus, they float as they must have floated in the seas of the Ordovician. At
once they can be studied in a different way from Lapworth's hieroglyphs:
under a microscope they reveal wondrous details of colony growth and construction. This shows that the tubes are put together from increments which
meet on a zigzag line. Under even higher magnification, on the electron
microscope, another tissue can be seen on the outside layer of the colonies,
composed of crisscross strips, smothered on like bandages in plaster when
applied to a broken limb.

A pioneer in studying isolated graptolites was Professor Bulman at Cambridge University. When I was a student there Bulman had already retired,
but his aloof figure was always about, bearing a tray of graptolite material in
one hand, and a cigarette in the other. The former fuelled him, and the latter

eventually killed him. He was a slender man, with a slight professorial stoop, which exaggerated his prominent nose. Bulman was so diffident that he often failed to acknowledge your existence. After I did some work on graptolites, I merited a certain recognition, and he would greet me with "Ah, Fortey," and a kind of bemused surprise, as if to indicate that my appearance in front of him had served only dimly to confirm my existence. For all I knew, I might have belonged somewhere on a classification table not far from the jellyfish and the flatworm.

The distinctive tissue construction of the graptolite colony proved to be the key to deciphering their place in the animal kingdom. In the Recent seas there are some humble, encrusting organisms known by the name of *Rhabdopleura*. They are neither hydroid nor bryozoan, although they are colonial animals, growing over shells and stones in little patches. They, too, construct their colonies in the fashion of graptolites, using the same material; indeed, similar kinds of cross-bandaged tissues that make the outside of the graptolite colonies have also been observed on *Rhabdopleura*. These inconspicuous animals belong to a group of organisms known as hemichordates, primitive creatures which fit into the tree of life somewhere below all the animals with backbones (ourselves included). Thus was another enigma of identity resolved. The few living hemichordate species are evidently the tail of a much more glorious history, when their graptolitic relatives swarmed in the oceans of the world in their role as the prime plankton feeders, and left their remains on the dark sea floor in tens of millions of abandoned skeletons. In this fashion, they also left behind one of the best natural timepieces in the fossil record, which served to calibrate Ordovician and Silurian events around the world.

Charles Lapworth was an early example of a scientist rising by virtue of his native ability and hard work. In the nineteenth century geology was often a gentleman's occupation. It even had a distinctly aristocratic flavour, since many of the most influential figures were well connected. An afternoon spent with hammer in hand was an appropriate, even a manly way for a gentleman of private means to occupy his time. While the ladies botanized, the gents geologized. Before the camera became commonplace, a notebook was the routine way in which to record observations. Since a skill in drawing was also part of the young gentleman's education, the field notebook was a primary record, and, in aristocratic circles, often finished up enshrined among the master's effects in the library. Letters from notable colleagues were likewise preserved. These circumstances combine to make the efforts of nineteenth-century geologists transparent to historians. We know about the scientific

disputes, the snobberies of the investigators, their obstinacy or their feeble-
ness. I doubt whether the electronic age will bequeath such an archive. Fizzy
notes on the electronic mail will be deleted as soon as delivered, and the living process of investigation will go unrecorded. Whether the education
received by the nineteenth-century scientists was superior to that delivered
today is disputable, but it can scarcely be denied that they wrote better letters.
Feisty, full of invective and observation, even the jottings of obscure naturalist vicars or forgotten academics strike the modern reader as written with
verve and confidence. Perhaps the assumption of a fixed social order served to
render debate a more intimate affair. People could let their literary hair
down, and had the skill to do it with colour. The upper middle class were
confident of their position. Disputes, of which there were many, were conducted between occupants of a predictable social *échelon*.

But it was not so for everybody. John W. Salter was the first English
expert on my own fossil animals, the trilobites. He was of comparatively
humble origins. He named many of the species of extinct animals of Cambrian to Silurian age which had been discovered in the ancient hills of
England and Wales by a dozen or so clerical and gentlemanly collectors in
the middle years of the nineteenth century. He owed his advancement to the
most powerful man in British geology, Sir Roderick Murchison.

Murchison was an aristocrat with a powerful network of connections
throughout the "county" class in England and Wales. He was endowed with
sufficient means to pursue the geology of the older sedimentary rocks without studying his bank balance too carefully. More importantly, he could
count on hospitality and intellectual succour as he made his grand progress
through the shires. Local parsons opened their personal cabinets of collections to him and felt privileged to help him to forge his scientific syntheses.
He wrote two great works, *The Silurian System* (1839) and *Siluria* (1854),
which summarized the ancient rocks of Wales, drawing upon all he had been
able to glean from his sojourns in a dozen country houses. He was doubtless
indebted to all the lords, squires and clergy he met along the way, but, with
the confidence of his class, had little hesitation in claiming as his own much
that had been hard won by others. He was unscrupulous, arrogant and over-
weaning, but the only man capable of summarizing the ancient history of the
strange land of Wales, a mysterious geological realm hitherto beyond the
reach of reason, and of bringing it to the attention of an avid public. As Lapworth later discovered in the Southern Uplands of Scotland, the fossils provided much of the evidence for new interpretations. John Salter was central
to this endeavour. He was gifted with a subtle eye, able to distinguish the fos-

sils of each rock formation one from another. He published prolifically. His contract at the Geological Survey kept his family, and ensured a wide reputation. But he yearned for a permanent post commensurate with his talents, a post that should be within Sir Roderick's gift. It was true that Murchison effected the advancement of those who had helped him. He engineered the appointment of Archibald (later Sir Archibald) Geikie into the post of Director of the Geological Survey. This was in no small part because Geikie had supported Sir Roderick in his interpretation of the structure of the Highlands of Scotland (an interpretation which was later proved to be mistaken, and which Charles Lapworth contrived to dismantle). But it may also have been because Geikie was socially presentable. Poor Salter was not of the same class at all. He was regarded by Murchison as little more than a superior servant, no doubt ordered in his estate by the deity that saw to such matters. In spite of his service, Salter never got the preferment that he felt he merited. Perhaps it was his frustrated ambition which turned his mind, or perhaps it was the frustration of an underdog who felt forever inferior to his master. He wrote increasingly imploring letters to Sir Roderick outlining his plight and demanding his patronage:

> Sir Roderick [he wrote on 4 October 1867]—you have a good opportunity to atone for past mistakes—Back up my application to the Treasury for a pension. Get me placed on the Coal Commission (you know I am fit for this)— Give me your set of Welch [sic] maps. Put my boy Willie to school (St. Paul's or any good Grammar school) and then I will ask you to send your latest photograph—
>
> I need only tell you that sorrow has *crazed* my wife; and she has left me— to let you know how much good yet remains to be done to
>
> J. W. SALTER

These pathetic pleas did not seem to move Sir Roderick. A supplicant who pursues his case too vigorously and hysterically may induce precisely the opposite effect to that he intends. John Salter succumbed to despair. On 2 August 1869 he took his own life by jumping off a steamer into the river Thames, where he drowned near Thames Haven. A few minutes earlier he had given his young son William the gold watch that he had been awarded many years before in recognition of his geological achievements.

The description of fossil faunas was an international endeavour. At about the same time as John Salter was working in Wales, in Bohemia (now the central part of the Czech Republic) the Lower Palaeozoic rocks and their fossils were made known to the world by Joachim Barrande. From the mid nine-

teenth century onwards he published a series of tomes, in French, beautifully
illustrated by lithographs of the highest quality, and each the size of a volume
of *Encyclopaedia Britannica*, which remain his extraordinary personal monument. There is a suburb of Prague which is known to this day as The Bar-
randium, now the site of the Czech National Film Studios. I was astounded
to find local people there who seemed to know all about the man who had
donated his name to the surrounding rolling countryside. Barrande's good
social connections did him no disservice, either. His wife's considerable fortune was sacrificed to the publications, and who now would dispute that this
was money well spent? So Barrande was, like Murchison, endowed socially
and financially as well as intellectually. The change which took place between
the mid nineteenth century—through the time of Lapworth—and Professor
Bulman's era, when natural ability could win both income and scientific
publication, was a great step: it took science into the public arena, where pro-
fessionals were paid for their skill and knowledge; this was a triumph of intellectual egalitarianism. It became somebody's *job* to describe the riches of the
distant past, because it was felt that some advantage would accrue to the common good. Geological surveys were founded: in the United States, inspired
by the great James Hall, and also in France, Ireland and Sweden. The resulting collections were housed in public museums, where they still remain for
study. Those of us who have worked in such museums become accustomed to
recognizing labels inscribed with the fine copperplate script of the pioneer
curators, who systematized and sorted the discoveries of their colleagues. It
all came about too late to save John Salter.

The growth of knowledge of extinct organisms throughout the world,
and the progressive refinement of the timescale that Lapworth and his successors introduced, are advances of which we see no end. The richness of
vanished oceans filled the drawers of museums and the shelves of libraries.
From lists and monographs it was natural to progress towards summing the
parts together, to turning to the ecology with which this chapter began.
When did the seas begin to be like our seas? When did coral reefs and clam
banks and rich swarms of plankton first divide the resources of the sea into
ecological compartments which we would recognize as familiar? The natural
systems of the seas which comprise its ecology assembled themselves through
the Ordovician, between about 490 and 430 million years ago, on a shifting
world of strange geography which I sought to clarify in Thailand in the baking hot rubber plantation.

The Ordovices were a tribe who inhabited the Welsh hills at the time of
the Roman Conquest. Lapworth gave their name to the Ordovician System,

following Murchison's example, who had named the Silurian after another tribe, the Silures. Ordovician rocks form the substance of many of those hills, where stone walls amble over rises, and sheep bleat dimly at the passing walker from behind damp tussocks. I am deeply fond of this Ordovician country, for all its concealed bogs and many incomprehensible gates. In a few places, like Snowdon and Cadair Idris, the Ordovician rocks make true mountains, where they are reinforced by flows of lava and pumice which once erupted angrily over a sea swarming with brachiopods, trilobites and conodont animals, over which lazy graptolites drifted in their incessant trawl for plankton. This was a land that Sir Roderick Murchison helped to unscramble, and claimed for his Silurian System. The equally redoubtable Woodwardian Professor at the University of Cambridge, Adam Sedgwick, likewise claimed much of it for his own Cambrian System. The tussle between the rightful compass of one system or the other provoked a celebrated, fiery dispute for several decades in the Geological Society of London. Charles Lapworth's famous compromise of 1879 was to assign the older Welsh rocks to the Cambrian, and the younger to the Silurian, while carving out a middle time period between the two, which he named the Ordovician, thereby settling a dispute that might otherwise have continued to the crack of doom. It was some time before the world accepted the compromise, for all its eminent good sense. I have a Russian monograph published about the time of the Second World War which still used "Silurian" in the old, Murchisonian sense. But graptolites, and the universal clock derived from them, stimulated the recognition of this huge slab of geological time all around the globe. Everyone now knows the Ordovician for what it is, the meat in the sandwich between Cambrian and Silurian, and every time the word is used it is a nod towards Charles Lapworth and his good sense.

Now, the Ordovician seems not merely a clever compromise, but also an entirely natural division of Earth's time. The Cambrian oddities were replaced by shells we know. A questing geologist patiently breaking open Ordovician sedimentary rocks to find their organic remains would discover hints of a world half-familiar. He might discover that reefs had acquired corals, and such reefs might not, at first glance, have looked so different from those around Bikini atoll today. Clams were suddenly there in the kinds of shore sands where we might gather cockles today, for it was at this time that they learned to burrow. Other organisms, which left no record of their bodies, churned up sediments to depths of some metres, just as happens at the present day where the sea floor is soft and well oxygenated. The geologist would find their activities recorded by mottling of the rocks produced when these vanished "worms" made their burrows. Then there were the remains of large predators in the distinctive guise of nautiloids, some of them curved like rams' horns and obvious cousins of the living pearly nautilus; others straight and huge, longer than a man. In Sweden these extinct nautiloids make small cliffs; in China, they make up much of the rock that was traditionally used in building pagodas. As we saw previously, there *were* predators in the Cambrian, too, but in the Ordovician it looks as if they swarmed in every habitat. If the nautiloid did not get you, the fearful sea scorpion would have another attempt. This was an early, subaqueous relative of the true scorpion, larger than a lobster, equipped with great pincers, the purpose of which does not have to be guessed at, and which provoked the evolution of protective devices. Many shellfish became larger, and acquired thicker shells. Their remains started to accumulate in beds like the "reefs" of oyster shells that are found now at the edges of estuaries. Many kinds of animals appeared for the first time, and some of these are still with us. There were the first sea urchins, second cousins to those that now bristle against the white limestone rocks in the Mediterranean Sea; there were starfish which looked very like species that a curious child will meet in a rock pool; and there were graceful sea lilies. The sea swarmed with crustaceans; there were fish, too, but these were diminutive forms without real jaws. There were sea mats, and sea snails and sea cucumbers, all the inconspicuous creatures that hide under seaweeds in rock pools. In short, if we had been able to take a dive in the Ordovician sea it would not have looked so different from today's. When we examined it more closely we would have seen that there were trilobites and shells that we could not readily place in our gallery of living animals, but the way in which these animals scuttled over the sea floor—some burrowing into the sediments, others clearly in the business of eating whatever they could find—would not be so different from what we find now in the seas off Cornwall or Nantucket or Sydney Harbour. This is where the repertoire of behaviour was fully rehearsed, and, although many crises would shake the performance, the underlying script was written.

THE FIRST OF THESE CRISES was an ice age.

In the Anti-Atlas Mountains of Morocco there are bleak hillsides where rocks are perfectly exposed, almost unencumbered by vegetation. Goats nibble away at the few spiky bushes that remain. There is no need to strip away a covering of cow pastures to get back to the geological past here, for the rock formations stretch way across to the horizon like so many layer-cakes stacked one on the other: the passing palaeontologist has only to pick out the cherries. There, at the end of the Ordovician part of the local geological column, there is a strange rock formation, unlike anything that underlies it. It seems no more than a jumble of boulders just heaped together, but a closer look shows that there are smaller pebbles between the larger boulders; in fact, the whole curious confection "floats" in a matrix of fine clay. The rock resembles an old-fashioned bag pudding, a duff, in which meat and vegetables are dotted throughout. This rock is the legacy of an ice age, a little-known ice age, 430 million years older than the one which our ancestors negotiated in northern Europe when woolly mammoths roamed the tundra. The strange rock was dumped by the ice as it retreated: as the glaciers melted they dropped whatever cargo they had once carried, from big boulders to impalpably fine mud, and deposited it with the abandon of a tired traveller suddenly laying down his load.

This was a great event. There is evidence around the world of a massive deterioration of climate as the Ordovician period drew to a close. It became colder and colder. Few areas were left that were warm enough to deposit limestones. Trilobites that had, until then, only lived in the polar regions suddenly moved into lower latitudes. I even discovered some in Thailand, now lurking among the rubber trees. The reefs that had burst gloriously into Ordovician life retreated to who knows where. The great ice sheet centred upon what is now Africa, for the then South Pole lay somewhere in the north of that continent, from which glaciers spread outwards in all directions. There must have been a thick ice cover, like that on the Antarctic, because as it grew, the sea level elsewhere in the world dropped drastically. As a result of the climatic change, many kinds of animals became extinct; in fact, well over half of all the species previously living. This is a more serious extinction than took place as a result of the last, Pleistocene ice age. Some of my favourite trilobites, those that I had collected in Spitsbergen, died out in this refrigeration, which affected the whole world. The sea was poisoned, possibly as a result of massive quantities of nutrients suddenly released into the oceans after the ice began to melt, encouraging poisonous algal "blooms" which used up all the oxygen. Graptolites were reduced to one or two survivors, which shortly gave rise to all the Silurian species. The end of the Ordovician was a punctuation mark in the history of life, providing a natural end to the first great phase of diversification and reorganization of marine life. Those animals and plants that survived went on to make the modern world; had the list of survivors been one jot different, then so would the world today.

It may come as a surprise to discover such an ancient ice age. It was not the first, nor will it be the last, and nobody knows its cause. I have mentioned that in the late Precambrian there is good evidence of more major glaciations. I shall describe yet another in the Carboniferous to Permian times. It seems that every couple of hundred million years the Earth refrigerates. It is a periodic crisis. On our shifting planet, where continents are not fixed, but can slowly wheel through the tropics to the pole, there is also a periodic fluctuation of climate, hot to cold, cold to hot; it seems that the world is for ever in the grip of mutability. There are also many minor climatic cycles between the great crises of iciness, and there is no chance for life to become complacent, for how can life be unchanging when the world changes around it with such frequency? The relationship of animals to one another and to the environment—their ecology—operated on a series of designs which became re-established once again when the climate ameliorated after the Ordovician ice age. Reefs grew once more in the Silurian; tropical abundance again guaranteed a cornucopia of life in shallow, warm seas. But the great crisis at the end of the Ordovician reminds us how great events shape the world, and that after such events nothing is ever the same again. Were it not for such crises, would life not just slouch onwards in an unenterprising way, like peasant farmers whose practices seem to have changed not at all since the dawn of recorded history? The ice ages force all life through a kind of lottery, where, if you have adaptations that prove useful, you survive, and if not, you perish. It is not even possible to predict which qualitites might be useful to survival and subsequent profit. Who could have predicted that small, ungainly, armoured fish would survive to give rise to animals that walked on land, and one day learned to talk and reason when ultimately transmuted to human form, even as many species of beautiful trilobites would crawl to their final grave before the dawn of the Silurian?

Chapter 6

Landwards

T HERE CANNOT BE a more important event than the greening of the world, for it prepared the way for everything that happened on land thereafter in the evolutionary theatre. A love of green is not just a sentimental attachment to rural holidays remembered from youthful days, *green* days. It runs deeper than that. In desert countries the rich sheikh celebrates his fortune with a garden sequestered away from the sun. We admire grandeur in wild scenery, mountains, canyons, deserts and glaciers. In such territory eventually this majesty begins to pall; a vague sense of dissatisfaction creeps in. Something is missing. But in greenness there is repose. It has been proved that the green wavelengths are least irritating to the retina. Red is angry, blue is cold, but green is restful. In *The Golden Bough* Sir James Frazer catalogued endless human ritual variations on the theme of welcoming the return of spring, of rain, of foliage and growth. There is the medieval Green Man, an atavistic figure to be found carved in stone or wood (rarely, coloured) in European churches and cathedrals. On his wild face tendrils grow from nose and brow. As they branch out they bifurcate, and give off simple leaves. They might be a representation of an early plant, like *Cooksonia* or *Rhynia*. All terrestrial animal life grew up alongside green plants, and we are still bound to them. Green still represents fertility and fecundity, and even amid the sophistication of city life, there is still some urge that drives people to parks and gardens. In China I joined a queue of hundreds of quiet Chinese in order to admire the plants in the great garden of Hangzhou. This was in the time before self-expression was encouraged, and the comrades lined up patiently in their blue serge suits. In a recess of the unconscious mind there may still be an awareness of the greening of the landscape in Silurian and Devonian times, that first Eden. Andrew Marvell's wonderful lines are worth recalling:

> The mind, that Ocean where each kind
> Does straight its own resemblance find;
> Yet it creates, transcending these,
> Far other worlds, and other seas;
> Annihilating all that's made
> To a green thought in a green shade.

During the Cambrian one-third of the world was devoid of life. The barren area was the land surface away from the sea. There may well have been bright stains of bacteria around springs, and covering such rocks as were washed regularly by showers. But the landscape would have been devoid of any softening tones of green. It would have seemed, to our eyes, naked and harsh. Nothing would have been there to consolidate loose soils, to absorb the worst of the weather, so that every rainstorm would have prompted a small flood, and stones and pebbles cascaded down slopes and tumbled freely into the choked beds of rivers. This was a world of erosion, one that left the elements free to wreak their worst. We might catch a glimpse of such a world in the remoter parts of the Arabian Peninsula today, where no trees break the hard horizon. Sandstorms whistle across the plains, or slowly build dunes. The rock is sculpted by wind and sand, which finds out every weakness. When there is a storm the flood carries great quantities of rock away; the waters gather finally into basins, where they evaporate into glistening salt pans. There is no sound save the wind, nor any sense that it was not always thus from the beginning of time. In mountainous regions in the Cambrian we must imagine a high and unclothed Himalaya, where frost prised slabs of rock away, and rain carried off its stony booty almost as fast as tectonics could uplift the ranges. Great fans of scattered rocks would have carpeted the plains beyond. There might have been a certain stark beauty to the landscape, but our eyes would soon have sought the comfort of an oasis, or a green valley.

    The greening of the land was an extraordinary transformation to the beauty of the world, not merely an opportunity for a great expansion in the compass of life. I like to imagine the breakthrough happening as a dramatic change of scene, so that plains and mountain slopes alike were verdured in a geological instant, and all the bleakness of the ancient landscape was suddenly painted over with a thin carpet of photosynthetic green. In fact it was probably not like this, neither so sudden nor so complete. The colonization of the land by plants entailed surmounting several important and tricky physiological obstacles. If one plant species made such a breakthrough away from water it must have had a hundred precursors which had solved these difficulties before it. It is unquestionable that algal ancestors lay below the first land plants. It was once thought that the first truly terrestrial plants were Devonian in age, but then examples were found from Silurian rocks. Now, students of this early history are looking for the kind of spores which might have been dispersed not through water but through the atmosphere, a sure indicator of terrestrial habits, at least at some stage in development. These spores are more easily preserved than the delicate fronds of whole plants. They can be extracted from sedimentary rocks by dissolving them in hydrofluoric acid. Spores* with the right kind of design and ornament have now been found in rocks of Ordovician age. This confirms, sadly, that my dramatic vision was incorrect, and that the greening happened over tens of millions of years.

There was not merely the business of moving from an environment supported by water to one of thin air. Tissues are delicate, and desiccation is the worst kind of destruction that can happen to them. So it was necessary for plants to develop a thin, waxy coat on the outside of their green fronds, which could cut out, or at least much reduce, water loss. Surely the first plants would have hovered tentatively at the water's edge, ever vulnerable to just too much roasting by the sun, crisping when the conditions deteriorated, plumping up with rain or dew. Breathing through such a waxy covering then presented a different set of problems—after all, what retains moisture also excludes air. The compromise is a special arrangement of cells in the surface of the plant to make *stomata*, minute holes guarded by sausage-like cells which can admit air but which expand to close the apertures when conditions get too drying—a natural air-conditioning system at a microscopic level. Once armed with waxy coat and stomata, a flat shoot could creep over mud, and put up shoots into the air, extending short roots to absorb nutrients even as the green surface photosynthesized and grew, expanding into a new world of air and light.

If you kneel down by a small stream bank where the atmosphere is moist and dank, listening to the tumbling water, brush aside overhanging ferns and

*They are termed *cryptospores* from the Greek, meaning "hidden."

on the bare bank above the water line you will very likely see such a prostrate green leaf. Dark green, and perhaps slightly crimped, this plant will make little effort to lift itself off the muddy bank. It forms a verdant crust. It seems to revel in the dim light, soaked, and skulking where nothing else grows. This is a liverwort. It may be as close as we can get in the living flora to those ancient plants that first made the break from water, though there is little enough in a liverwort that can be preserved as a fossil. It is a plant so primitive that it does not even have stomata. Now, liverworts are confined to the edges of water and to dim caves, once more occupying only the niches in which they may have started. Photosynthetic pads like these have a two-dimensional spread over mud and soil. Exposed thus to the sun's heat and radiation, like a prostrate sun-bather on the beach, they must resist the effects of harmful ultraviolet rays. It will be recalled that these rays are partly filtered by the ozone layer, which is formed high in the atmosphere by transformation of oxygen molecules. Presumably, there must have been enough oxygen in the atmosphere of the Ordovician to start such a protective process, the long, patient result of 3,000 million years of algae and "blue-green" bacteria releasing oxygen into the atmosphere. This reminds us of a link with the earliest days of the Earth: in a sense, leaves were conceived in the first, tacky microbial mats.

The task of almost any other plant you care to name is to grow upwards, towards the light, aspiring to the sky. This at once creates a whole environment: protected, sheltered places beneath stems and foliage. But prostrate mats require some stiffening before they can rise above the mud, and they need points of growth where shoots can be concentrated so that they do not grow haphazardly. As shoots get longer they need a system of transporting water from the roots to the shoots, to keep the whole structure turgid; contrariwise, the products of photosynthesis need to be distributed to the rest of the plant from the leaves. The plant, in short, is little more than a photosynthetic factory with problems of distribution and supply like any manufacturer. I wonder if it is coincidence that complex and large factories are referred to as "plants." Lest this seems a drearily reductionist view of the wonders of plants, consider what happens when any part of the system fails: that wilting pot plant deprived of water; the etiolated and spindly specimen deprived of light; the floppy failure forced to grow too fast. Plants are what late-twentieth-century systems analysts would call integrated feedback systems, and if the feedback fails they let you know by wilting, and eventually by dying. The technical problems that need to be overcome to make a viable, upright plant are problems of plumbing and engineering, chemistry and aerodynamics.

In the first place, there is support. Some plant organs achieved a certain measure of rigidity by their cells clubbing together, leaning on one another for mutual support. When I was a child there were still a few farms on which the wheat was harvested using equipment less sophisticated than the combine harvester. The cut sheaves were stacked into standing bunches which dotted the fields. They were known as "stooks." It is odd to reflect that this is a truly obsolete word, for other than in the background of a painting by Constable or Brueghel I cannot imagine where a stook will be seen in the future. But the principle is not so different from that behind the construction of the stem of a plant. In cross section most stems are composed of stacked, long tubes, which associate together for strength. The tubes conduct fluids—sap, the "blood" of plants. The cell walls acquired a thickening, often making a characteristic spiral pattern. The tough, organic building material lignin further reinforced the structure. Plumbing and engineering combined to allow upward growth. This full complement of features is typical of vascular plants that clothe the world today.

Shoots may have come before leaves. Some of the earliest plants in the Silurian were small and creeping, sending up upright shoots that bifurcated, once or twice, like candelabras. None was identical with any plant still alive, although there was a passing resemblance to some of the simplest mosses. Leaves were often nothing more than small scales that clothed the shoots. It is known that the earliest fossils of land plants are associated with sediments that accumulated in rivers and estuaries. It was around some ancient lake or river that shoots first poked their heads above water. But why did they leave the sheltering water? What prompts life to nudge its way across such thresholds?

The question can be asked as appropriately here as at any other point in this narrative. Some writers, especially geneticists, invoke the imperative drive for reproductive success, the urgent bidding of DNA to propagate itself until it subdues the very stars. But then the time was also undoubtedly ripe. If the planet had an evolutionary history which was both part of, and engendered by, the organisms that shaped it, then this was the moment when new things almost *had* to happen. The atmosphere was ready, the climate was ready. But chance alone could equally be cited as the prime mover. Ponds and lakes are notoriously unreliable habitats, drying out at some times, overflowing at others. The crucial engineering changes possibly happened as a way of coping with unpredictability. How much better to weather a dry period than to die out completely, and what better incentive to favour plants which could reproduce by means of spores, dispersed through the air? The earliest stages of the story are indeed recorded only by the spores that spread beyond the habitats of the first land plants and were left behind in shallow seas and deltas. The microscopic spores themselves had to be specially constructed to survive out of water: they have a tough coat, the exine. This enveloping cover is also impalpably thin, but still serves to protect delicate reproductive contents. Its toughness is confirmed by the survival of these tiny objects for hundreds of millions of years. Maybe one day a fortunate hammer blow will reveal the *Archaeopteryx* or *Australopithecus* of plants.

Early plants are preserved as black films on shales; often they look little more than dark scribbles on the rock surface. Their fine detail has to be teased out by means of preparations which clear their ancient tissues. Diane Edwards, professor at the University of Wales, waxes ecstatic over a dark smear conjured from the early Devonian rocks that crop out along the mouth of the river Towy in South Wales, because she can guess what details they may yet reveal. She has contributed more than anyone else to our knowledge of the vital steps taken by humble plants in ancient times.

Plant fossil hunters particularly like to find sporangia, the vessels that held the spores. These little sacs can be picked apart and the spores they contain extracted. In this way we can attempt to link the common fossils known *only* from spores with actual species of extinct, primitive plants. Sadly, many fossil shoots are sterile. But the cases that are reliably connected show that there were progressive adaptations to wind dispersal through early plant history, as proved by smaller spores. Occasionally, petrifactions are discovered which allow for a complete reconstruction of plants, down to the last cell. In the most famous of these, the Devonian Rhynie Chert, a hard, greyish rock discovered in the early years of the twentieth century in Scotland near the village that gives it its name, each cell has been filled immediately after death by silica, so that when the chert is sliced thinly, the minute structure of the plants can be examined. In some of these plants the vascular tissue is well developed, and it is not so difficult to imagine them being related to ferns. Ferns are presumed to have a history almost as long as that of liverworts. Living ferns, too, belong to the realm of green shade, although some species are tough and can endure drought. I was astonished to find them growing on arid slopes in the middle of Australia, in the company of *Spinifex* and *Eucalyptus*. On the underside of the fronds the little spore sacs appear as brown dots. If they are dusted on to moist peat the spores will germinate; but the little plant that arises is not the spore producer. Instead, specialized portions of this reproductive generation produce male sperm; eggs are produced in another. Only when the one has been fertilized by the other can the mature fern grow. This fertilization happens as sperm swim through the medium of water, be it ever such a thin film. Hence the fern still has a link to its algal forebears, and an obligation to the water from which it crept, and, appropriately, ferns are still at their most extravagant in places like the forests of New Zealand, where there is abundant and continuous rainfall. There, tree ferns erupt skywards like fireworks; epiphytic ferns hang from branches, as delicate as lace; creeping ferns cover every patch of untenanted soil. There is nothing obsolete about ferns, even though their fossil record stretches back to the days of the first tenancy of the land.

There are some places which are holy sites for palaeontologists. The town of Ludlow in the Welsh Borderland is one of these shrines: it was in the vicinity of Ludlow that many of the important stories concerning the conquest of dry land by plants and animals were constructed from clues found in the Late Silurian rocks. The town retains a character that neither one-way systems nor superstores can obliterate. Timber-framed houses still lean out over the streets, surprising in their top-heaviness, and showing off their oak timbers with an ostentation that signifies centuries of comfortable and peaceful affluence. The Feathers is an old and fine hotel decked out in this black-and-white livery. Geologists stay there, and have done so since the early days. You can mention Ludlow to Czech or Estonian geologists and they will grin with recognition, pat you on the shoulder and say: "Ah . . . the Feathers! Very *good.*" There are rooms with eccentric angles, and watercolours that are difficult to hang straight. There is good ale, without which no scientist can draw any conclusions of any worth or lasting value. People still assume that you are honest there, and do not demand twelve kinds of identification before they will cash a cheque.

There is a tract of land running from Church Stretton to Ludlow in the east, to Bishop's Castle in the west, along the old border with Wales. It is not grand country, if by grandeur you mean spectacle. But I find that the Long Mynd, an ancient and wild inlier of Precambrian rocks that stands high above the surrounding low hills, is all that is needed in the way of spectacle. I am always trying to find excuses to return to Bishop's Castle, where Miss Sheckleton used to provide the perfect bed and breakfast in a Tudor house with flag floors and improbable staircases. Her breakfast was neither too much nor too little, the egg yolks were soft and the whites were cooked, the toast warm and cut into triangles on which to spread thick, home-made marmalade, and pots of tea arrived unbidden. At such a breakfast time nobody could be more content than a geological researcher looking forward to the day's discoveries. In Shropshire, wandering up streams, hammer in hand, you could be forgiven for thinking that little had changed since the Devonian period, for there are still ferns and liverworts galore: male ferns stacked against the banks, polypodies sprouting from rock surfaces like so many feathers, and all along the stream sides there is still a crust of richly green liverworts.

At the end of the Silurian period, and in the following Devonian, this part of the Welsh Borderland lay to the south of a growing Caledonian mountain range which stretched away to where the Highlands of Scotland now spread the northern part of Britain with wild, boggy moorland and glens. The peaks that remain there now are but ground-down relics of what once extended as far as northern Norway. Towering like the Alps, snow-clad peaks fed streams which carved away the mountains even as they grew, carrying material away to be deposited on plains beyond and, ultimately, to the sea. Gradually, the sea retreated southwards as the apron of debris, mostly sand and silt, accumulated from the slow demise of the Caledonian Mountains. In many areas the iron that weathered away under the ancient tropical Sun colours the rocks a rich, deep red, or even purple. These rocks have long been known as the Old Red Sandstone. In South Wales it was these rocks that yielded to Diane Edwards some of her favourite fossil plants. Where cows and sheep now graze in the peaceful fields of Shropshire there was then a maze of estuaries, streams and lakes, over which the sea stepped from time to time, although with less and less frequency. The deposits of the riverbank, or the floors of estuaries, have yielded many of the remains that record the very transition from sea to land, by way of intermediate habitats. What is now a domestic landscape was witness to the most profound events in the history of life. It somehow seems too tame, this comfortable countryside: one almost demands a spectacular geological feature to match the profundity of the events recorded in the rocks. For here there are cadavers of early fish, some still so incompletely known as to be mysteries. There were fish the size of herrings without hinged jaws, just a simple opening for a mouth, and many, like *Cephalaspis*, were more or less protected by hard, bony plates on the outside, and were almost certainly feeding on algae and other organic materials grubbed from the mud. During Silurian times fish must have made a transition from marine habitats to being capable of coping with both fresh and brackish water. Once they had overcome the physiological problems of making this transition, of swapping salt water for fresh, one can visualize these sluggish animals grubbing their way upstream, following the nutritious mud, tapping willingly into a resource that hitherto had been exploited by nothing more advanced than bacteria. For a brief heyday, it was a good time to be a simple fish.

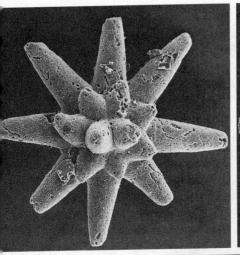

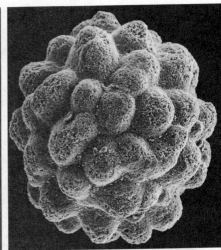

20. Skeletal fossils of animals from the early Cambrian of South Australia (photos courtesy Stefan Bengtson); enlarged between 50 and 150 times. Small shelly fossils show a great variety of shapes, but can be problems to interpret in terms of living relatives.

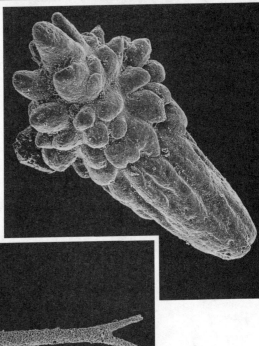

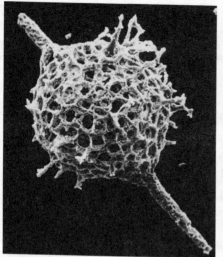
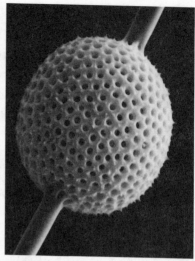

21. ABOVE: Radiolarians—
a protistan group with a long
history and siliceous skeletons.
The example on the left, a
millimetre across, is from
Ordovician rocks in Spitsbergen;
on the right is a living radiolarian
from Barbados.

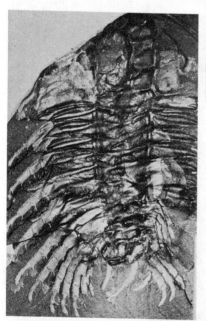

22. Burgess Shale (Cambrian) arthropods
preserving amazing details of their limbs:
Olenoides (trilobite), left, and *Marrella*, above

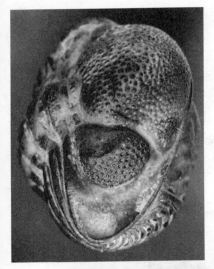

23. A trilobite (*Phacops* from the Devonian) showing its sophisticated eye—note the individual lenses exquisitely preserved for hundreds of millions of years.

24. Reefs ancient and modern: the structure of the Silurian reef in Shropshire is very like that of the modern reef (left) even though the organisms of which it is composed are only distantly related.

25. Conodonts—an assemblage of tooth-like conodonts, each a millimetre or so long. For many years their precise nature remained an enigma. Now they are thought to be allied to vertebrates.

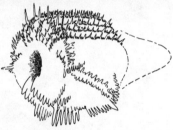

26. Professor Lindström's reconstruction (far left) and the *real* conodont animal as recently reconstructed by Mark Purnell (left).

27. Graptolites "writing in the rocks." The saw-like edges were the tubes inhabited by small living animals. Ordovician *Didymograptus* on black shale (natural size)

28. Charles Lapworth, namer of the Ordovician System

29. Tree ferns still form an exuberant canopy in New Zealand.

30. The Feathers in Ludlow, a well-known rendezvous for Silurian palaeontologists

31. A thin section through Devonian Rhynie Chert reveals the minute details of each plant cell—a snapshot of the early colonization of land. This is a section through *Rhynia*, magnified forty times.

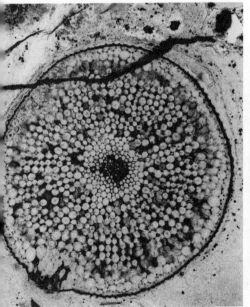

32. Willi Hennig, the originator of cladistics

33. An early and distant relative of the spiders, a *trigonotarbid*, painstakingly reconstructed from sections by Jason Dunlop of Manchester University

34. BELOW: Part of the skeleton of "Boris" from the late Devonian of Greenland, the many-fingered early land vertebrate. Its scientific name is *Acanthostega gunnari*.

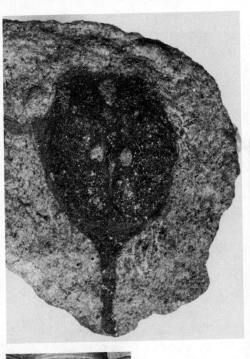

35. Jawless fish, *Spizbergaspis*, from the Devonian of Spitsbergen. The head of this animal clearly shows the paired eyes behind the forward nose.

36. A spiny trilobite, *Ceratarges*, from the marine Devonian rocks

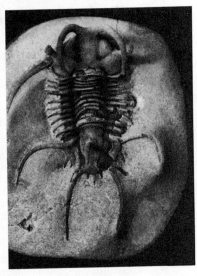

37. LEFT: The wonderfully regular bark of the Carboniferous tree *Lepidodendron*

38. FAR LEFT: One of the ornamental columns in the Natural History Museum, London, based on exactly the same species of *Lepidodendron*—a case of craft imitating nature

39. The living *Selaginella*—
a humble survivor from the days of
the Carboniferous giant trees

40. Carboniferous cockroach
Aphthoroblattina johnsoni. This
specimen comes from Coseley,
Staffordshire.

41. The "Bolsover dragonfly"
Tupus diluculum—a giant of its
kind—flying in a Carboniferous
swamp

It is interesting to reflect on how historical accidents played their part in the story of life. Were it not for the convulsions of the Earth which threw up Caledonia and the Appalachian chain* there would not have been the myriad lakes and streams in which nurseries for new species could be established. Species are produced by isolation, and in the Devonian mountain chains these unique historical circumstances allowed for little sequestered pockets which in turn permitted animals to go their own way. To take an appropriate linguistic analogy, there are more extreme dialects in Norway among a few million people than there are in the whole of the vast United States. These dialects developed along the thin, but impassable Caledonian chain that follows the present Atlantic coast of Norway. Glacial erosion has sculpted this coast into a series of spectacular and deep fiords. But contact between settlements on the shores of adjacent fiords is limited, other than by sea. Before modern communications had become established, local Norwegian populations developed their own ways of speaking, almost their own languages, in isolation from their neighbours.

It is not difficult to imagine that, tucked into equally isolated crannies of the ancient Caledonian range, there were places where adventurous, innovative species could thrive away from too much direct competition. Land life was born and bred at the edge of dying mountain chains. These vanished Alps and Andes themselves were thrown up by what Charles Lapworth called the "great Earth engine" of tectonics. The great breakthrough to land was stimulated by the restless crust, upon which life was carried as a flea may be carried by a bison. It was the disappearance through subduction of oceans which had been wide in the Ordovician that brought continent into collision with continent and threw up mountains that stretched through the Appalachians of the southern United States, through Vermont to New Brunswick and Newfoundland, and ultimately through highland Scotland to Norway. Genesis was as much about geography as genetics.

EVEN AS THE PLANTS WERE CREEPING over the land, animals were leaving the water, too. Their early history is, if anything, even more obscure than that of the plants. It will come as no surprise to learn that the arthropods were among the vanguard. Some have even claimed that terrestrial tracks showing the whiskery or pleated series of marks typical of arthropods indicate

---

*I should add that there were other and similar mountain chains developed at the same time in other parts of the world, and that these further enriched the process.

that they may have accompanied—even predated—the landward march of plants. Such tracks are certainly known from the Silurian rocks of Western Australia. But it is probably special pleading to claim such precocity for the arthropods, because it will be recalled that evidence of land plants in the form of their cryptospores has been found in rocks as old as Ordovician. But it is doubtless true that there are fossil footprints of arthropods in localities where the arthropods themselves are reluctant to put in an appearance. The animals that made these footprints may have been too delicate, easily washed away, their skeletons dispersed and broken. But their tracks were evidently rather easily preserved on the sandy flats that surrounded temporary lakes, or on the levees that flanked the rivers. Thus may the most casual of actions in the life of an animal be destined for immortality. The transient may be linked to the eternal:

> To see a World in a grain of sand,
> And a Heaven in a wild flower,
> Hold Infinity in the palm of your hand,
> And eternity in an hour

as the poem by William Blake described it.

Arthropods enjoyed a brief residence at the apex of the pyramid of life during the period of transition from water to land. Water scorpions—eurypterids—haunted the estuaries, lakes and rivers of the Silurian and Devonian. They were the giants of their jointed-legged kind. The longest species of the genus *Slimonia* was the size of a man. At its front were pincers; the flattened body was carried on spindly legs; there were eyes on the top surface of the head; at the back end there was a leaf-like "paddle." Other species had a spike at their posterior end which probably functioned as a sting. The appellation of these animals as scorpions is no coincidence—they were indeed relatives of those most poisonous predators in hot climates. In Mexico scorpions are still one of the most frequent causes of death, far outstripping snakes; the notion of a two-metre subaqueous scorpion is one that chills the blood. The largest specimen of *Slimonia* in the Natural History Museum in London is in a six-foot-long slab of sandstone, a mighty animal by any standards. It once fell on the foot of my assistant, turning his big toe blue for weeks, and proving that it could still do damage nearly 400 million years after its demise. These animals must have terrified jawless fishes, lurking in the shallows until some unfortunate creature crawled into the compass of their arms. The lungs lay on the underside of the body, oxygen-absorbing tissue tucked away, and folded and crimped like a napkin in a posh restaurant. *Slimonia* needed to husband its oxygen. I imagine it lying motionless for hours or even days at a time, much as some species of crocodile do today, until a misguided creature approached close enough to be grasped. Then . . . a sudden lunge, fast as the snap of a mousetrap.

There are limits to arthropodan horror, but they are imposed by the laws of physics rather than the constraints of imagination. The arthropod skeleton is a wonderful construction—but it works best at small size. The arthropods have an *exo*skeleton, a series of tubes that house the flesh and muscles inside. That is why we have to pick at lobster legs with delicate forks to get out the last of the meat. But as legs built on this plan get bigger—and this applies especially to support in air and on land—so the efficiency of such tubes as support dwindles; they need to have thicker and thicker walls, and then the operation of any kind of joint becomes progressively difficult. So does breathing. As in plants, oxygen has to be absorbed into the animal through an impervious skin. "Lungs" may be no more than small tubes—trachea—connecting with the surface of the body, through which air may pass and be absorbed by way of soft membranes into the body cavity. At very small sizes, its subsequent diffusion is sufficient. But as arthropods get larger the surface area of the animal increases in proportion to the square of its dimensions, but its *volume* requiring to be serviced by breathed oxygen increases as the cube. The arithmetic is obvious. A doubling in size requires an eightfold increase in oxygen absorption; there has to be an upper size limit beyond which breathing becomes almost impossible. So it seems likely that our eurypterid was torpid most of the time, lying in wait in some muddy ditch or stream, biding its time, harbouring its strength for a swift kill. (It will be recalled that the living parasitic arthropods known as bed bugs are capable of living without food for long periods of time until a meal presents itself, when they spring, or rather crawl into action with surprising alacrity.) Venom also helps to reduce the struggling-time of prey, a technique that many snakes still employ to advantage. The sea scorpions were soon outclassed as predators by jawed fish, but I like to think that the *frisson* of terror we still feel when we disturb a scorpion is akin to that felt by a Devonian fish disturbed from sucking up Devonian mud by the claws that lurked beneath it.

The Rhynie Chert preserved more than just early plants. The extraordinary seal from decay provided by siliceous water could entrap the most inconspicuous of animals. Plants and animals alike were petrified—turned to stone—by a hot spring, one that dribbled siliceous fluids around stems before they could decay. The silica replicated each cell with absolute fidelity. The plants probably grew around warm pools, fed by nutrients that accompanied the waters welling up from deep within the Earth. Beneath them there would have been a humid microclimate, and plant debris would soon have rotted to form the first humus-rich soil. This was a habitat, a haven from the hardships of a young world; a green arbour of opportunities to earn a living. In the Rhynie Chert there are spider-like creatures called trigonotarbids that were certainly equipped with massive jaws, and must surely have been predators. They did not spin webs, but probably simply crawled over their food and chomped it. But they prove that already by the Devonian there must have been a food chain. Then there are mites. These are both the most numerous and least conspicuous of arthropods, being at most a millimetre or two long. Today they teem in the soil, they clean up small particles, they feed on human skin and impinge on our awareness only because the droppings of house mites cause asthma. They are a food source for thousands of other creatures of the loamy dark. They share a distant common ancestor with the spiders. There is another tiny animal, a springtail, which has been claimed as the most primitive of insects. It has no wings, but like all insects it has only three pairs of legs. Living springtails can easily be observed as tiny, dark "bugs" around the edges of ponds; disturb the dank vegetation there and you will see little insects "hopping." The "spring" in the tail allows for this powerful escape reaction. The Rhynie Chert specimen comes complete with its "spring." They have almost no other fossil record, but their ubiquity today probably indicates that they have been pursuing their inconspicuous life habits since very early times. In another Devonian age deposit, from Gilboa, in the eastern USA, there are other small animals, including a strange creature called a pseudoscorpion. These miniature predators are still common in almost all damp places today, where they cull other tiny animals that live in soil. Their arms seem hardly to have changed since the Devonian.

What is extraordinary about these early communities of small arthropods is not how different they are from their equivalents in the living world, but how remarkably similar. It can be concluded that once animals and plants left the water they set up a mutual dependency which has endured ever since. This is the decomposing community of animals that live in the soil, breaking down organic matter, recycling it, returning it to the soil once again. In its own way it is an unspectacular little ecosystem, conservative, even dull. But it is vital to the health of the planet. If the springtails were to undergo a mysterious demise, together with the mites that live in soil and the minute fungi upon which they feed, quite soon there would be an ecological crisis of a magnitude we can scarcely imagine. Nutrients would become locked away, the soil would become progressively impoverished, larger plants would die, and soon animals would follow suit. Novels that seek to portray post-holocaust worlds always seem to assume that the soil will magically survive, and that a bean cast into seared soil will quietly proceed to a successful crop. But the soil is not a passive medium; it is alive. I doubt whether there would be many readers for a post-holocaust novel that was concerned with the hero's desperate search for a mite. But alas for the world if the mites and their diminutive allies failed to prosper!

There are other survivors. After all, liverworts, mosses, clubmosses and ferns are still found in many places around the world, surviving messengers from the days of the great greening. The meek may, after all, inherit the Earth. But they remain meek—through hundreds of millions of years. It is as if lack of ambition somehow secured longevity. Live and let liverwort! These organisms are reminiscent of the Good Soldier Schweik, the soldier who survived by dint of always being somewhere away from the front line. Stolidly avoid conflict, sit on the sidelines, look for the uncompetitive option. Become a liverwort. If persistence has to be purchased at the price of lack of adventure, so be it.

So the backboned animals followed shortly after the plants made their excursions on to land. By the Devonian, there were fish equipped both with backbones and with jaws alongside their jawless relatives, swimming through freshwater lakes, probing and exploring. These jawed fish included one species that was the ancestor of all land animals, be it lizard, bat, bird or dinosaur. Genetic similarities between all land-living animals equipped with backbones tell us that it was indeed likely that the step was made but once. This is not to say that other animals may not have lingered about the edge of the two worlds, trying their luck. In mangrove swamps in the tropics curious little fish called mud skippers can waddle out of water on their fins, but somehow they do not seem to be contenders to usurp the place of lions. It is difficult now to imagine the circumstances of that swampy exodus that changed the face of the Earth in so many respects. The hand that types this page was born in a limping, stumbling appendage that triumphed only by virtue of being first. I am reminded of the temptation, that any biographer has to resist, of making the tale seem too pat, the steps—in this case the most literal steps—too logically arranged. The first vertebrate limb was probably a poor thing; it was natural selection that refined it. We can imagine a whole range of flippers and fins that *might* be transformed into legs and hands by shifting this bone or that. All land vertebrates are termed tetrapods—literally "four feet"—and it is obvious that most amphibians, reptiles and mammals do indeed have four limbs. In creatures where they are lost, as in snakes, this loss is an evolutionary change from ancestors that originally also had four limbs. Where a pair of limbs is transformed, as wings are in birds, it is equally apparent that the wing is the *equivalent* of a limb, but changed for a different function—wings were not sprouted anew. Glance at a struggling, naked chick in a nest, and the bones—the arms—that go to make the wing are only too clear. A biologist would say that the wing and the limb are *homologous* structures—that is, they share a deep evolutionary identity. Limbs may be lost, but they cannot be reinvented, because the blueprint is fixed. Further, cats and humans alike have five toes. So do nearly all the tetrapods: where the number has changed—as it has, for example, in the horse, which has but a single toe—this is because of a proven reduction and loss of toes; the primitive number was still five. But why five? After all, it is well-known that there are genetic "sports" among humans which may result in whole villages with six digits. It does not seem to be a lethal mutation. It poses no major difficulties other than making trickier the early learning of the decimal system. The number five may seem to be the "right" number for digits in some indefinable way, but this may be no more than wayward chance, the fickle selection of one individual rather than another on a Devonian shore. I remember the circumstances which made me a palaeontologist and how they depended on one, crucial examination result; there could so easily have been a different slip of paper, another outcome.

It is now known that some of the early tetrapods had seven toes. The fossils of these ichthyostegids—an awkward name perhaps not altogether inappropriate for animals which probably waddled rather uncomfortably across the Devonian landscape—were recovered from sandstones and siltstones in east Greenland, rocks which were not so different from those in South Wales or the Welsh Borderland, being another part of the great Caledonian chain. The climate in Greenland was warm then. Ichthyostegids have been claimed to be little more than "walking fish," and it is true that they retained from their aqueous ancestors both a fish-like tail and long jaws with rather undifferentiated teeth. But their limbs had become at least workable legs, even if they operated akimbo and spreadeagled. Their toes were the surprise. Early workers may have assumed that there *had* to be five toes because we tend to see what we expect to see, and, after all, everybody knows that tetrapods have five toes. Jenny Clack and her colleagues at the University of Cambridge reexamined some of the early collections of these animals, as well as making elegant new finds of their own. What they discovered was a surprising variation in digit numbers. In this lumbering world seven was as good as five. Per Ahlberg tells me that he is studying a species with eight toes. At some stage one of these fish-like land-dwellers with five toes pulled ahead of the rest, and this was the founding father (or mother) of five-ness in all subsequent terrestrial vertebrates. If one of its other waddling contemporaries had been the successful stock for all evolutionary lineages that followed we might now be struggling with seven-fingered gloves, or be able to play an octave with one hand without crossing our fingers. Or else villages with six digits might be the norm, and whispered remarks would be made about those freakish places in which the inhabitants were born with only five.

Lungs were necessary, too. Terrestrial breathing presents some of the same problems that stomata solved in plants—it is a question of taking the process of oxygen absorption *inside* the organism. External gills would soon dry out, even on a mud flat. Small insects and millipedes solved the problem by incorporating small tubes called tracheae into their bodies, but as we have seen this could not work at large sizes. Among the vertebrates, lungfish had already developed in early times the necessary air-breathing organ which gives them their name. Oxygen is absorbed through the moist, sequestered walls of lungs and tracheae alike. Some lungfish live in Africa, and when the rivers dry out, as they do in the dry season, these types can wrap themselves in cool mud and breathe air until the weather improves again. Was some such ruse the prompt that helped natural selection to improve the first lung? After all, we know that the Caledonian climate was hot, and the rocks are full of evidence of drying-out pools, leaving behind the tessellated testimony of mud cracks. In living animals the lung is crimped into folds to help oxygen absorption, and it is quite easy to see how, when this extraordinary organ had been developed, improvements in its design would have been at a premium. Once equipped with a scaly skin that would not dry out, with four waddling legs (regardless of toe number), and with lungs, the tetrapod vertebrate that was our ancestor—and ichthyostegids are a good approximation to it—could lumber away from the water. It was probably no herbivore, in which case it must have eaten other animals washed up on the shore, or caught the arthropods that had made the same journey away from water. It would, presumably, have returned to the water to breed, because a larval stage under water would have been retained from its piscine past. Not every change has to be made at the same time.

In the Mall, not far from St. James's Park, there is an imposing square building at the end of a Regency terrace, which houses the Athenaeum Club, favoured by bishops and professors and senior civil servants. But there is another club that meets there from time to time in the dining room in the basement. This club is called the Tetrapods. It was founded by Julian Huxley and his friends in the 1930s. Biological matters of the day are addressed by a visiting speaker, and the chairman of the night bangs the table for order with the pubic bone of a whale. The fortunes of the dining club, like those of many other tetrapods, have gone up and down. At its low ebb, it met in a pub called the Goat. Some of the members have been coming for a very long time—in fact, nobody remembers who some of the oldest members are. But from such places ideas do diffuse through the scientific community, and I remember hearing about the important discoveries of the multiple digits on one of those very evenings. The chairman afterwards made some poor jokes (also a tradition) about some of the members returning to primitive tetrapodous modes of gait by the end of the evening.

THE LANDWARD TRANSFORMATIONS WERE probably already underway in the Silurian, but were completed in the Devonian period—that is, between about 410 and 360 million years ago. I should briefly describe the life and times of this crucial time period. The Devonian was named after the county of Devon, in southwest England. The countryside is singularly lovely, and much of it owes its character to the Devonian rocks. They underlie most of Exmoor, where even the constant inroads made by tourists cannot dull the charm of the Barle Valley, or overwhelm the friendliness of Dulverton. Tougher sandstones make for moorland, where sheep and heather flourish. Modern farming methods still seek to nibble away at the untamed stretches, but enough remains to help us to understand how wild it must once have been. Much of the charm resides in the contrast between the open moors and the sequestered valleys with their ancient cottages, many of them thatched. Westwards again, marine Devonian shales make up many of the sculpted cliffs of Cornwall. In south Devon, around Torquay, there is a completely different set of Devonian rocks, which contain marine fossils; little of the story of the colonization of land is preserved in this area. It presents instead a continuation of the riches of the Silurian seas, and there is the testament of corals, clams, brachiopods and trilobites to prove it. The Devonian rocks in parts of Ohio, New York State and Ontario are similar in character, but even more prolific. There is a famous trilobite called *Phacops rana* which teems in certain localities, and which has great eyes with spherical lenses and a knobbly head. There are dozens more such animals in great areas of Brazil, or Morocco, or Turkey. Looking at these creatures you would never guess that elsewhere during their lifetime a great greening was in the process of softening the face of the Earth.

The two different faces of the Devonian—marine *versus* the deposits of lakes and mountain basins—are a kind of temporal schizophrenia. The non-marine rocks were lumped together as Old Red Sandstone, and it was some time before it was proved to everybody's satisfaction that these richly coloured red rocks recording life's greatest adventure were the exact contemporaries of unremarkable pale limestones and dark shales. They looked so different, and had so few fossils in common. There was precious little to link the plants and fish of the Caledonian basins with the marine rocks which contained trilobites and brachiopods galore. This led to a grand controversy in the nineteenth century as to the identity and age of the Old Red Sandstone. Martin Rudwick has described how the Old Red problem was settled, using as his source prolific contemporary letters and documents to bring alive a debate every bit as fiercely contested at the time as the origin of species a little later or the discovery of the AIDS virus today. At its end, one of England's most peaceful counties was attached to one of the most revolutionary periods in life's long history.

One of the problems any thoughtful person has in understanding this momentous breakthrough is how to account for such a profound shift by a series of small, stepwise changes in body design. Yet the more evidence that accumulates, the more species are discovered as fossils which show one or another of the features associated with making the change to life on land. Clever anatomists can show that the bones in a reptilian arm can be compared closely with another set of bones in an ichthyostegid fossil. Molecular biologists discover more and more deep similarities in the genes which attest to the common heritage of tetrapods. No, we can take shared descent now as a *fact*, at once as commonplace as our own parentage and as miraculous as birth itself. The only satisfactory explanation as to *why* plants and animals pushed further and further into hostile territory is that their daring was rewarded by overwhelming success in reproduction in this virgin habitat. They had more babies, which prospered more, and in turn bred more freely. Such species were, in a word, fitter. This is natural selection at work. If the cradle of these changes was some isolated basin at the edge of a vanished Himalaya, some place where the first ill-engineered limbs could try out their new function, their subsequent reproductive success is attested to by the appearance of tetrapods practically *everywhere* in the Carboniferous period that followed the Devonian. There are other reasons for this, too, which we shall come to.

Somehow the colonization of land is a more graspable breakthrough than the evolution of the cell, vastly important though that was. It is easier to appreciate the problems that had to be faced, to imagine ourselves as that first doddering tetrapod, to feel the need for breath, or imagine ourselves sliding gratefully beneath dark, damp vegetation to escape the worst of the Sun's attentions. But just as we can empathize with the animals at the moment of invasion of the land, so we almost inevitably throw our own human values into the scene. We incorporate the language of ambition. We talk, as I have just done several times in the previous paragraph, of achievement and attainment and success. It is easy to turn this story into a drama of the race to be first. I myself find it almost impossible to filter out the idea of the will to succeed from my narrative of life's changes. Science itself is so often motivated by the will to be first, to attach one's name to a discovery. The language is strikingly the same. A scientific achievement is described as a "breakthrough." It allows for the development of a new field, the exploration of a new landscape of discovery. The race to be first is what motivates teams of hollow-eyed experimenters to sit up all night watching the bubble chamber in the hope of catching the fleeting evidence of a fundamental particle. The race to be first pushes hardy palaeontologists into the bleak Gobi Desert in search of new dinosaurs. Teams of medical researchers fight to beat the other team to be the first to publish the structure of a gene, or the properties of a chemical which can suppress the growth of cancerous cells. It is no wonder that I have conflated evolutionary achievement with the process of scientific advance. Of course, there are important differences from natural selection; after all, scientific research deals with things other than reproductive success (unless you stretch a point and regard the successful promulgation of an idea as a kind of fecundity). Temporarily at least, some scientists have even got there first by usurping the credit from somebody else, or by claiming the work of a clever student as their own, or by pulling strings in the Academy of Sciences.

In Nature, on the contrary, there is no sin. Nor can Nature cheat. While there are areas where the morality of scientific discovery is murky, and many a bright mind who has been wronged by a greedy professor, there remains one golden rule: *you must not fake your results*. Palaeontology is particularly vulnerable to false claims, or to deliberate attempts to mislead. The trust that is axiomatic to science means that if a fossil is reported as coming from a site, and being genuine, well, then you do not examine its bona fides. The Piltdown Man scandal is the best-known case where this trust was abused. The fabrication of a "missing link" between man and ape was a bold conceit, to say the least, and in retrospect it is astonishing that the hoax was not discovered sooner. But the history of palaeontology has been marred, and in some ways enlivened, by other cases of fabrication which are less well-known to the public. There are even several at the stage we have reached in our history.

The case of Professor V. J. Gupta emerged during the 1980s. Gupta's name was a familiar one to those concerned with the geology of the Himalaya. He had written a textbook on the subject, and had perfectly respectable academic qualifications from the University of Wales, at Aberystwyth, where his former colleagues will vouch for his easy charm of manner and intelligence. When he returned to India and attained high standing at the University of Chandrigar, it may well have seemed he had been given just reward for a valuable career. He published many scientific papers. Of these, a number were concerned with establishing the ages of rocks in remote areas in the Himalaya. Clearly, this bore closely upon the understanding of the geology of the highest mountain chain in the world. The publications described fossils collected by Gupta himself, at some personal risk, because the fossils were snatched boldly from politically difficult areas, or from very inaccessible mountains. The majority of these publications were written with joint authors known for their expertise in one or another group of fossil animals. There were papers on trilobites among them. What happened was that V. J. would send a parcel to the appropriate professor including the fossil material, indicating its importance. The delighted professor would recognize many of the fossils without much difficulty; this was very satisfactory, because the age of the rocks in question was thus established. A fairly effortless paper would be the result, one more for the professor's list, and another for V. J., and so everyone was happy. The only problem was that some of the fossils did *not* originate from where they were said to have been found. There were combinations of fossils unique to other, well-known areas—such as Bohemia—and it looked as if they were even preserved in the identical way, as shown, for example, by a distinctive colour or chemistry. It began to appear suspiciously as if the "Himalayan" fossils might have been collected (or even purchased) in other countries during the course of V. J.'s travels. After all, he had spent time in what was then Czechoslovakia, and nothing would have been easier than to pocket a few fossils from one of the well-known localities there. John Talent, an Australian academic, collected together a number of examples of such jiggery-pokery and published a scathing *exposé* in *Nature*. Some of the allegedly Himalayan fossils were pinned down to other, well-known localities far from Asia. It seemed that success had even made V. J. careless, because some of the same photographs of specimens appeared more than once, though allegedly from different localities! When you have got away with improbable feats of duplicity then it begins to feel as if you have

total immunity; you can get away with anything. So it was with V. J. Some
other facts fell into place. The trilobite co-author, Fred Shaw from Lehman
College in the Bronx, had been sufficiently interested in the trilobites he had
described with Gupta that he wished, not unreasonably, to do some fieldwork
in the Himalaya in order to collect more and better ones. He travelled all the
way to Chandrigar, on the promise of doing this, only to find that the professor was not available. He had mysteriously disappeared. Poor Fred was rather
miffed to make this discovery, but even then he did not immediately assume
that he had been the victim of a ruse, such is the belief in the probity of the
scientific community. But after Talent's case had been made, all the work carried out by Gupta was tainted by uncertainty, and if there was good there, it
was condemned along with the bad. There was, of course, a scandal. V. J. was
stripped of certain of his titles. His regular attendance on the international
conference circuit stopped. Certainly, his work in future will be quoted only
in the context of doubt. But India is a large country, and Chandrigar is a long
way away from London and New York. I am told that V. J. Gupta still draws
his salary and still has a position at the university.

     As to motive, advancement was probably the most important. Academics
rarely become rich, but they may at least become comfortable enough, and
influential:

> Fame is the spur that the clear spirit doth raise
> (That last infirmity of noble mind)
> To scorn delights, and live laborious days . . .

For some, the fame is tempting, but the laborious days are too exhausting,
and the delights irresistible.

SUCH DIVERSIONS DO not help the efforts of hundreds of honest workers
who are attempting to unravel the story of life. But the Gupta case brings up
the question of evidence, and it is an interesting one. How do we know when
we are nearing the truth about history? It is quite clear that we know a lot
more about the landward move of life than we did fifty years ago. New discoveries in the field are part of it. So are new perceptions of old facts. But
even facts are slightly nebulous. After all, it was a "fact" that five toes were
universal until seven toes were discovered by looking again at an "old" fossil.
Without impugning the honesty of any individuals, all palaeontologists know
that some observers are more reliable than others. Some scientists, you might

say, are better at recognizing facts. We are becoming used to the idea in particle physics that the observer is part of the equation. Perhaps this is a more general phenomenon than we admit. In my own history, I can reinterpret the events of my life in retrospect, as I have doubtless modified the autobiographical facts interwoven within this book, if only in the interests of improving the story. Perhaps our fakers lie at the extreme end of a spectrum of reliability in the face of the facts.

If facts are treacherous things, how can we know that we have arrived at the truth? I suspect that we will never know enough about the colonization of land. Year after year there have been surprises, and there is no sign of an end. The next Devonian plant is never quite what we would have predicted from the last one. Every new early land vertebrate springs some sort of surprise. There are good reasons for this. Serendipity is one: the mere chances of finding a fossil are low, and poorer still for early animals that were rare. The rivers in Wales or the glaciers in Greenland carve this way and that at the behest of the elements, not to accommodate scientific researchers. So each discovery is some kind of small miracle, a jackpot on the wheel of chance. Facts may be slippery things, but one thing we know is that we do not have enough of them. Then there is preservation: delicate animals like insects and spiders require a Rhynie Chert to preserve them. The most primitive insects were without doubt wingless, and we know one or two fossils that prove it. They were also very small, and size alone militates against survival as a fossil. So when wings appeared in the fossil record they did so as fully developed structures. A specialist can point to various primitive features, but a non-specialist can recognize them for what they are without prompting. There are no linking fossils yet known, and the chances of finding them now must be slight. So in this case we can say that we have a shortage of facts; we *know* what we are missing. This leaves the way open for speculation as to the origin of wings, which there has already been in abundance, as we shall see. In other cases, it might not be so clear what still remains to be discovered, such as what plants might have lived around strange pools that may yet serve to link mosses and liverworts or ferns with their ancestors. I am forced back on to a thoroughly practical and rather unsophisticated definition of the limits of ignorance. When continued and persistent collecting reveals only more of what we already know, the searcher eventually gives up through sheer exhaustion. Facts still unknown may be hidden from sight, but it becomes unproductive to wish for what can be sought but not found. The complete reconstruction of history will for ever elude our grasp. There will be little chance of ever discovering direct evidence of insubstantial and tiny worms like nematodes, even if the search were carried on to the end of time. When land was colonized, an invisible phalanx of small animals and plants moved in alongside *Ichthyostega* and *Rhynia*. It is salutary to remember that there was always, and will always be, this hidden history.

There are ways of coping with the impossibility of ever knowing history, or completely reconstructing a narrative of the past; a way round being an eternal supplicant for more facts. The solution involves turning speculation about evolutionary history away from arguments about actual descent to theories about *relationships*. This is relativity rather than narrative. This practice is embodied in the discipline of cladistics. I have spent much of my working life near the centre of cladistic theory. I have been berated for not being a pure enough practitioner of the method by some of the more demanding zealots. Equally, I have been condemned by other traditionalists as a "damned cladist." Passions have been stirred about the manner in which we gain knowledge of the past that are astonishing in their intensity. Some scientists have not spoken to other colleagues for years, other than in grumbling undertones. Yet cladistics is in no way offensive or unpalatable.

We are all, perhaps, used to seeing evolutionary trees in accounts of the evolution of life. These are similar in intention and design to our own family trees, where it is known whom Uncle Cuthbert fathered, and Aunt Mildred's sons sired. These are literal, historical narratives of descent, of genealogy. They comprise the unexciting parts at the front of the minor books in the Bible where who begat whom is laboriously chronicled. With ancient history, prior to written records, genealogy is much more problematical. So it is with species. But what we do know is that if evolution happened, then evolutionary changes—whether they affect limbs, such as described in this chapter, or genes, such as periodically emerge throughout this book—will be handed down to future species which shared the same common ancestor. These derived features are the basis of cladistics. Instead of trying to reconstruct actual trees of descent, *cladograms* are drawn which map out the relationships between species as a series of regularly splitting branches based upon the characteristics of the animals themselves. Every branching point is where a new characteristic was acquired, which is handed down to any species lying above it on the diagram (*see* page 162). The animals or plants that descended from a common ancestor comprise a "clade." Names of groups of animals or plants are given to clades that share a set of characteristics. The groups can be very large—a good fraction of the animal kingdom—or very small, such as a genus comprising a handful of species, but the principle is the same. We have already met the tetrapods, those animals that share four limbs, a vast group if ever there was one. On the other hand, *Homo* is a genus within the tetrapods comprising intelligent apes sharing a capacity to make tools, walk upright and copulate at all times, not only at oestrus. To construct a cladogram one does not need *all* the species because the only concern is relative common ancestry. One can easily construct a cladogram showing that humankind is more closely related to pigs than to fish, without implying that men descended from pigs, or that pigs are more human than we might have thought. Such a cladogram would be based on our shared characteristics: with pigs we share (*inter alia*) our warm blood, four limbs, ear structure, giving birth live and suckling our young. These characteristics are sufficient to show that man and pig belong together in a clade—we call this group the mammals. With fish *and* pigs we share backbones and paired eyes, and these are just two of the characteristics that show that all three animals belong in a still more inclusive clade—of vertebrates.

If the principle is no more than just stated, then the practice is a good deal more complicated. The more species that are treated on the cladogram, the more potential diagrams of relationships there are, especially because not all characteristics point the same way. For example, we know that both bats and birds have wings; yet bats are not birds, they are mammals. We recognize this because bats share more characteristics with other mammals (suckling their young, fur, and so on) than they do with birds (flight). Our common sense appreciates that bats are flying mammals rather than furry birds. What this natural nous does intuitively is to apply Occam's razor to the question of the ancestry of bats. We recognize more characteristics linking bats with other mammals than bats with birds; it seems a simpler and preferable arrangement to assume that Nature would have to make fewer alterations to turn an insect-eating mammal into a bat, than to turn a bird into a bat. This simple guiding principle is what William of Occam (*fl.* early fourteenth century) is said to have determined: given the choice, we prefer an explanation that is simpler. As cladograms get more complicated because of the inclusion of more species and more characters a computer will be required to determine which of many hundreds of possibilities is the simplest arrangement of species.

This might seem to be pretty esoteric stuff, and hardly likely to stimulate fury and controversy. Cladistics, however, examines cherished notions about descent from a new standpoint. Quite often new answers emerge, and those answers are not necessarily popular with people who have championed another, traditional view. The origin of terrestrial habits is an example. The tetrapods had been regarded as descending from a curious group of lobe-finned fishes, a view often repeated in popular books. The most well-known lobe-fins have fins with a fleshy central axis which look superficially leg-like, and it seemed a reasonable suggestion that these were the basis for the evolution of that most necessary of terrestrial organs. The living lobe-finned fish, the coelacanth *Latimeria*, is possibly the most famous of "living fossils," and its nickname, "old four legs," speaks clearly of its traditional place in the history of life. It is, indeed, a most lugubrious survivor from a distant age, sluggishly plying its trade in a few places in the Indian Ocean. But there were other kinds of fish living in the streams and lakes of the Devonian. Among them were the earliest lungfishes. There are several species of lungfish still living today, of which the most primitive is *Neoceratodus*, a slow-moving animal clothed in large scales which now lives only in Australia and copes with low oxygen conditions in water by breathing air, an adaptation that may have seen it through many hard times since the Permian period when its fossil relatives lived. *Neoceratodus* lacks limb-like fins altogether. The cladists got to work, analysing *all* the characteristics of lungfish, including subtle features of the bones of the head as well as obvious ones like lungs. They compared these features with what they shared with lobe-finned fish on the one hand, and tetrapods on the other. A heterodox answer was returned: it was the lungfishes, not the classical lobe-fins, that were more closely related to the tetrapods. Our coelacanth was shuffled sideways off the main line of descent: "old would-be four legs." Naturally, those who had espoused the lobe-fin theory for most of their lives were disbelieving; there were arguments, there were tantrums. But the cladists were probably right. Now that the molecular sequences of molecules can be elucidated (providing yet another list of characteristics) there is additional evidence supporting their view. A return to the fossils showed that *early* fish like *Eusthenopteron* had fins which might, after all, be candidates for conversion into limbs. Later descendants of the Devonian lungfish, like *Neoceratodus*, had lost them over the passage of millions of years. In sum, it was a revolution about the understanding of those seminal days on Devonian shores.

The early protagonists of cladistic theory in the 1970s were united against what was largely a hostile "establishment." Opposition threw them together, convinced (rightly, in my view) of the value of their perceptions. The founding father of cladistics, the German entomologist Willi Hennig, was elevated to the status of secular saint. His works, I might say, are not an easy read. The new discipline acquired a language of its own, which further alienated those who had not been converted to the methods. (I have avoided all such language here, but the inelegance of it—symplesiomorphy, synapomorphy, semaphoront, to give a few examples—may have had not a little to do with the alienation some have felt.) Cladists talked to other cladists, and those who disbelieved the canon grumbled or raged among themselves. It was a fine example of what the philosophers call a change in paradigm.

A change in a way of seeing the world follows a kind of trajectory. The early, heroic days unite a number of dedicated scientists who follow the developments of the new theory. The invention of a language for the new concepts serves both to reinforce their own identity and to exclude heretics. The inner group of leading practitioners then becomes a kind of high priesthood. Acolytes are admitted if they embrace the canon. Research students are always the most fervent of converts, because they have the energy of youth combined with a desire to please their new masters. They like to be part of the coterie. The Natural History Museum in London, and the American Museum of Natural History in New York, became the centres for theoretical developments of cladistics in the heroic days. Colleagues would enquire *sotto voce* of a friend, "Are you or aren't you?" You became a cladist the way you might have become a Buddhist. Some of the most trenchant of the high priesthood were almost preternaturally designed for the role. Fish experts, in particular, dominated; Gareth Nelson, from the American Museum, radiated the fervour of a slightly demented prophet. His lectures had a kind of insouciant brilliance that made you feel really stupid for not understanding what on earth he was on about. Surely, you felt, anything so incomprehensible must be worth unscrambling. The opposite approach was embodied by the fish expert in the museum in London, Colin Patterson, whose presentations combined clarity and charisma in equal measure. Then there was Dick Jefferies, who has esoteric theories of his own about the origin of all vertebrates. Dick is the very embodiment of professordom, with an extensive knowledge of almost everything. He is given to quoting slabs of Chaucer, and can read embryology textbooks in the original German. For added spice there was David Hull, a philosopher of science from New York, who was following the cladistic debate as a kind of living experiment in the way ideas are worked through the scientific community, and who stalked around the department in London listening to everything that was being said.

Some of the antagonists were more colourful than the protagonists. Beverley Halstead was a professor at the University of Reading with a knack for publicity and a capacity for making mischief. He was also strangely likeable considering the trouble he caused. He retained a quirky boyishness and great energy into middle age; he seemed to appeal to women. He had worked

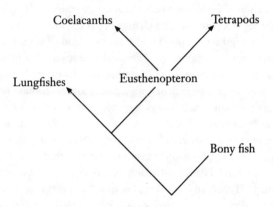

Traditional Evolutionary Tree

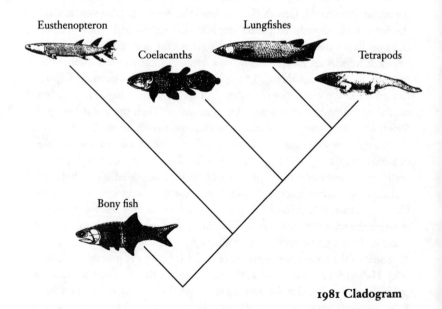

1981 Cladogram

extensively on Silurian and Devonian fossil fish, and was thus directly concerned with the implications of cladistics. He did not approve of them. I recall his outburst at a meeting at which Dick Jefferies outlined his theories about the origins of all vertebrates. "Stop this madness, Jefferies, before it's too late!" he cried. His mischief-making impacted on the Natural History Museum in London after an exhibition opened in 1981, which included representations of the evolution of mankind illustrated by cladistic diagrams. MARXISTS TAKE OVER AT THE NATURAL HISTORY MUSEUM was the kind of headlines that Halstead conjured out of his critique of the exhibition. He had rather ingeniously combined his own two personal bugaboos (cladistics and Marxism) in a rant against those in charge of the exhibits at the museum, comparing the intellectual grip of these people with the kind of implacable orthodoxy which was associated with the then Soviet regime. It was an absurd charge, but it made good publicity. Another charge that hit the papers was that DARWIN WAS DEAD at the Natural History Museum. This was a different misreading of the use of cladograms—the point being that you do not have to *assume* that evolution has taken place to construct them. Equally, of course,

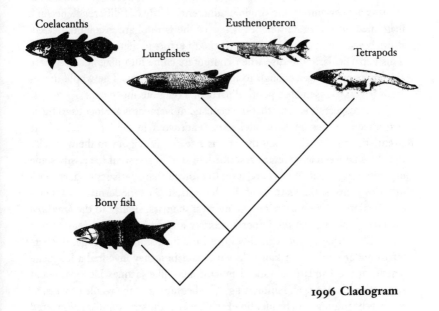

1996 Cladogram

the shared characteristics on which cladograms are based argue for a shared common ancestry; that is, that evolution *has* taken place. But the debate ran for several months. I sought out the cuttings file, which is quite a handful. In the end, the exhibition was modified somewhat, which no doubt Bev Halstead regarded as a triumph. He would have been nearer the mark if he had noted that cladograms make very *dull* exhibits, however interesting they may be to the scientists who made them. Bev had an endearing habit of turning up in the Palaeontology Department after one of his salvos looking as pleased as punch, like a naughty but confident schoolboy who had just been caught doing something rude in the school gym. It was a loss to the richness of our intellectual entertainment when he was tragically killed by a lorry a few years ago.

The next phase in the trajectory of scientific change begins as the new ideas become part of the mainstream. They become incorporated into undergraduate courses. Cladograms illustrating the relationships between fish and early tetrapods become the norm; those who once fulminated against the ideas retire or go mad, or, as in the case of poor Bev, die early. It is not long before almost everyone is using the new methods. At about the same time, schisms appear among the original adherents. Points of difference become magnified into separate schools. A few of the faithful are excommunicated. Rows break out among people convinced of errors in methodological detail. This is particularly the case when computers come into play because there are few people clever enough to understand algorithms. The whole movement begins to resemble a political party or a religious movement.

In retrospect, it is clear that the cladistic movement is analogous to those that affected other fields of intellectual endeavour in the latter half of the twentieth century. The main thrust was a search for rigour in the way relationships between animals are described: in sum, if you want to classify some organism you should do so in relation to evolved characteristics it shares with other organisms, the features of the thing itself. The modernists in literary criticism attracted an aversion among rival scholars, similar to the *brouhaha* directed at cladists, in the importance they attached to the primacy of text itself. They tried to unbuckle literature from its historical baggage, the better to understand its workings. As with cladistics, they invented a language which, in the hands of advanced practitioners like Jacques Derrida, could seem at once opaque and infuriating. The beginning of the whole process, in so far as beginnings can be identified at all, was in the structuralism pioneered by the anthropologist Claude Lévi-Strauss. Descriptive anthropology was previously rich with case studies of tribal habits and systems, which, being almost infinite in diversity, required a reductionist view, a discerned set of interpretative principles, to render them more than a mere testimony to mankind's inventiveness. The change from water to land, this unscrambling of history I have described, is not merely about the discovery of facts; it is also an anthropological study, every bit as human as the intention of a novel.

Discovery does not stop. Some of the more extreme cladists seemed at one time to think that their new methods might render almost obsolete the need to discover new extinct animals. This is dangerous nonsense. There can be no better illustration of the possibility and importance of new discovery than the application of X-rays to the Devonian Hunsrück Shale of Germany. This dark slate contains slightly smeared examples of various animals—trilobites, sea stars, crinoids—as fossils go, not especially appealing. It was known that "soft parts" of animals were preserved in the Hunsrück, apparently coated or replaced by iron pyrites at a very early stage, which painted over the parts that normally decay to invisibility. In the 1970s Wilhelm Stürmer reasoned that the pyrites should register on an X-ray of the slabs. He could scarcely have anticipated the parade of wonderful details his X-rays revealed. These animals were as if frozen on the point of death: starfish petrified with their minutest details conserved; sea lilies with their most delicate, feathery pinnules stretched out to gather seafood, more graceful than a peacock feather; and, just as showy, trilobites with limbs and antennae still scrabbling; and other, strange arthropods that had no record since the Burgess Shale of the Cambrian. These were sea-dwellers, but how much could a discovery of like kind add to our knowledge of the invasion of the land? Whose hammer may yet uncover further secrets?

Chapter 7

Silent Forests, Crowded Oceans

TREE-TRUNKS RISE to more than thirty metres. They are sheer, virtually unbranched up to their crowns. The air is so humid that the moisture congeals upon your shoulders into great drops which trickle away down your back. The canopy of the trees far above is cutting out much of the light; in some directions the way is dark, but where a tree has come crashing down there are bright shafts of sunlight, picked out in wisps of mist, which illuminate a thousand ferns, sprouting like green fountains from the forest floor. The upturned tree has shallow roots that bifurcate regularly, held up like a great hand blocking the way. It is hard to move through the forest because everywhere roots are tangled, and there are rotting logs across the path, some crusted with the satiny green of liverworts, others slick with damp moss. The browns and greens are unrelieved by splashes of colour; there is a sense of dimness, of almost oppressive growth. Everywhere, too, there is a warm smell of decay. Touch the bark of one of the great trees and it feels rough: its surface is tessellated with lozenge-shaped protuberances. There is no sound—no distant howlings, no birdsong. Somewhere ahead, a glimpse of a dark stream, or maybe a pool, snaking between the trees. Suddenly there is a splash, a distinct sound of movement. An animal swims lazily near the edge of the pool—a small crocodile, perhaps? Look more closely and you see other evidence of life. Two very large cockroaches scuttle rapidly away beneath the

166

rotting stump of a tree-like fern. They flee the light, their antennae tucked away in the darkness. Flitting slowly through the moist atmosphere there are flying insects, too, some of them large, fleshy and ungainly. None resembles the butterflies you half expect to see in this swampy forest. When one lands on you—probably mistaking you for a tree—you instinctively brush it off, and it falls, struggling, into a pool. Perhaps you have disturbed something by your sudden movements, for from inside a log there are subtle scraping noises. It could be a scorpion. You shudder and walk on.

This is a walk through a Carboniferous coal forest, 330 million years ago. The world is full of trees, and hidden creatures. There have been many changes since the greening of the Earth in the Devonian. The vegetation which previously clothed the land only tentatively now triumphantly smothers it. It is easy to see a resemblance to today's tropical rain forests in the soaring trunks and the humid atmosphere, a place in which decay and regeneration are so intimately entwined. There are even liana-like herbs that twist through and over the trunks of the great trees, and through all the silence there is a kind of expectant stillness that comes only from fecundity. But in other ways it is so different. It is silent because there is nothing that knows how to make a noise. The only sounds are generated by the scrape of insect limbs, or maybe a low amphibious hiss. There is little colour, because there are no plants that have yet developed the flamboyance of flowers. Dark greens and browns are the colours made by lycopods and tree ferns; here and there the fresher green of a shoot, or a flash of primary colour produced by the bright fruit body of a fungus. But there are insects everywhere. Some of them are familiar, like the cockroaches. There are others that resemble dragonflies, but fly more sluggishly than those which dart around our village ponds. All the insects have wings that jut out from the body in the same fashion as those of dragonflies. The crocodile-like animal was actually an amphibian, rather than a reptile. There is an unparalleled variety of these animals, crawling through the undergrowth, hiding inside rotting logs, or swimming through the pools. One of the latter snatches up the insect that you had brushed away, and swallows it in the deliberate, convulsive amphibian style.

It is surprising how quickly trees developed from lowly herbs. There was a drive upwards. Competition for light must have been part of the reason for stronger and taller herbs, which could spread their fans towards the sun and shade out rivals below. As a solution to an engineering problem, a tree is remarkable. As it grows so it must support the canopy that gives it life. Large leaves are heavy. If you have ever tried to wave a banana leaf you will appreciate the truth of this. One way of lightening the load is to divide the leaf into leaflets, which may even become feathery. Another way is to produce smaller pads. Both strategies were adopted many times. The more successful tree will produce more leaves, but this will do it little good if the whole structure then collapses. The support of this considerable load is the responsibility of the trunk. Cells with walls thickened by lignin are strong, and when these cells are packed together the whole is reinforced many times. Recall how concrete is often reinforced with iron rods that render it slightly flexible. In the same way, a trunk must have some flexibility, or the first gale will snap it as easily as you can break a twig between your fingers. As the trunk lifts the whole photosynthetic apparatus aloft it introduces a vertical obstacle between root and crown, where the real work of absorbing water and nutrients, or of turning light into organic substance, takes place. Immediately a problem is posed: how to supply vital water and nutrients to the leaves so far above the ground. The trunk is in a sense just dead space, a conduit. The cells of the trunk include hollow "tubes" through which water can pass upwards. Without this water, wilting of the canopy follows immediately. These tubes (tracheids) have reinforced walls which contribute to the strength of the trunk, and are arranged in bundles. But then there is the problem of pumping the water, lifting it against gravity from soil to sky. This requires energy, yet plants have no moving parts, like the heart of an animal, that can supply the pumping energy needed. By a natural consequence of the structure of leaves, and with an ingenuity that a modern engineer would have trouble contriving, the force is produced instead by evaporation. As water evaporates from the surface of the leaves under the warm sun, capillary action *draws* water upwards from the roots. The hotter the day, the greater this transpiration, and the faster water is lifted. Wilting occurs only during drought, when the water supply runs short. The trunk is thus both strut and hosepipe. But this is not all. It must also be able to grow in *girth*, as well as upwards, for a spindly tree is doomed to fall. This is accomplished in many trees by having a girdle of proliferating cells running near the perimeter of the trunk. In detail the trunk structure of tree ferns, lycopods and other Carboniferous trees differed greatly from one to another. Some tree ferns were little more than bundles of separate stems; other trees had carefully differentiated tissues for conduction of fluids and nutrients. All were masterpieces of hydraulic engineering. I make no apology for invoking such design principles in describing trees, for they are truly marvels. It is almost impossible to describe the sequence of changes that happened to early plants without using the language of "improvement." This is not the same as positing a designer—the guiding hand of Mother Nature, or some other metaphysical abstraction—for the principles on which the design works are almost a logical consequence of striving for light, coupled with the evolution of leaves equipped with stomata. If it seems soulless to attribute the glories of arboreal form to natural design alone this underestimates an appropriate sense of wonder at the extraordinary and creative inventiveness which life has repeatedly shown.

The evidence for the glorious flowering of forest life in the Carboniferous is often humble in the extreme. Twenty years ago I used to climb the cliffs around Saundersfoot, in South Wales. The Coal Measures come to the coast there, where they are eroded by the winter storms. Ribs of sandstone are prominent, frequently showing ripples and other evidence of the water in which they were originally deposited as part of ancient streams or deltas. Between the sandstones there are soft, crumbly shales, often dark grey. Occasional thin, still darker lines are diminutive coal seams. I used to tap gently at the shales with a geological hammer, and they would split into flakes. On the surface of these flakes there were often fossil leaves, dark as soot, but clearly showing leaflets, and carrying the fine veins which we know well from living leaves. These represent one small, leafy part of an ancient tree, carried away in flood to a silty grave. On the surface of the thin coals there could be a sheet of lozenge-shaped impressions, the bark of *Lepidodendron*, a common lycopod tree of the Carboniferous. Coal, as everybody knows, is composed of carbon produced from ancient trees—trees which did not decay to powder because they were preserved in boggy ground away from oxygen. The very name of the time period, Carboniferous,* hints at the main product of this geological age. Thick, commercial coals were the product of generations of trees, but there are thinner ones too, like those I studied on the south Welsh coast, that may represent a single drift of logs, rendered down to a sliver through time and compression. It is in these, in the guise of bark impressions, that direct evidence of trees often abounds. However, the details of the trees' constituent cells—from which the history of their fluid engineering might be reconstructed—are much more rarely preserved, being captured especially by "coal balls," petrifactions in silica, which soaked the original tissues early enough to preserve every fine detail. Coal balls are extremely dull to look at from the outside, but they reveal their wonders when they are cut and polished. Many were retrieved by fossil collectors from tip heaps (their silica

---

*In North America it is customary to divide the Carboniferous into two. The earlier time period is called Mississippian, after the fine deposits of this age that flank the Mississippi Valley; the later time period is called Pennsylvanian, after one of the great coal-producing areas of the northern hemisphere.

makes them useless for burning), the waste from commerce paradoxically yielding unanticipated riches to knowledge.

At the moment coal is in decline in the West as a major economic resource, demonized as a source of global warming and sulphur pollution, and as the progenitor of acid rain. It began in the Industrial Revolution, which was driven by coal power, when the solar energy sealed away in the Carboniferous was released again into the grinding cogs and blazing furnaces of 10,000 factories. The carbon dioxide so freed, let alone its sulphurous contaminants, might better have remained imprisoned. For we are exploiting the sunlight that filtered in shafts through *Lepidodendron* trunks, danced upon the filmy ferns and caught the dark-loving cockroaches unawares. This sunlight cannot be revived now except through the destruction of the things it made. If global warming is a real phenomenon, fuelled by a greenhouse effect caused by carbon dioxide released into the atmosphere, then it is no more than the consequence of our meddling.

Reconstruction of ancient fossil trees is a difficult art. Such huge objects do not fossilize whole. The branches drifted away from their roots, leaves away from the branches. Spore cases are rarely preserved—but the spores (and related organs) themselves are common fossils; the only problem is to link them with the mother plant, since they were redistributed into seas and lakes. It is like trying to complete a jigsaw puzzle without any clear idea of what the finished picture should resemble. You try to find corners that link together—maybe a fossil leaf attached to a twig, or a spore-containing organ like a cone dangling from a branch preserved in a dark shale. Many of the links have not yet been made, which leads to the intriguing possibility of several different Latin names being attached to different parts of the *same* extinct plant. Spores might have one name, bark another, roots a third. *Stigmaria* is a common kind of root fossil which attaches to more than one kind of Carboniferous tree. In many sandstone formations *Stigmaria* is all you find. These rather pure, white rocks were sometimes the "seat earths" upon which coal trees grew. The roots, which are adorned on their surface with regular depressions, are often as thick as your arm, and may extend horizontally for several metres. The depressions on the surface gave rise to rootlets. Linking *Stigmaria* to lycopod trees required discoveries of rare fossils that bridged roots and bark—missing links in the most pedantic sense.

If stature was the most obvious change in vegetation from the early Devonian into the Pennsylvanian part of the Carboniferous period, subtler changes were happening with the reproductive system, as the limitations of spores were overcome. Seeds, female reproductive parts, became larger and full of food; while male pollen stayed small. A seed is a way of shortening the odds on successful germination. It can remain dormant through hard times until good times come again; it can survive cold. A well-stocked larder can give the seedling a start against its competitors. Some seeds can float. Pollen stayed small because its main job of fertilization does not benefit from larger size, and wind dispersal is a great boon to the cross-fertilization of the species. The male and female organs were born on the same plants in many species, but on different plants in others. It is clear that seeds were a real advantage to terrestrial life because so many different plants developed them. Seed-ferns (pteridosperms) were abundant in the Carboniferous; their foliage was not so different from that of their fern contemporaries but their seeds could be as big as dates. The conifers, ginkgos and, much later, flowering plants all prospered because of their seeds. Our own lives would be as nothing were it not for the seed habit, for arable farming is all about gleaning seeds, and upon the cultivation of seeds rests civilization itself. Surely it is no coincidence that semen is "seed" in the Bible, an analogy attesting to an instinctive appreciation of their position in the propagation of all the things that matter.

There is a compelling parallel between seeds and eggs. The amphibians were tied to water, because they had to return to it to breed, just as they still do. Tiny, soft, glutinous eggs hatched into tadpoles, naked and simple animals, compelled to live in fresh water. Only later in the life cycle could the problems of land living be addressed. Paradoxically, many living amphibians have turned this schizophrenic life cycle to advantage, because tadpoles can survive in protected sites. There are species of tropical frogs whose tadpoles only develop inside the water trapped in the crown of leaves of bromeliads, plants which live like hanging baskets attached to the branches of trees. These tadpoles are protected aloft, above the attentions of all predators other than snakes. Until very recently frogs, newts and toads seemed to be holding their own very well against the inroads of mankind (they are currently declining mysteriously and globally). They were not, evolutionarily speaking, second-class citizens. But water was, and remains, a tie. Like liverworts, amphibians find it difficult to cope with drought. Just as the appearance of the seed "inflated" earlier spores to a new size, so the reptile egg was comparatively gigantic and nutritious. Like a fleshy seed, it developed a tough casing which served to protect its fluid contents from drying out as development proceeded. This is far from simple, as the enclosing membrane needs to be strong, but must also allow the embryo to "breathe": the membranous skins of reptilian eggs perform this difficult task. Most important of all, large egg size permitted the comparatively full development of the baby before hatching. It is marvellous to see newly hatched turtles scrabbling to the sea, with all their senses nourished in the egg. Baby crocodiles are crocodiles red in tooth and jaw. Later, much later, the shells of eggs became toughened with calcium carbonate against dehydration. This is the final answer to that old conundrum: "Which came first—the chicken or the egg?" It was the egg. The evolution of the leathery case of the yolky reptile egg was a most profound change. Exactly when the change happened is difficult to say (soft eggs don't preserve well as fossils), but it must have followed the appearance of giant dragonflies and great trees propagated by seeds. The biological world changed in a kind of sympathetic harmony, as if enmeshed in a dance for which no excuses were accepted.

None of this should be taken as a kind of onward-and-upward description of life's story. It is equally important to recall that mites and ferns and amphibians did not *die out* as each of these new strategies evolved. Rather, they continued to do new things within their own limitations. In the Royal Botanic Gardens at Kew, in west London, there was a glass house for the filmy ferns, the most evanescent and delicate of all, tucked away in a corner of the gardens, where not many visitors found their way. The greenhouse was a relic of a nineteenth-century obsession with ferns, and an almost symphonic tribute to the power of green patterns to delight the eye. There are those who scoff at the taste of high Victorians, but I find entirely sympathetic this appreciation of richness of design without superfluity of ornament. Fern houses were popular then as never since. Another, equally delightful, celebration of ferns is in the old botanical garden in Glasgow, where they seem to fit perfectly with the mahogany staircases and filigree ironwork and exuberant plaster ceilings. Fashions have changed again in the plants popular for cultivation and display. Nowadays garden centres cringe under the illumination of brilliant geraniums or florid busy lizzies. Ferns are tucked away in some corner or other under the general label of "houseplants." Their horticultural history serves as a convenient symbol of the changing fortunes of some of the plants that thrived in the Carboniferous.

Lycopods formed great trees in the coal forests, superficially resembling huge palm trees (to which they are totally unrelated). Today, the creeping herb *Selaginella* survives to provide a living link with the past. I saw this plant growing near the edges of paddy fields in Thailand, a creeping little thing decorated with delicate, tiny leaves, skulking on the edges of the forest on bare soil where little else could survive, a durable bystander, a commoner surviving from the lineage of a royal house. Horsetails (arthrophytes) are much more conspicuous in the living flora, though rarely welcome. They are feathery, jointed plants, which look rather like miniature, leafless Christmas trees, easily recognized when they produce their reproductive "cones" which are raised up like so many brownish fingers among the plants. They often creep from a common rootstock which produces many plants, and they can be an implacable weed. If you try to pull them up they break off at one or another of their numerous joints. The gardener is left with a handful of green stuff and the sure knowledge that the plant will sprout again, more vigorously than ever. Horsetails can grow in swamps or dry fields, and seem to require little nourishment. Their history goes as far back as the Devonian, and they, too, formed arbours in the Carboniferous. Some clambered, others again grew into shrubs with quite large leaves. Perhaps in those moist forests the horsetails were as various as ferns, but few of these several designs have survived from the early days. Why this should be is as mysterious as the changing fashions in horticultural taste. For horsetails thrive vigorously enough today to compete with any of the more "modern" plants. If some catastrophe swept across the Earth, reducing civilization to dust and the works of humankind to nothing, when the rains returned to moisten the ruins, the first shoots that would sprout from the ravaged plains would probably belong to a horsetail. It may be, perhaps, that those which survived from Carboniferous times were the only species blessed with that kind of adaptability; their tree-like relatives may have been mightier, but they required a pampered world of damp forests. The ascetics outlasted them.

IT WAS IN the Carboniferous forest that the last physical threshold was crossed: from the ground to the air.

A recurrent dream haunts many sleepers. By dint of frantic waving of the arms the dreamer rises into the air—he is flying! Soon, he looks down upon Rome or New York from a great height, master of the insubstantial air. Like Puck, he'll put a girdle round about the earth in forty minutes. Would that it were so easy. Flying is governed by laws of physics as implacable as those that govern the upward growth of plants. Flapping of the arms would never produce lift sufficient for flight. Sceptics have long since explained how angels would need impossibly large pectoral muscles to power flight from Earth to the heavens. None of the elegant figures of Annunciation painted by quattrocento artists could have succeeded in lifting themselves from the ground, at least without divine intervention. But the smaller the body size the easier it is to be carried aloft. In a sense, the first spores that dispersed through the air had conquered that element for the first time. For spores, all that is needed is to be extremely light, and to be dispersed into the air where they can be wafted away on the merest breeze. But it was the insects that led the way for other animals. There are still many times more species of flying insects than of all the flying animals from the rest of the animal kingdom added together. There is no difficulty in imagining the advantages of flight to an insect. Such species could get from food plant to food plant with greater efficiency than their walking relatives. They could find a mate more easily (especially one from a different gene pool). They could divide a feeding (grub) phase from a flying, dispersive, mating stage thereby concentrating time on the serious business of feeding and saving sex for the briefest of liaisons. So the advantages of flight are obvious. The problem remains of how to start the process of "growing" a wing. It is known that the early flying insects had rather stiff wings that stuck out from the body, rather like dragonflies today. Wings are clearly made of the same stuff as the cuticle that covers the rest of the insect body, although they are supported by struts and engineered to a high specification. The supportive veins were net-like on the earliest species. One favoured theory has the wings developed initially as membranous outgrowths of the body—perhaps evolved for some quite other purpose—which then, almost accidentally, acquired a use in gliding from one plant to another. Once associated with this use any improvement in "wing" size would be favoured by natural selection. Then there would have been a subsequent mutation changing rigid fixtures, like those of a glider, into articulated ones that might allow for something approaching aerodynamics. This is perhaps not so implausible considering the many articulations on the limbs and antennae elsewhere in the insect body. One of the common mutations which happen in the experimental fruit fly *Drosophila* is like this scenario in reverse; it is called "wingless" and hardly requires further elaboration. Even so, I find it easier to conceive of the genesis of trees than the genesis of wings, although both are fundamentally problems of engineering. This is a case where even one fortuitously preserved insect fossil displaying an early stage in the evolution of flight would almost certainly furnish new and surprising facts bearing on the history of the conquest of the air.

It was not long before flight became spectacular. By the time that the lycopod trees were reaching gigantic proportions so, too, were some of the insects. There was a dragonfly as big as a seagull. It achieved brief fame after its discovery in the 1970s as the "Bolsover dragonfly," the biggest insect that ever lived, named for the English town where its colossal wings were first hammered out of the coal measures. I imagine it flitting lazily among the massive tree-trunks, coming and going through the misty vapours. It was not designed for great speed. But it was proof that by the later Carboniferous the air was colonized, full of life.

Dragonflies were not the only giants in the Carboniferous. There were also huge millipedes. They have left their footprints behind on sandy surfaces—paired prints testifying to a diameter of some centimetres. They could have been as long as a man. What a terrible creature this must have been, a kind of marching army of legs wending its way through the damp vegetation. Perhaps it was this you could hear scraping away inside the hollow logs? Some of its living relatives are highly venomous, and if this extinct animal had been poisonous in proportion to its size, it would have been a thing to avoid at all costs.

Giant trees, giant insects, giant millipedes . . . there seems to have been an urge to become colossal in the Carboniferous. Perhaps there was some special reason to suspend the normal rules. Ingeniously, it has been suggested that the Carboniferous was an unusual time when oxygen production from photosynthesis outstripped the capacity for oxidation to remove it again, so that it formed a higher proportion of the atmosphere than it does now, or had previously. This rich, heady air allowed the arthropods to grow into monsters. This also neatly explains the special types of coal that are found frequently in Carboniferous rocks, which appear to be the product of charcoal baked in forest fires. High oxygen would have produced a heat of enormous intensity; from time to time the great forests must have been swept by forest fires which would have been even more destructive than those that ravage woodlands in Mediterranean and tropical latitudes today. The trees would have had to cope with this; maybe, like eucalypts and *Banksia* in the Australian outback, they would have developed special adaptations to resist the fiery heat, and sprout again after it had passed through. In ravaged forests the heat actually encourages the germination of seeds, which bide their time before a conflagration signals them into life. Perhaps coal measure seeds were developed as a paradoxical consequence of fire.

Not every forest animal was a giant. There were scorpions. There were relatives of the present-day horseshoe crab (*Limulus*) crawling around muddy channels. Some of these animals did not look very different from their living counterparts; the horseshoe crabs had a circular headshield which covered clawed legs. A spike on the back end served to right the animal if it was turned upside down. Little compound eyes on the headshield observed the Carboniferous scene. They look a little like trilobites; indeed, *Limulus* includes their closest living relatives. Not much meat on them, one might think, but in southern Thailand I was astonished to find them on the menu. Apparently, this was a rare seasonal dish. Where, I wondered, could the succulent pieces be hiding? To eat one would be the nearest I might ever get to tasting my favourite fossils, and I could not resist asking for one to be cooked for me. The animal arrived, poached, its circular headshield as large as a dinner plate. The mystery of the meat was solved when the headshield was opened up. Inside, near the front, was a mass of eggs the size and shape of lentils. These are big, yolky eggs for any arthropod. Nothing else on this tough animal was worth considering as food, and so limuloids were only on the menu while they carried their eggs. They had a most peculiar, strong, oily, fishy taste. They really had to be mixed with lots of rice to be palatable. If trilobites had the same flavour it might go some way to explaining their long history in the fossil record.

Both scorpions and *Limulus* might have claims to be "living fossils." This is a contradiction in terms, but a useful conceit. Of course, living animals are not actually *identical* to their fossil forebears, and what is meant is that these are survivors from earlier worlds. Often it is implied that there used to be many more of their kind, and now they alone linger on, the last of their persuasion. There is a covert implication that these survivors are like members of the Flat Earth Society—they once had their day, but are now to be cherished chiefly for their eccentricities. Did these animals survive by luck or by virtue? Like the coelacanth fish, did they have some quiet backwater where they could potter along, oblivious to the modernization of the world? It is all too easy to look at the survivors and concoct stories about their longevity. I am reminded of interviews with very old people in which they are asked the secret of long life. One will swear by abstemiousness, another by whisky; one will swear by celibacy, another by a happy marriage. Few will attribute their good fortune to luck alone. Well, it does seem reasonable to attribute the survival of scorpions to their capacity to thrive in deserts, and perhaps the limuloids survived when the trilobites did not because they could tolerate brackish water conditions, another specialized habitat. But it is also interesting that all these long-distance animals have acquired the habit of having relatively *few* babies, of larger size, which have a greater individual chance of survival. But, after all, maybe the less satisfying explanation is the true one: they were just lucky, and virtue had nothing to do with it.

Within the record of the rocks of the Coal Measures, with their seams of carbonized forest trees and their legacy of fires, their fossil soils ("seat earths") in which the giant trees grew, and their sandstones produced in strands and streams, there are occasional thin, dark beds of rock which record a different kind of life altogether. Marine fossils—ammonites, trilobites, even corals—can be found in these rocks. It is obvious that the sea must have advanced over the Carboniferous coal swamps when the dark rocks were formed. This drowning must have killed trees, amphibians and scorpions alike. But it did not do permanent damage, because the plants and trees return into the rock section a metre or two above each of these "marine bands." The sea advanced to drown the ancient forests, but the trees and their animals could not tolerate the inundation, withdrawing to drier highlands for the duration of the flood. Why this happened is not agreed. At certain times it is obviously connected with the ebb and flow of an ice cap; at others it may be associated with especially active periods of tectonics. But the marine bands are yet another testimony to the mutable Earth. For it was recognized several decades ago from the study of the fossils they contain that the bands were exactly contemporaneous across many of the coal basins. If separate basins have identical marine bands only one explanation fits the facts: the bands had to be formed when sea level rose relative to the land surface. This happened globally, a so-called "eustatic" sea-level rise, and must have been responsible for the sea encroaching upon the land.

When sea level rose sharply, seawater spilled over the lush coal swamps, burying all the evidence, imposing a uniform grey mud upon the landscape. A worldwide rise in sea level, provocative of such faunal changes, is known as a "marine transgression." When the sea withdraws again, the geologist will describe it as a marine "regression." The fact is that particular marine bands can be recognized all over Europe, and many can be followed into America, or into Asia. This immersion of the coal swamps halted the abundance of photosynthesis for a short while, reinstating the ancient dominance of the sea, and such changes in sea level are under eustatic control. Throughout the whole fossil record the sea ebbed and flowed, and graphs can now plot the level of the sea through the geological column—slopping over the continents at some times, draining off at others to leave the slopes exposed. The phenomenon is particularly clear in the ancient coal swamps because its effects are so dramatic.

However, in spite of the slop-and-drain of the sea, the revolution set in train by the conquest of land was never reversed. Amphibians could move away into the safety of streams. Their tadpoles were not threatened. Insects, now air-borne, could fly to other plants in other places. Many of the amphibians were sluggish, salamander-like animals. Some grew to large sizes unmatched in any of their living descendants. They were fulfilling roles as predators then which are now occupied by reptiles and mammals equipped with greater speed and sharper senses. Yet they were fully equal to the demands of the times. But the commonplace description of the Carboniferous as the "age of amphibians" does less than justice to the vigour of the group. After all, frogs and toads survive and prosper, but they are not like *Limulus* and its allies; no "living fossils" these. Rather, they have turned their limitations to advantage: their tadpoles can prosper where little else will grow, their highly specialized tongue is better at catching insects than almost any other of nature's fly-traps, and their back legs provide a wonderful method of escape. My cats routinely bring frogs into the house as trophies, carrying them gingerly in their mouths. Unlike mice or birds, they usually survive the experience. Their glutinous and foul-tasting skin discourages Felix from taking a big bite; they play dead; then they emit a loud squeak like an exploding whoopee cushion that causes the cat to drop them in alarm. Toads are too well protected to even get across the threshold. I do not wonder that Kenneth Grahame made Mr. Toad such a self-confident figure in *The Wind in the Willows* (although his obsession with speed must have been intended ironically). Living amphibians prove that endurance in a group of animals with an early origin in the history of life is not the same thing at all as being primitive. It all depends on what happened in the meantime, and some organisms like frogs and toads have been very busy, evolutionarily speaking, long after the late Carboniferous.

IT MAY HAVE BEEN in the thick of some Carboniferous forest that the first amniote egg evolved, one protected from desiccation by a tough surrounding membrane; and thus the first reptile was born. Like many other early histories in this book this one is also obscure. But a Carboniferous species of reptile (recognized from its bones) was discovered ten years ago in Scotland by Stan Wood. It is known, in the trade, as "Lizzie." To look at, perhaps, it is not too impressive—a black thing on a black rock. But its importance is vital in calibrating the conquest of land. Stan Wood is a canny Glaswegian who has made finding rare fossils his business. He has the raffish candour of the natural salesman, the business sense of a Getty, and a miraculous eye for a fossil. If he had not discovered a Carboniferous livelihood in fossil-hunting he could undoubtedly have made a living selling coals to Newcastle. His charm and persistence are prodigious. By mining and splitting slabs of rock from the right levels in quarries in the Midland Valley of Scotland he has discovered a succession of strange new sharks, giant scorpions, amphibians, and "Lizzie." Many of these discoveries have been purchased by the Royal Scottish Museum in Edinburgh, and thus reside permanently not many miles from where they were first discovered. A few, like "Lizzie," have gone to museums in Germany with longer pockets than Edinburgh. After spending a dinner sitting next to Stan Wood at the Tetrapods Club I would have been prepared to buy "Lizzie" myself.

Glasgow has a tradition of taking geology seriously. In Victoria Park, right in the middle of town, there is a museum built in Victorian times over a fossilized forest. It is one of the unsung wonders of Britain. You go into a fairly standard park in Kelvingrove surrounded by arrowed railings, an open space amid the houses, like hundreds of others in the suburbs of British cities. There is a path that leads into a rock garden, but it is a garden that reveals something of the geological foundations of the city, because it has been cut downwards into the subsurface. Massive sandstones lead you on. You enter a low building, and inside there is the surprise. The floor of a Carboniferous forest has been unearthed. Laid out before you are several massive trunks of *Lepidodendron*, a tree of impressive size, with trunks that would match any oak in diameter. Only the bases of the trunks are preserved, but the roots of the tree spread outwards over the sandstone that was once a forest floor. The trees are spaced apart as they would be in any dignified forest today. Here is a place in which to allow your imagination to take you back into the dark mysteries of those vanished swamps. Pause a while to appreciate the passage of 300 million years between the time these giants flourished and your own intrusion upon the disinterred past. Reflect upon the continuity these trees illustrate with the first photosynthesizers in the dim recesses of the Precambrian. This is no sentimental maundering, but a spectacular slab of the real and tangible past.

It is interesting how the Victorians preserved such monuments for the edification of the whole population. The belief that knowledge itself was good, and led to moral improvement, was also doubtless tinged with a measure of sentimental wishfulness, but as a goal it can scarcely be faulted. The things displayed are almost always *the real thing*. There is no notion of virtual reality, or of spectacular shows in which a little explanation is slipped in furtively, as a gangster might have dealt out a Mickey Finn. Rather, the objects are allowed their own eloquence. As I watched some of the indifferent bystanders in Kelvingrove failing, alas, to connect with the past I was driven to reflect that a welter of twentieth-century explaining may have blunted our capacity to respond directly to marvellous objects. We would rather watch the video that tells us what we ought to feel and think. It is possible to reconstruct these vanished forests in electronic ways that make them seem more real, more alive than grey, cold fossils sheltering beneath a Victorian canopy.

But, like all such images, when the power is switched off the memory fades; what is fact gets blended in with all manner of fiction. This is why we should return to obstinate facts, forged in stone, like Kelvingrove.

Coal mines seek out the seams that *Lepidodendron* and its allies left behind. When I was a schoolboy the geography master would quiz us with: "Well, boy, how would you describe coal?" To which the correct response was: "Black diamonds, sir!" Some of the lustre has gone now. The legacy of its sulphur has blighted forests and sterilized lakes. It would be difficult to overestimate coal's contribution, for better or worse, to the culture of the industrial age. In the northern hemisphere the Carboniferous forests powered the change from a rural to an industrial culture. The invention of the coking process, which removed many of native coal's impurities, allowed the commonplace manufacture of steel, and permitted Isambard Kingdom Brunel's steamships and single-span bridges, together with railroads and mighty beams. It was coal that powered steam engines that in turn powered looms and lathes. Cloth from Bradford, steel from Philadelphia—fortunes were made, and industrial dynasties founded. The whole edifice of manufacturing was built upon black mineral foundations hacked from mines by the toughest of all labourers. These same workers were prominent among those who fought for the rights of trades unions, who defined the political power of the common man, and who ultimately altered the relationship between governors and governed. The triumph of natural engineering that created the first tree entailed, hundreds of millions of years later, the acme of *artificial* engineering in the hands of practical men of power, who believed in building character as much as they believed in building bridges.

In the oceans, some of the animals that we first met in the Ordovician had already died out in the Devonian; planktonic graptolites, for example, had disappeared for ever, and so had many of the more exciting trilobites. There was a series of extinctions in the Devonian which have only recently been recognized, and whose cause is still debated. While the plants and bolder animals completed their hold on the land, other kinds of forests grew up in the sea. There were forests of sea lilies (crinoids), sweeping this way and that with the restless currents. The sea lilies are relatives of sea urchins, and have nothing at all to do with plants. Their "flowers" are cups fringed with arms, and tethered to the sea floor by long stems. The arms filtered out planktonic food which was fed to a mouth in the centre of the cup. In the warm, well-nourished seas of the Carboniferous, sea lilies thrived as at no other time in geological history. Whole limestone cliffs can be composed of nothing but the remains of sea lilies broken up and then re-deposited by storms. Experts can distinguish hundreds of different species, based especially upon the details of the plates that made up the cup and arms. Some species evolved very small, and lost their stalks, becoming part of the plankton. Others were giants. The living sea lilies still form "gardens" in the deep sea. When the pioneering research vessel *Challenger* dredged the depths of the ocean for the first time in the early years of the twentieth century the scientists were astonished to find the whole deck awash with crinoids. Surely this was a glimpse back to the Carboniferous, when the crinoids swarmed both in shallow seas and in the continental shelves. Through much of Europe and North America the Mississippian rocks are massive limestones that owe much of their constitution to these stalked plankton feeders; in thin section the characteristic, spongy-looking tissues of echinoderms can be seen even if the fossils are not obvious on the surface. Polished slabs of this rock can be made into an ornamental "marble."

Some of the crinoids lived among corals and contributed to reefs, which could attain dimensions similar to those of reefs today. The corals themselves are a delight when they are cut and polished. Net-like corallites as delicate as spider's webs are revealed, sometimes tinted a beautiful colour—rose or pink—by mineral changes to the calcite of which they were composed. It is not surprising that these coral limestones, too, have long been used as ornamental stones. Not all corals lived in reefs: there were other kinds of deeper water corals that formed cups or small colonies rather different from the skeletons of the massive reef-dwellers. Sponges, bryozoans and algae were other contributors to Carboniferous reefs.

The Carboniferous was the heyday of brachiopods. These rather unassuming shelly fossils teemed in the shallow seas in numbers we might associate with oyster beds in the living sea. The shells often accumulated in banks or drifts. The brachiopods known as productids had a passing resemblance to oysters. They grew very large: the name of one of the largest—*Gigantoproductus giganteus*—describes it appropriately enough. Certainly, they were as big as scallops. They are unlikely to have held so much meat, however, because the valves of the giants nestled inside one another in the same way that you can cup one hand inside another, leaving little room for flesh. The greatest of the brachiopods, like the humblest, were filter feeders. Some curious flat forms invite comparison with the kind of clams that encourage symbiotic algae to reside in their tissues, and it seems possible that they lived in the same way. But there were also small, fat brachiopods, and smooth ones, as well as ribbed ones, light ones, heavy ones, communal ones and solitary ones. And alongside the brachiopods there were snails and clams and early relatives of the coiled ammonites termed goniatites, and nautiloids distantly related to the living pearly *Nautilus*. The seas thronged with life of all kinds.

This richness requires scholars to describe it, to record and preserve it in the academic pages of monographs. To devote your life to the recording of vanished life might seem an arcane activity in this age devoted to the short term and the fast buck. I have known some of the greatest of the monographers. G. Arthur Cooper has probably described and named more brachiopods than anyone else in history. From his position in the Smithsonian Institution in Washington he operated a kind of archivistic factory, recording the history of a great phylum of animals by publishing book after book on the species that populated vanished seas. He started work in the 1920s and was continuing his studies in the 1980s, before retiring, eventually, at the age of ninety. In the course of a life of extraordinary single-mindedness he named brachiopod fossils from the earliest history of the group, and ranged all the way to those still living. He is a small man, but slow in manner, and only the sharpness of his expression gives a hint of his reserves of energy. During the Great Depression he was saved from absolute penury only by the kindness of another scientist from Yale University who hired him on a minimal salary from his own pocket. Visitors to the Smithsonian Institution during the 1960s often used to lunch with Cooper in the FBI building across the way from the museum, crossing The Mall to get there. On one such occasion when I was with him there was a dime in the gutter, and I was astonished to see the great man leap upon it. He looked at me with his canny eyes. "Anyone who grew up in the Depression knows the value of a dime," he said.

There is a difference between scholarship and pedantry, although the two sometimes get confused in popular perception. The difference is this. Scholarship at its best is about adding to knowledge, and the scholar remains for ever a student, even as he or she acquires more knowledge of their field than anyone else alive. Such people nearly always remain humble in the face of what they still do not know, for above all they understand that the task of comprehending history, or even enumerating its facts, is altogether impossibly huge. G. A. Cooper is an example of such a scholar. The pedant, by contrast, acquires a detailed knowledge of some small thing, and then proceeds to round upon any person bold enough to infringe his small area. Rather than being in awe of history, he denies the magnitude of the task to hand by small-mindedness, by guarding some small piece of the past with pettifogging pickiness, ardent to assert his command of every last footnote and detail. The pedant is immediately recognizable at scientific meetings as he advances to correct the latest speaker on a detail in his address. His sentences always begin with a clause sounding something like an apology—"I fear you neglected to mention . . ." To the casual observer, the pedant sometimes *seems* more knowledgeable than the scholar because he is more assertive, and more in command of the stray detail. The reverse is the case, and the academic world suffers on account of it. In a field like palaeontology, where what there is to know is almost beyond reckoning, it is comparatively easy to corner a small area of expertise: some little-known fossil snail, perhaps, a neglected fern, a forgotten nautiloid—there are easy pickings for those inclined to pedantry. Like Robert Browning's *Grammarian* they might master "learning's crabbed text," but beware, for "this man decided not to Live but to Know."

It is tempting to emphasize the novelties in this history of life and, in order to restore some kind of balance, I shall finish this chapter with a contrasting case, a story dominated by continuity. It was in the Devonian and Carboniferous that the distant relatives of sharks began cruising the oceans. They have stayed near the top of the marine food chain ever since. The shark is self-propelled appetite, voracity made mobile, a hunter refined through hundreds of millions of years. The leading predators on land have changed several times: *Tyrannosaurus* has been replaced by the tiger, *Velociraptor* by the wolf. But in the oceans the shark has cruised on, seeking prey, picking out sickly fish for a quick death. There have been events that wiped out most of the species on land and in the sea, but sharks swam onwards, improving their design by stealth. Sharks much like those in modern seas were present 170 million years ago. In fairness, it must be remembered that not all sharks are predators. The largest of living species, the whale shark, is a plankton feeder, amiably cruising the oceans and sucking in krill. Some of their relatives, the rays, have similar habits. But their anatomy tells us that this peaceability was a later development, and that hunting is fundamental to being a shark. In some ways they remain a primitive group of fish, commensurate with their antiquity. The lopsided form of the tail was more generally present in fish in the Devonian and Carboniferous. They have a series of slits on the sides of the head to irrigate the gills, when more advanced fish have a broad flap covering the gills. These primitive features have been turned to good account by the sharks. Their backbones are not bones at all, but are made of tough cartilage. It used to be thought that sharks were primitive in this characteristic also, because cartilage is comparatively primitive, and ossification of the vertebrae is an advanced characteristic of animals with backbones. But it is now considered that the flexible cartilage skeleton is yet another specialization towards the relentless pursuit of food. Sharks were perfectly capable of producing bone, which they have in the form of hard tubercles in the skin. They also have that projecting "nose" which sticks out in advance of the rest of the animal. Not only is it designed exactly to cut the water, with the precision of (and much the same shape as) the front of a submarine, but it is also equipped with the most sensitive nostrils, which can smell blood in minute traces across many kilometres of empty sea. There are special organs, the ampullae of Lorenzini, which look like a pepper pot, underneath the snout: these have the most exquisite sensitivity to vibrations. The struggling of a wounded fish can be detected far away, or the very breathing of a crab. No vulture has a more finely tuned appreciation of a meal. With the poise of a lawyer detecting a fee in the offing, the shark cruises the ocean and misses nothing.

The teeth of living sharks are produced on a kind of conveyor belt that constantly ensures that new, ferocious weapons are available. They are triangular, and flattened, and their ranks are directed backwards into the throat, thus ensuring that any animal once ensnared will find escape impossible. The cast-off teeth of sharks are common fossils in rocks of Triassic age or younger. The mouth is located on the underside of the head, which lends the shark its wicked smile. Few species approach the size of the Great White Shark, the monster of *Jaws*. There are many more shark-like fish that have the dimensions of a dogfish, and most of the early ones were rather small. Apart from the cruel teeth, the remains of sharks do not fossilize readily—cartilage is not as easily preserved in the rock as bone—so the record of whole animals is poor. Some of the early species which are relatives of the sharks had more bony plates in the head region and have a comparatively good fossil record. Stan Wood found the most fantastical shark fossil (*Stethacanthus*) in one of his Carboniferous localities. It carried what looked like an anvil on its back. Astonishingly, the upper surface of this structure was covered in small bony plates that look for all the world like *teeth*. Imagine a shark plated on the outside with teeth! It was as if some genetic dental instruction had been taken over by a mad scientist and recruited to churn out teeth willy-nilly, everywhere. The "teeth" are preserved, burnished black on the rock, so that the creature truly looks an emissary of darkness.

The sinister personality attached to the shark is different from that of any other of Nature's "bogeymen." This must be more than a matter of jaws; after all, scorpions kill more people each year than sharks do. Perhaps it has something to do with their antiquity—they are dark mementoes of a forgotten time. Or could it be the sudden appearance of the animal, like an avenging angel, almost out of nowhere, without warning, to wreak terrible havoc? Until recently it was thought that the larger sharks never slept, being continuously in motion until death. At least some species have now been filmed in a state of "suspended animation," but it remains true of the large sharks that no other kind of animal is so wide-ranging in pursuit of food. They have an inexorable quality about them. Somehow, the fact that they are so much less intelligent than, say, tigers does nothing to allay our fears: rather the reverse, because we know that they cannot be deflected by guile or reason. Sharks just continue to attack as long as their appropriate sensory organs are being stimulated; there is something of the machine about them. We never see pictures of sharks being kind to their young, like tigers. Loan sharks metaphorically borrow their voracity. It would be too pious to point out what service sharks render to the hygiene of the oceans, or to natural selection, or to maintaining the health of natural populations by weeding out sickly individuals. Fear does not respect the notions of evolutionary biologists, nor do nightmares yield to statistics.

Chapter 8

The Great Continent

A TOYOTA LAND CRUISER sped across the sands in the desert of Oman. There is an art to getting across sand in a heavy vehicle. Although the terrain seems to invite caution, if you go too slowly the wheels dig in. Desert sand is tremendously lacking in substance: one error of judgement and you are suddenly down to your axles in impalpably fine and unbearably hot sand. Then there is nothing for it but to dig yourself out. There is little pleasure in wielding a spade under the implacable sun. As fast as you dig them out, so the wretched yellow grains slide back down into the hole. Eventually you can see the bottom of the tyres, and then, and only then, you can ram sand sheets under them. Sand sheets are portable, corrugated metal runways. If you are very lucky, you can whizz away over the sheets in your four-wheel drive and escape the clutches of the dune. If you are less fortunate you buck and slide and grind for a few yards, throwing up dense sprays of sand, and then lurch once again to a halt. So there is only one way to go at sand, and that is steadily but as fast as you dare, so that there is more than a little skiing, and a certain proportion of sliding, and just enough traction to keep you moving. Then you might escape from the sand on to stony desert, which is often astonishingly flat and seemingly endless, and on which it is possible to drive as recklessly as anyone could wish. Not least this is because there is nothing to hit in any direction, nor any landmark, and you must guide yourself by compass. Once in a while you will see the tracks of some other vehicle, which may have passed your way years before, for no track fades in such a barren place. The ground is burnished copper with "desert varnish," a shiny patina on the surface of stones which have been buffed by the relentless wind. The Arabian Peninsula is where true desert still spreads out inviolate. The occasional vehicle like mine seems an impertinent intrusion upon a land of perfect emptiness.

I had a purpose for being there, which provides a justification, if not an excuse. The peninsula has a rich past, and was not always so punished by merciless hot winds. Beyond the stony desert there are low hills, where sedimentary rocks attest to a time when this barren land was covered by a fertile sea, and snails and sea urchins lived side by side in the shallow waters. Earlier still, there were trilobites crawling where now only the occasional hardy scorpion emerges at dusk on spindly limbs. I was being taken to a small wadi— a valley cut by waters that flowed more prolifically only a few thousand years ago. Even now, the wadi can suddenly be filled with a flash flood which could drown the unwary traveller in minutes. We climbed down steep sides past a few scruffy bushes with perfectly spherical, succulent leaves. The bottom was quite flat and sandy, a natural path. As I picked my way along this path I started to notice the rocks to either side, which formed steep cliffs thirty metres or so in height. They were a curious jumble. In places, there were big boulders, irregular in shape, sticking out of the cliffs, but the boulders were set in a greyish clay. Looking more closely I could see that not only were there boulders; there were also small pebbles—in fact, rocks of all manner of shapes and sizes, all stuck into the cliffs like plums in a pudding. There seemed to be various *kinds* of rock types as well: white pebbles of limestone, big sandstone blocks, even pieces of granite and other rocks that had cooled from hot magma. Yet so loosely were they held by the clay that some of them could be plucked out of the cliffs. I took a small piece of some kind of fine-grained igneous rock in my hand and inspected it at close quarters. It appeared to be flattened on one side. On this same side there was something of a polish—in the sunlight it reflected slightly—but the polish was disfigured by a whole series of scratches. There could be no doubt that something had gouged a series of grooves across the surface. The majority were no more than a millimetre or two deep, and they were mostly parallel to one another. But this was obviously a tough rock, so whatever had done the scratching and polishing must have exercised some considerable pressure.

Further along the wadi was a most extraordinary sight. The whole valley opened up into a kind of natural amphitheatre, surrounded by steep cliffs evidently composed of the curious, medley rock. But the floor of the amphitheatre was another kind of substance altogether, a hard, solid rock, in complete contrast to the clay-and-boulders, which could now be seen to lie as a great blanket immediately on top of this floor. What was extraordinary was that this hard rock, too, had been polished and gouged, but on a colossal scale. The whole valley floor had been scraped: great grooves scored the surface in a set of lines that ran virtually parallel from one side of the wadi to the other. Some of the grooves deepened, in places, as if they had been scraped by the fingernails of some Titan clawing the ground in rage.

The truth is more astounding still. What I was looking at was the evidence of an ice age. Here, where now there beats a sun that melts ice as fast as a hot frying pan melts butter, there had once been rivers of ice: glaciers. Coarse boulders which had been included in these glaciers and carried along like blunt weapons had been responsible for scraping the hard floor of the wadi; finer rock material had produced the polish. Many tonnes of ice had provided the necessary pressure to produce what could now be recognized as a glacial pavement. The scratches on the pebble I had examined were explained. And then, when the ice sheet which had once covered the Arabian Peninsula finally melted away, all the material it had formerly carried was dropped. Boulders and stones which may have originated many kilometres away and been carried on the back of the ice were dumped together with the finest of rock that had been ground into flour and mud. The whole together accounted for that odd, plum pudding of a rock formation which lay atop the glacial pavement. If you want to see identical rocks today you must go to Norway or Canada, where permanent ice still lingers. There, among dwarf willow trees and oozing bogs, the rocks still show the scratches produced beneath ice that has hardly retreated. Glacial boulder clays abound, with their promiscuous selection of scratched pebbles. The match is perfect. I had seen these unmistakable rocks high in the Arctic many years before; the only possible conclusion was that a great ice age must have scoured the ancient Arabian Shield.

My visit to the Arabian Peninsula impressed on me the perceptiveness of those geologists in the early years of this century, who recognized the signature of ice in several places in Africa, Arabia and South America. It must have required a leap of the imagination to be certain of what might at first glance seem improbable under a tropical sun. There was no general perception at that time of the mutability of the world, that the shape of continents and the distribution of climate could change and change again. Rather, the continents were conceived as fixed since the beginning. To infer such a startling difference in climate in the remote past was an extraordinary step of reasoning. But to take it further, to recognize that Africa, South America and India as well as the Arabian Peninsula, and even Australia, were formerly united as a single continental entity, well, that was boldness of a different order. The continents could be fitted together in a kind of crude mosaic: that much was obvious to any child with a penchant for jigsaw puzzles who saw how complementary in outline were the west coast of Africa and the east coast of South America. Fit them back together and suddenly the glacial deposits of the kind I had inspected were attached to a single ice mass; leave them apart and the ice would have had to grow at different latitudes and from different centres. The age of the glacial deposits was given by fossils to either side of the distinctive unit. These showed that it was likely that the ancient continent, if that is what it was, existed in late Carboniferous and/or the succeeding geological period, the Permian. If that former continent had truly existed, then it followed that it must have fragmented, and those fragments drifted apart, over the last 160 million years or so, to give us our present continents in their present positions. The concept of "continental drift" was born.

There were antecedents of this theory as early as the 1850s, but the father of the modern ideas was a German meteorologist, Alfred Wegener, who published his famous book *Our Wandering Continents* on the subject in 1915, while he was a teacher at the University of Marburg. The implications for the history of the Earth were so profound that most scientists at the time simply reeled away from them in perplexity. The ice deposits alone, for example, showed that the South Pole could not have been steadily located where it is today. Southern Africa, India and South America would all have been within the polar sphere of influence, as the South African geologist Alex du Toit had recognized by 1927. Evidently the poles, too, changed position as the continents moved. I have no idea where I would have stood in the great debate that followed (though of course I like to believe that I would have been on the winning side). The story of how the greatest "anti-drifters" included some of the most distinguished physicists of their day is familiar; these distinguished mathematicians could find no acceptable mechanism to account for it. Much of the evidence *for* drift was accumulated piece by piece by field geologists and palaeontologists, who discovered distinctive late Palaeozoic and Triassic fossil evidence like the *Glossopteris* flora—a typical cool-climate terrestrial flora whose distribution would have been difficult to understand without uniting the southern hemisphere as one continent, as the British botanist A. C. Seward was quick to appreciate; Wegener himself mentioned *Mesosaurus*, a small, distinctive marine reptile found in rocks of the same age in South America and Africa. These geologists and palaeontologists were concerned less about mechanisms than recording what they regarded as inescapable facts. The India + Africa + South America + Australia assembly was called Gondwanaland by the Austrian geologist Eduard Suess at the end of the nineteenth century (the name comes from the Gonds, inhabitants of central India). There is a still greater assembly that includes Eurasia and North America, fused not only to one another, but also to Gondwanaland, so that *all* the world's continents were once united as a single entity in the late Permian: this is reflected in the name Wegener gave it, literally "all the Earth"— Pangaea.

There is a pecking order in science, as in so many human activities. This order is scarcely a formal one, but it informs most scientists' perception of where they fit into the scheme of things. The most rarefied and clever scientists, sitting at the top of the hierarchy, are the theoretical physicists, invariably mathematicians. Isaac Newton set the pattern, with his almost supernatural appreciation of physical laws which had hitherto been obscure. Albert Einstein has been the archetype in modern times, and Richard Feynman seemed an appropriate descendant in the line. Perhaps Stephen Hawking embodies the notion of pure brain in this tradition better than anyone else now alive, for, released from his paralysed body, his capacious brain roams free among inconceivable things, nodding to eternity as to a neighbour. Such great men cruise in abstract worlds where lesser souls seek the consolation of metaphor, or the mundane comfort of an analogy. Somewhere not far below them are experimentalists, who might be physicists and chemists (and, these days, more often than not also the biochemists). These scientists translate the dreams of the theoreticians into experimental tests. In the past such experimental wizards might have been arcane figures like the aristocratic chemist Lavoisier squirrelling away in a forgotten wing of some crumbling stately home, while the estate rotted and mad aunts raved in the east wing. Nowadays they tend to be businesslike and decisive, clever scientific administrators running teams of white-coated assistants, spending millions of dollars on the latest technology and driving teams of research students to get there before their rivals. They search for fundamental particles, or genes, or viral structures, or whatever is timely. Both the theoreticians and the leading, inventive experimentalists are likely candidates for Nobel Prizes and full-page obituaries in *The Times*. Few field geologists are likely to achieve such prominence.

The story of "drifters" versus "non-drifters" records one of the few triumphs that the field scientist has won over the theoretical physicist.* It is one to be notched up for the little man—or at any rate the slightly smaller man. The evidence of the distribution of glacial rocks and diagnostic fossils continued to be ignored for several decades by some of the more influential physicists. It was not until the development of palaeomagnetic techniques in the 1950s—whereby the ancient pole positions could be fixed with some certainty—that proof of wandering poles and ever-changing continents could be adduced in a way that was satisfactory to most geophysicists. The ancient pole positions obtained from different, scattered continents today only made sense if they had once been conjoined. Here, at last, were "hard," experimental data that supported the drift hypothesis. But there was soon much more. The exploration of the seas using modern research ships revealed trenches and mid-ocean ridges and many other features that painted the first detailed portrait of the oceanic floor. Its details were only explicable in terms of Pangaea and its subsequent break-up. The final marriage of observation with theory came with the development of plate tectonics in the late 1960s and 1970s. Then late Palaeozoic Pangaea finally lost the status of hypothesis and became, as it were, historical fact. Continental reconstruction became all the rage. In the richer universities, computers were developed which could hold sufficient data to produce sophisticated "best fits" between ancient continents, supplanting the drawing boards and guesswork used by their pragmatical predecessors. It all became very scientific.

It was at about the same time that the scientists who had finally confirmed the physical reality of the ancient "supercontinent" proceeded to play down the role of the field geologists and palaeontologists in unravelling the story. The physicists who despised history as not being sufficiently experimental had little problem in ignoring it. There was some reason for this. The further development of plate tectonic theory particularly required information on both the nature of the upper mantle and the deep structure of the Earth. This was obviously inaccessible to a mere geologist with his hammer and hand lens, but open to those who used seismic information to probe to depths where no man could ever venture. The incontrovertible evidence for the subduction of plates was derived from profiles generated in the laboratory by teams of analysts. In the earthquake zones—that is, the active margins of the plates—at the edges of oceans, the deep reflection profiles reproduced the very image of the oceanic plate dipping away under the adjacent continent as

---

*Lest the palaeontologist be painted entirely as the "good guy" I should recall that some of the last, and most recalcitrant anti-drifters were also of their number, like the late Professor Curt Teichert of Yale University.

it was subducted to oblivion. The discovery of "stripes" of basaltic crust with normal or reversed magnetization on the ocean floor proved how the plates had migrated away from the mid-ocean ridges. In short, the burden of evidence slipped away from fieldwork and into the laboratory, or research vessel, the domain of mathematician and "black box." Computer techniques soon became more important than a sharp eye and an unfettered imagination. Pangaea was proved. Mathematics was once again the master.

Despite its incorporation into "hard" science there remains something romantic in the idea of an ancient land where all was one: a continent where unity reigned. Pangaea is the real lost land—a fact, to be sure, but it reminds us that the idea of lost lands has a longer cultural tradition. Maybe the ghosts of these other, fabulous lands instinctively made the hard-headed physicists in the 1920s like Harold Jeffreys originally recoil from Wegener's Pangaea, or Suess's Gondwanaland. The story of Atlantis was retailed by Plato in *Timaeus*. This land lay west of the Pillars of Hercules. It was an island renowned for its prosperity and beauty, and was powerful enough to have an empire that included part of Africa. Its impious and arrogant ways led to it being engulfed by the sea, and drowned. Appropriately enough, its end was tectonic, and not such an implausible one at that. The Mediterranean Sea is a tectonic jigsaw, and seismically very active, as Plato and his contemporaries would have appreciated. Greek fishermen may well have seen the dramatic landward rush of the sea that accompanies even the most minor submarine earthquake. I saw something of the kind while I was lying on the beach in a narrow bay in northern Yugoslavia (as it then was) in 1967. The sea suddenly vanished. I suppose that I was far enough away from the epicentre of whatever caused it not to feel anything on the ground. But I was able to walk towards rock pools with fish and clams and seaweeds that normally lived well below low tide. You could even pick up bloated sea cucumbers that were stranded out of their element. But not for long. The waters returned very fast, like a tidal race, and soon encroached far further on to the shore than the normal high-tide mark. Startled tourists were forced to grab their bottles of suntan lotion and retreat. If this was the effect of a tiny tremor one could imagine how dramatic would be the onslaught of the sea upon a low island following a major quake. Lands really do rise and fall in this region, as stranded beaches testify on many Greek islands.

For all its fabulousness, Atlantis does offer a kind of mythic parallel to Gondwana and Pangaea. The idea of great lands vanished beneath the sea was a part of serious scientific thought until well into the nineteenth century. T. Vernon Wollaston was a biologist now chiefly remembered for having published one of the first critical reviews of Darwin's *Origin of Species*. He had worked, quite seriously, on the faunas of the Canary Islands and Cape Verde, places where there are numerous endemic species living in similar circumstances to those which Darwin made famous in the Galapagos Islands. He reached entirely the opposite conclusions to Darwin, claiming rather that the species he studied were separately created and immutable. He attributed some of their diversity to the former existence of a larger continent, parts of which had foundered in the manner of Atlantis, leaving only islands behind. He is revealed in his attachment to, and derivation of, this theory in a book of his verse, *Lyra Devoniensis* (1868), part of which reads:

> O blest Atlantis! can the legend be
> Built on wild fancies which thy name surround?
> Or doth the story of thy classic ground
> With the stern facts of Nature's face agree?

Physicists, the true keepers of "stern facts," would neither then nor now want anything to do with such a poeticizing lover of Atlantis. But its appeal has not faded, even today. A year or two ago there was a headline on one of the American down-market tabloids which blared ATLANTIS DISCOVERED! Accompanying this startling revelation was a map showing the lost land lying in the North Sea just off the coast of East Anglia, not far from Ipswich. It is doubtful whether Plato would have recognized it.

The biological implications of having all the major land masses conjoined in the late Permian are profound. We are so used to our dispersed and scattered continents that we take the facts of isolation for granted. Imagine for a moment that all the continents were united together today. There would be a free-for-all as animals and plants dispersed in all directions; some would prosper, without doubt, but many others would not—consider, for example, how mankind's comparatively recent introduction of cats and foxes into Australia has had a devastating effect on dozens of the endemic marsupial species. There, the rabbit-sized potoroo was once common; it now survives mostly in cat-free enclaves. If everything could compete, there would be more losers than winners, because a habitat can only support so many species in a single ecological niche. Overall, therefore, one can imagine a loss of diversity. In this respect, a world joined would be a world reduced. Of course, the climate would not be the same everywhere, so that climatically determined ecological zones would still be of great importance, but the single continent would also modify the climate in unsuspected ways. The supercontinent did indeed stretch from the poles to the equator, and much of the variation in the biology of the time was related to where on that climatic profile an animal or plant found itself. In truth, *total* unification happened only briefly, if at all, and as soon as it was annealed the great continent began to come apart again.

When we left the Carboniferous it was a world of plenty. As the various parts of Pangaea converged, eventually to be pieced together into the super-continent, so the climate deteriorated. The rearrangement of the continents ensured that the South Pole was situated within the continental interior, and this in turn predisposed the growth of an ice cap. It will be recalled how the north polar ice cap today is comparatively thin compared with the vast and merciless ice sheets that surround the true continent of Antarctica. Ice sheets grow thicker on large land masses, away from the emollient effect of the sea. As it was, the greater part of southern Gondwana was eventually affected by the glaciation that left such impressive evidence of its passage in the Omani wadi with which this chapter began. In South Africa, in Cape Province and Natal, there are deposits of glacial origin that thoroughly dwarf anything I saw on the Arabian Peninsula; so also in New South Wales and Queensland in Australia, and in India and Antarctica. Glaciation was under way by the end of the Carboniferous and extended into the Permian; it was the first time that the world had suffered the advance of continental glaciers since the extinction which had ended the Ordovician. The detailed evidence shows a complex series of ice sheets that waxed and waned by turn. Could it have been the wide spread of cool climates that encouraged the first development of an animal metabolism that could maintain its own temperature, a kind of central heating of the body? If the Carboniferous reptiles' grade of organization estab-lished them alongside the amphibians, cold-blooded both, was there before the end of the Permian perhaps a small, cold-tolerant animal that had fur, was warm-blooded and gave birth to its young alive? No fossil of anything that we would recognize as a mammal is known until the Triassic, long after the gla-ciers had waned, so the time is out of joint. But, one wonders, will a lucky hammer one day strike the evidence?

LET US TAKE A TOUR around Permian Pangaea. First, to either side of the equator there were colossal deserts where little but scorpions could thrive. These deserts ranged through much of the centre of what is now North America and onwards through Europe. Where lush Carboniferous swamps had sweated a few millions of years before, and great dragonflies skittered through the mists, now there were crescent-shaped barchan dunes and seas of sand. You can see the legacy of these great dunes in massive, yellow sandstones, like those in the north-country English town of Penrith, in which every grain has been polished as smooth as a grain of millet by the incessant jostling of the Permian winds. Much of what is now South America was similarly arid. It is hardly necessary to say that fossils are exceedingly rare in such rocks, which buried all the older rocks beneath them. But you can find fossil footprints in places. I was astonished to see tiny footprints every morning running over the sand dunes in the Arabian Peninsula, left behind by who knows what nocturnal interlopers in the cool of the night. Elsewhere in the Permian desert the sun dried up whatever pools and lagoons there were, and evaporation left a legacy of deposits of salt and gypsum. Through present-day Africa, India, Antarctica and much of Australia you could say that the glaciers scoured out another kind of lifeless desert, blasted frigid regions which a few specialized plants that could cope with cold (*Gangamopteris*) occasionally colonized. Skirting the coldest, polar areas there were forests of the tree *Glossopteris*, which formed endless groves through Gondwanaland in the southern hemisphere. In the northern hemisphere, in what is now Siberia and Mongolia, the Angara flora is the equivalent habitat, but with different species. The *Glossopteris* tree has dense wood with growth rings, a sign of seasonal growth, and almost willow-like leaves, which must have carpeted the forest floor in the cool period. It is not difficult to imagine this tree having much the same grip on this ancient landscape as conifer woods do upon Canada and the taiga. In the tropics, however, as always in the tropics, there were places which were neither harshly cold nor arid and baking. Rich rain forests flourished over what is now northern and southern China, and similar forests probably grew in the Malay Peninsula and in Venezuela. Some species even extended into North America. This is where the descendants of the Carboniferous forests grew on. Huge lycopods and horsetails, tree ferns and seed ferns still continued to form coals in places.

In the shallow seas that surrounded the huge continent there were many marine animals that continued to evolve from their Carboniferous forebears. The tropical seaway which ran between the northern Gondwana and Laurasia (that is, Asia plus Europe) was named the Tethys, after the wife of the god Oceanus, by the great geologist Eduard Suess (1831–1914), ultimately professor of geology at the University of Vienna and author of one of the masterpieces on the physical history of our planet, *The Face of the Earth*. In one form or another, Tethys persisted for millions of years. According to Homer, Oceanus is the beginning of all creation, and Tethys the mother of all things, antedating even the gods themselves. How appropriate it is that the rocks that were laid down in Tethys are so often full of the remains of past life—although not of the birth of the gods. The Tethys extended eastwards through the Mediterranean region to the Crimea, east again into the Pamir Mountains, and thence to Timor and to southern China; westwards it extended to what is now the state of Texas. It was a hot sea, but full of life— thousands of different kinds of corals and bryozoans and brachiopods and snails and ammonites. In Texas and New Mexico, in the Guadeloupe Mountains and the Glass Mountains there were giant mounds made of sponges, bryozoans and brachiopods in tumultuous profusion. The very last of the trilobites lived in the Tethys, along with many of the echinoderms, like cystoids, that had lingered on from the Ordovician and Silurian. Many of these animals were to die out in the extinction at the end of the Permian.

There were some strange animals that lived in the shallow, tropical seas of the Tethys. Among them, there were small, conical animals propped up with tangled spines that lived together in clots. At first glance they resembled corals. But unlike corals they were not subdivided into fine divisions, or septae, on the inside of their shells. Furthermore, they had "lids" that fitted rather closely on to the cones. The anatomy of the "lids" reveals that these odd creatures were, after all, aberrant brachiopods. They had taken to living together in makeshift colonies, often mixed up with sponges or bryozoans. They were named richthofeniids after the German sinologist, Baron von Richthofen (rather than, I believe, the Red Baron, First World War flying ace). The resemblance to corals may not have been entirely fortuitous because, like corals, they probably extracted nutrients from the surrounding sea, and a conical shape is a sensible one for this kind of living, expanding as it does upwards and outwards into the seawater. Like corals also, these curious brachiopods may have grown together for mutual support, although coral specialists always become very annoyed when you describe them as "reefs." Martin Rudwick concluded that the lids may have flapped to create feeding currents. Since brachiopods feed by extracting small, edible particles from the seawater, such feeding currents would have helped to bring them nutrients. Some species had remarkable perforated "lids"—like a colander. They may have fed passively, taking what the currents brought them, if they lived where the seas were well irrigated by currents. It has even been suggested that, like other coral look-alikes, they could have had symbiotic algae in their tissues, which provided many of the modest supply of nutrients they needed. Brachiopods are not greedy animals. They may have lain out for years under the tropical sun quietly growing, heedless of the brief time over which they would make their mark upon the Earth.

Something about the tropical seas of Tethys encouraged gigantism. Could it have been the result of the unusually hot climate? Even single-celled foraminiferans became colossal by their normal standards. From their beginnings in the Carboniferous there was a progressive diversification of a remarkable group of these normally inconspicuous organisms. These relative giants are known as fusulines. They are spindle-shaped fossils, which can be several centimetres across. Apart from provoking a reflection that this is remarkably large for a single cell, they do not look very notable. From the outside they are usually greyish, more resembling seeds than any kind of animal, but they had limy skeletons unlike any seed. Inside, they are wonders of engineering. Often they are abundant and welded together to form a hard limestone, and the limestone must be cut and polished and then examined in section to observe them properly. Then it can be seen that fusulines are divided into numerous chambers. Their walls are perforated, or spongy-looking. They look as if they had been filled with bubbles, so numerous are their internal partitions. A closer inspection shows that these chambers are not randomly arranged. Rather, they are in layers. The fusuline has grown by wrapping layer after layer around itself, and each layer is itself subdivided into the tiny chambers that give it such a frothy appearance. Since the animal grew from an initially tiny cell it is clear that the wrapping must have spread outwards from a small start to make the spindle shape. The best comparison is with one of those Austrian confections spun out of strudel pastry, or even the cake we know as a Swiss roll (I do not know what they call it in Switzerland). This extraordinarily complex technique permitted a single cell to grow outrageously. It grew its own buttresses, and manufactured calcite living rooms and extensions until it had contrived to make a protoplasmic palace.

These are but two examples from the teeming Permian seas of the former tropics, but they serve to show how marine life continued to do new things. Even the single-celled protist, with its pedigree of more than 1,000 million years, was as inventive as if newly minted. Just because some type of organization has been on the Earth for a long time does not imply that its capacity for change is obsolete. One of the glories of the fossil record is that it continually surprises. Some organisms seem to vanish, or disappear into obscurity, only to reappear in a new guise. Rather like the last concert performances of Frank Sinatra, there always seems to be another appearance in a different place. The ammonites nearly disappeared on several occasions, only to burst forth again into prolific life. However, eventually they gave what was truly a last performance, and no curtain-calls.

In northern Europe there was another sea that spread from Russia, through most of Europe and touched briefly upon Durham and Yorkshire in northern England. This was the Zechstein Sea. There were different species living here from those that swarmed in Tethys. There were clams and ammonites, and especially bryozoans and brachiopods. My favourite fossil name is that of one of the brachiopods—*Horridonia horrida*. Poor brachiopod, what can it have done to deserve such a horribly horrid moniker? It must have been a purblind brachiopod worker who named it, what with ferocious dinosaurs, fearsome sharks and sabre-toothed mammals in contention for the title. The horriblest thing about it was a set of spines attached to the valves, hardly a match for the teeth of *Tyrannosaurus*.

The Zechstein Sea was subject to the aridity that was so typical of the Permian. Parts of it evaporated, baked dry by the relentless sun. Seawater is salty, and most of the salt is sodium chloride, the salt of the cellar and the cook. But there are other chemicals dissolved in seawater, and these will crystallize out if sufficient quantity is evaporated under the right conditions. Collectively, these minerals are known as evaporites. A cut-off lagoon may be supplied with seawater at a lip that connects with the open sea, and this can act like a kind of cauldron continually topped up with liquor. The precipitation of minerals from solution is dictated by their relative solubility. The first mineral to come out of solution is the hydrated sulphate of calcium, the mineral known as gypsum. Brine has to be further concentrated to more than one-tenth of its original volume before rock salt, sodium chloride, begins to be deposited; but once this precipitation begins, huge deposits may form quite quickly. The anhydrous sulphate of calcium, anhydrite, is another commercially important mineral which crystallizes from solution at about this stage. Then, once the brine has been concentrated almost to dryness, there are different salts of potassium and magnesium that crystallize out in smaller quantities. If the evaporites are then sealed in by clay or marl they are protected from redissolving even if the climate changes and becomes wet again, and may finish up at considerable depths below the surface. The minerals that precipitate out are all of industrial significance: calcium sulphate is plaster of Paris, and gypsum and anhydrite are involved in the manufacture of plasterboard and many other industrial materials; hydrochloric acid, sulphuric acid, potash and ammonia were among the reagents that empowered the later phases of the Industrial Revolution and which were produced in commercial quantities in factories around the evaporite sources. The chemical industries that grew up all over Europe in the late nineteenth and early twentieth centuries were ultimately based upon these evaporites. Mighty Imperial Chemical Industries grew up in humble circumstances in Middlesbrough, above the Permian rocks. Railway systems were developed specifically to transport the wondrous chemical products of the industrial age to ports where they could be exported around the world. One can argue, usually inconclusively, as to whether or not this exploitation was of benefit either to mankind or to the world, but one cannot argue about its importance in the development of industrial culture. Ultimately, the chemical edifice all rested upon the existence of the supercontinent, Pangaea, and the climatic patterns that allowed for evaporation of seawater on a massive scale. These were foundations built not upon rock, or upon sand, but upon salt.

LAND VERTEBRATES ESCHEWED evaporitic environments. They are grisly places for life. I have seen something like their modern equivalents along the Gulf coast and in parts of the Great Salt Lake in Utah. The heat is so oppressive that it seems to be a malevolent presence, almost as if it were alive. Everything shimmers and wobbles in the distance. The mirage does indeed look like a shining lake, and only its shifting and brilliant unreliability give its illusoriness away. Mountains at some remove suddenly heave themselves above the horizon and float on the mirage, as if they were icebergs in Hades. Salt and gypsum crystals glisten on the flats with a spurious lustre. This is where the minerals are crystallizing out from solution under the warmth of the sun. Underneath this treacherous *sabhka* there are hot muds that can entrap a vehicle more effectively than any dune. It is no place for an animal hoping for a chance to perpetuate its genes.

Away from the evaporite basins there were species of reptile that could cope with dry conditions. Leathery skins and cold-blooded metabolisms revelled in the heat. Indeed, they still do. Semi-deserts are still populated by reptiles by day. The Pangaea episode of the geography of the world was a crucial one in the development of life on land. At the beginning of this period there were reptiles which were starting to have a certain sophistication, but by the end there were dinosaurs and sea lizards and mammals, and every major backboned group other than the birds had made its appearance. The subsequent story of terrestrial life concerns the juggling between dominance of one group or another.

Palaeontologists charged with unscrambling this narrative must, above all, love bones. Apart from this quirk of affection, many of these scientists seem quite normal. They appreciate that the story is written in bones, and in the changes that happened to bones. Bones provide a route map for evolution. There are very rare cases when something other than bones—skin or hair, for example—is preserved as a fossil. Vertebrate specialists have to be able to remember an extensive anatomical lexicon of bone names that twist the tongue almost as much as they trouble the memory. It is a language as specialized as Basque, and spoken by far fewer people. But it is a scientific language which is necessary to be able to chart and describe the fusion, loss and exaggerations that bones are heir to; it establishes, if you will forgive me, the bona fides of the researcher. The bones of the skull and the jaw are particularly eloquent, for it is here that the character of the animal is most readily revealed—both its diet and its ancestry. Teeth alone are richly rewarding. Sharp, stabbing teeth testify to predatory habits, and sharp teeth with saw edges belong to meat-eaters. Flat-crowned teeth, especially those equipped with ridges on the surface, are typical of herbivores. Small, pointed teeth (in small animals) often belong to insectivores. The skull is not merely a tough house to contain and protect the brain; it is the site of the insertion of muscles for the jaws and the seat of the sensory faculties: sight, taste, smell and hearing. There are advantages to interpreting the life habits of Pangaean vertebrates compared with those of, say, conodont animals. We can, with a little effort of the imagination, put ourselves in the place of a Permian reptile. We can search for animals still living which can provide interesting analogues. We can understand the engineering of its jaws. We are conscious of a biological affinity with animals with which we are familiar—if not as to a cousin, at least to a distant relation appearing in some side branch of the family tree.

The skulls of the early vertebrates were solid affairs: they were literally bone-headed. The earliest reptiles were lizard-shaped animals with skulls of this kind. Many of the transformations which happened in skulls were towards a more open structure. Holes appear. They are termed windows, and this may be an appropriate designation, since they have helped to cast so much light upon the main lines of vertebrate evolution. In some cases these windows allowed for the freedom of movement of stronger and larger muscles. It is a curious fact that much of the early evolution of land animals was about biting more effectively. Early land vertebrates like *Ichthyostega* could do little more than snap their jaws shut. Clamping the jaws together, or chewing, requires altogether more subtle and flexible musculature. Rearrangements of the muscles entailed rearrangements of the bones. Today, only tortoises and turtles are reptiles with no major skull openings, apart from those in the eye sockets. These animals, one might say, adopted the bony alternative, sealing themselves thoroughly in. As a survival mechanism it seems to have worked rather well, a moral already appreciated by Aesop in his fable on the tortoise and the hare. At the other extreme are those with two pairs of openings (diapsids) in the skull besides the eye sockets—a group including many lizard-like animals, and crocodiles, together with all the dinosaurs, and birds. Why this last group belongs with the dinosaurs we shall discover later. Synapsids had only a single opening on either side of the skull, positioned rather low down. One set of reptiles had this structure, a group of extinct animals which is accepted by most palaeontologists as including the closest relatives of the mammals. This is reflected in their vernacular name: the "mammal-like reptiles."*

The oldest fossil of an egg with a shell was discovered in the Permian rocks of Texas, and may well have been laid by one of these animals. They were probably the commonest and most conspicuous reptiles that roamed over Pangaea. They are particularly well known from magnificent fossils from the Karroo Series of South Africa, which were laid down within a cool temperate environment, except when interrupted by the great glaciation. Although they started small, mammal-like reptiles had evolved by the Permian, and in the Triassic some comparatively massive species appeared which could be three metres or more in length. Their teeth show that they had also evolved several different life habits. Some were herbivores. Others in the same fossil deposits had massive heads with powerful muscles; their teeth include what are obviously fangs which could exert a powerful grip, and then slice meat sufficiently well to swallow it in large chunks. The "sail-backed" reptiles included the well-known form *Dimetrodon*, which often appears labelled quite incorrectly as a "dinosaur" in reconstructions of the Permian world. The "sail" runs along the back, supported by fantastically elongated spines arising from the vertebrae, and is supposed to have acted as a kind of solar panel to warm up the animal early in the morning—a cold-blooded animal's answer to being the early bird that catches the worm (or rather, since *Dimetrodon* was a predator, the sleeping reptile). Since these animals lived at comparatively high latitudes for reptiles, this talent may have been of considerable importance in their survival. Much more compact and muscular mammal-like reptiles were known as therapsids, and included some fearsome predators, like *Anteosaurus*. These were able to cull herds of their herbivorous relatives, which included dicynodonts, very abundant animals looking like nothing so much as a reptilian hippopotamus, with stocky bodies and short-toed feet. Dicynodonts had a pair of long tusks, while the rest of the jaw was

---

*I should also mention the euryapsids, with one opening like the synapsids, but high on the skull. This group included important marine reptiles such as ichthyosaurs and plesiosaurs—and is completely extinct.

modified into a kind of beak that could shred and grind plants. Some of the plants were probably tough, like the leaves of cycads and ginkgos, and it must have been hard work making a meal of them. But any animal that learned to do so was bound to prosper. At the same time as these dietary changes, the waddling posture which had been typical of the early reptiles was being replaced by a more upright gait that allowed the legs to be tucked more under the body. This led both to greater speed and to greater load-bearing capacity.

By the Triassic, there were some reptiles that were able to exploit their flexibility of diet and habit to prosper virtually all over Pangaea. There was never again a time when the world was so available to the wandering tetrapod—at least until the invention of the boat. The longest journey we can undertake on foot today is probably not much longer than the journey that Marco Polo completed in the thirteenth century from Venice to China. We have to imagine what it must have been like to be able to walk from Australia to China, by way of Africa, as did *Kannemeyeria*. Nobody knows how long it would have taken a reptile to have made such a perfectly pandemic perambulation of the known world, although it must have been many hundred times longer than Phileas Fogg's eighty days. Even 80,000 years might be within the limits of accuracy of correlation of rocks over such considerable distances. Before the recognition of Pangaea, such feats were explained by means of land bridges that were thought to have once connected the existing continents. Imagine a thin isthmus linking South Africa and South America, much as Panama links the Americas today. Cumbrous reptiles were presumed to have scuttled across these bridges, thereby explaining their global spread. Other theorists preferred to think of reptiles clinging to floating logs, referring to evidence given by mariners of sightings of geckos hanging on to uprooted trees drifting in the middle of the Pacific Ocean. It was alleged that even something longer than a man might, if flattered sufficiently by chance, make it across an ocean. Before we deride such curious notions it is worth recalling that it is assuredly easier to imagine a drifting log than a drifting continent. The sceptic would ask: can the fossil remains of a handful of reptiles really be enough to break the continents loose from their eternal moorings?

It will be clear by now that by the time of Pangaea there was in existence on land something which can be called a food chain. There were reptilian herbivores that moved in herds; there were small, lizard-like creatures that are very likely to have been insectivores; there were predators of several kinds that preyed upon these herbivores—but also, no doubt, scavenged when there was a chance. This structure—a "food pyramid" with a few large and serious carnivores sitting on top—is one which has been a constant and recurrent theme in terrestrial life ever since. Although hardly one example of the original *kinds* of animals survives today, the *structure* lives on, with different animals playing out the behavioural roles that were once adopted by the mammal-like reptiles and their contemporaries. It is another case of the ecology having an endurance which outstrips that of any individual kind of animal within it. This principle has been met with before—for example, in the marine reef and in the forest. As was the case with these other enduring ecologies, it is difficult to avoid the language of improvement, of adaptations getting better and better still, to describe the changes that happened in the ecosystem through time; but use of such language is misleading. The fairly obvious improvements in the musculature of the jaw, or in posture, among the reptiles are not a matter for the ecology, which works rather as a system. If one of the early reptilian predators had suddenly acquired the speed and efficiency of a tiger, the whole Pangaean system might have collapsed immediately. Changes in the ecology happen by animals interacting with one another; a better and faster grazer will be balanced by a more effective hunter, but not one that is so efficient that it eliminates its own food supply, as did the seamen who once feasted on dodos. The best analogy I can devise is with the evolution of the symphony. Improvements in the instruments are real enough. The cello replaced the viola da gamba for good reason. Trumpets got better valves. French horns changed their key (and their design). Harpsichord continuos mostly disappeared. Serpents became extinct. The oldest violins and violas continued to be the best. But the changes in instruments did not necessarily signal a change in the quality of the music. It was possible to write a wonderful symphony for the older instruments, which worked together in harmony perfectly well. Some of the later symphonies became more complex, to be sure, and more interactions between different parts of the orchestra were possible as it became larger in the nineteenth century, but a rise in the quality of the music was not a consequence that followed inevitably. If life is a succession of symphonies performed by different instrumental species playing together, then the orchestra is only the medium, not the composition. And from time to time there were major changes to both front and back decks.

The most significant of these changes were wrought by extinctions. Nor is there any question that the greatest of all extinctions happened at the time of Pangaea. This was probably a double event, with a massive loss towards the end of the Permian, about 250 million years ago, and another close to the end of the Triassic, about 10 million years later. I find it difficult to comprehend the magnitude of the change that happened through this interval. It reset evolutionary history. A narrative that had run from the end of the Ordovician was drastically halted. Somehow, it is difficult to take seriously the extinction of 500 brachiopods, or 100 clams. Would five times the number feel more "real"? The passing of the dinosaurs at the end of the Cretaceous, which we will come to in a later chapter, is what an extinction *ought* to be: it is about catastrophe and tragedy, it is about something spectacular going into sudden eclipse. The demise of quotidian species—no matter how many—seems curiously pedestrian. Yet the end of the Permian saw the reconstruction of the natural world. It was, truly, a mass extinction, a carnage of a magnitude that had never troubled the Earth before. As a corollary, the Permian extinctions subsequently prompted the appearance of the majority of animals that dominate the living world today. The transformation was evidently greatest in the sea, where it has been claimed that up to 96 per cent of all species died out (and nearly 60 per cent of the families, the higher units of classification). These mass extinctions were selective. Some groups of animals were affected more profoundly than others. Many did not survive at all. If an average small town were affected to the same extent we can imagine looking around deserted streets afterwards and recognizing no familiar faces among the few survivors on the pavements. Nobody would man the checkouts in the supermarkets. There might be crying in the night from a lost child. But mostly there would be emptiness.

The facts are these. Trilobites were never seen again. The last Permian trilobites seemed to be entirely successful creatures of the same general kind that had been thriving for at least 100 million years. But their longevity did not make any difference to their fate. The fusulinids, maestros of the single cell, completely disappeared. Among the corals, not one survived. The calcite corals that were such a feature of the Tethys, and which had been part of the reef ecosystem from the Ordovician period, were entirely eliminated. The corals that followed later on were still, obviously, corals, but they had a different symmetry and a different mineralogical construction of their skeletons; many authorities believe that they may not be closely related at all to the corals of the Palaeozoic. Among the brachiopods there was a mass mortality. Many of the great groups of the Palaeozoic did not survive, or were so diminished that two species continued where there had been 100 before. All the molluscs were affected: many types of clams died out, and as great a proportion of snails. The ammonites, most numerous of all the fossils belonging to the Mollusca, suffered more than one mortality. I have been told that all the later ammonites were recruited from no more than four or five survivor species. It is the same story with echinoderms: this was the end of the cystoids and blastoids, and Palaeozoic crinoids; many of the archaic kinds of sea urchins would never be seen again. The litany of disappearances runs on. But this did not take place as the result of one single catastrophe. Rather, it seems that last appearances were staggered through a long period of time within the late Permian, representing a decline rather than a quick assassination, with a *coup de grâce* only at the end.

Nobody disputes the facts, but the cause is quite another matter. There is a shortage of evidence. There are very few places in the world where rock sections show continuous deposition through from the Permian to the Triassic. It is, in its way, a kind of Dark Age of geological time, a period when evidence is hard to come by. Fortunately, it has not attracted as much attention from fabulists as that strange period in European history. Some of the best rock sections are in inaccessible places: in the Salt Range of India, in Kashmir, in northern Iran, in Greenland. There was a great draining (regression) of the sea from the coasts of Pangaea at just the critical time—which left a gap in the history of many areas: no sea, no sediments, no fossils. Who knows whether the last trilobites lingered on in some forgotten corner of the world, but were never preserved? Trying to collect fossils in late Permian sediments is a dispiriting experience. The rocks are clearly reluctant to yield anything. Perhaps—somewhere—there were refuges replete with life biding its time until the Mesozoic. But the overall impression is of a sea sadly depleted, drained of diversity, inhospitable, deficient.

The record of land vertebrates is more ambiguous. Many of the main kinds of reptile, like the mammal-like reptiles, survived from the Permian into the Triassic, but most of the species and genera did not. Much of the fossil evidence derives from the Karroo Basin of South Africa, but the rock successions there are known to be incomplete. Even names of fossils cause problems, because different names have on occasion been given to animals just *because* they were of different ages. Then various bits and bobs of the same animal have been dubbed with different names. It adds up to a rather confusing picture. None the less, the consensus now is that many of the land-based vertebrate animals suffered an extinction (but less marked) at the same time as the lowly clam and brachiopod.

There are a number of explanations for the great extinction. The evidence by now is distinctly cold. The would-be detective has no witnesses who are not mute and, worse, extinct. There are certain features of the case which are beyond dispute. The marine animals that suffered most comprehensively were those inhabitants of the Tethys that favoured life in warm, tropical waters. This kind of fauna accounted for much of the diversity of Permian life, and by the same token its removal must have accounted for much of its loss. Then there was something of a long-drawn-out effect: different animals apparently disappeared at different times. It seems to be different in kind from the extinction at the end of the Cretaceous associated with the disappearance of the dinosaurs, or the oft-forgotten extinction at the end of the Ordovician associated with a previous glaciation. There is no doubt that the Pangaean glaciation happened *before* the extinctions, and searches for evidence of meteorite impacts at the end of the Permian, which might have induced dramatic climatic change, have proved inconclusive. Next, there is the evidence of Pangaea itself. The existence of a huge supercontinent running from pole to pole must surely be part of the equation; it is simply too peculiar an arrangement of the land mass not to have had an influence on almost everything. Finally, there is good evidence of a regression of the sea, which must have restricted the widths of the continental shelves and extended the land margins of the supercontinent itself. This list of features may be long enough—in some combination—to have worked the destruction; but if so, how?

There have been some intriguing red herrings. I like the idea of mass poisoning, as if some cosmic Borgia was suddenly loosed upon the world. Vanadium has been fingered as one of the elements which might have been responsible (high concentrations are known in mineral deposits at this time). Sadly, though, calculations indicate that the sheer quantity of the rare elements required to do the dirty deed would be inconceivably large—even if the mechanism were plausible. Increased radiation—breeding infertile mutant degenerates—is another explanation that has passed out of favour.

Another, and very ingenious, plot invokes a drastic reduction in salinity. Most marine animals are intolerant of brackish conditions, and would certainly die out if such conditions spread to the ocean. Building on the undoubted conclusion that there was a global abundance of salt deposits of Permian age, the notion was advanced by N. D. Newell that *removal* of so much salt would leave the oceans too brackish for most marine animals to bear. They would be unable to reproduce. In the last decade evidence also came to light of drastically reduced oxygen in the seas at the critical time, derived from study of the isotopes preserved in several rock sections.

Then the distinguished palaeontologist S. J. Stanley noticed that there was some good evidence for a cooling of the global climate near the end of the Permian: and climate change is a familiar cause of extinction. Following this event, the early Triassic has a rather limited fossil fauna which is remarkably uniform over a great area. The same climatic deterioration rather neatly explains the disappearance of the coral reefs, which only thrive under tropical conditions in the late Permian. But what about the reduction (even if not so drastic) of the land animals? Why should they pay any attention to salinity or temperature changes in the sea? If the regression of the sea happened at the same time, then the face of the land would change in harmony with changes in global oceans. A widespread regression of the sea reduced the variety of habitats by exposing huge areas of low, marginal flatlands—it was as if, suddenly, the whole of Europe were dominated by mile after mile of rough grassland. This might be attractive enough to a handful of species adapted to grazing—but many others with different ecological preferences would have nowhere to go. The whole scenario was reinforced by the fact of Pangaea—for there was a shortage of refuges on such a vast and homogeneous continental monster upon which everything interconnected. Bad news spread universally. The idea has a plausible attraction.

The greatest interruption to the narrative of life is perhaps best interpreted as a fateful *combination* of these several misfortunes. If life had always to survive the vicissitudes of chance, this was the time when bad luck was heaped upon bad. Climate, sea change, geography—all dealt a succession of appallingly short hands. The marine inhabitants of the Tethys were used to a steady tropical swelter, and could not cope with a cooler world. There were reptiles unable to maintain a satisfactory level of activity when the climate cooled. Oceanic animals suffered an oxygen crisis that suffocated them. It seems like an impossible inventory of disasters combined. Yet we can understand such a concatenation of tragedies in a novel or a biography. I am probably not the only reader compelled to stop reading Thomas Hardy's *Jude the Obscure* because so much loss visited upon one character is impossible to bear. The death of all Jude's children is more than fiction should be allowed to carry, but it is the horror of it which makes our eyes stray from the page rather than a feeling that such misfortune is impossible. Luck, we know in our bones, sometimes runs out completely.

We like to think that virtue prospers. Much of the fiction written prior to Thomas Hardy allowed it to, even if the way to prosperity was uphill, and strewn with setbacks and misfortunes of a temporary kind. We dole out virtue to ourselves to explain most of the beneficial things that happen in our own lives. Luck? Of course there is, but secretly we believe we deserve the good things that happen; it is a tribute to our greater intelligence, or talent. Even being in the right place at the right time demands a level of unusual perspicacity. Such spiritual optimism used to run as a *Leitmotif* through accounts of the history of life. The idea of extinction was, in some subtle way, bound up with the notion of punishment for inadequacy. This idea lingers on in the derogatory journalistic use of "dinosaur," as trotted out in clichés like: "industrial dinosaurs doomed to extinction" or "trades union dinosaurs have had their day." What if luck were, after all, paramount? Tectonic plates bear us away to our fate. We are carried as inexorably as we would have been if the myth of the world being borne upon a tortoise's back had been true. We may be refrigerated or baked, drowned or drained, pampered or starved by turns, according to where the plates may carry us. The configuration of plates even largely controls the climatic pattern by guiding current and air masses. The end of the Permian was when all the physics and chemistry of water and climate combined to squeeze life through a colander on and about Pangaea. Even afterwards, survival was not a matter of superior quality. It was not, as Francis Bacon said, that "Prosperity doth best discover vice, but adversity doth best discover virtue." Could it be that there was no virtue in survival, but just a great measure of luck? Was it a lottery as mindless as plucking numbers from a hat?

It does seem to be true that the abilities possessed by survivors almost coincidentally contributed to their good fortune. There was no way of knowing in advance that these properties would one day be favoured, nor were survivors necessarily better suited than victims to carry the story of life onwards. In a cooling climate, the more compact mammal-like reptiles were better able to conserve heat, for example, than spindly reptiles. If any had already become "warm-blooded" like the living mammals this would have been even better (it is not known if they were). In the seas, it was an advantage to have a long-lived larva that was capable of spreading widely. Species tolerant of changes in temperature and salinity were some of the other survivors. Some of the most successful Permian marine animals that eventually died out were, on the contrary, quick growers without a prolonged larval development. *Autres temps, autres mœurs.* So survival was a matter of hidden assets. But for any animal or plant that *did* survive, the aftermath was opportunity, regardless of putative virtues, as we shall see.

The face of Pangaea had another side, like the dark side of the moon before satellites disturbed its unregarded privacy. If the continents were concentrated into one hemisphere then the *other* half of the world was an enormous ocean. It was an ocean that has entirely vanished; the subsequent movement of plates has consumed its last trace. This, truly, is a lost world, which can only be visited in the imagination. There would have been oceanic islands there, and where sea mounts reached towards the surface of the sea

Chapter 9

Monstrous and Modest

A COMMON PERCEPTION of geological history is that the world moved rapidly from a kind of greenish primeval soup through to the dinosaurs, which then (mysteriously) became extinct, leaving the world free to be populated by people wearing skins over one shoulder and carrying clubs. The least confused bit of this history is always the dinosaurs. They were dominant in Jurassic and Cretaceous periods (190–65 million years ago), and, with the possible exception of *Chrysanthemum*, the Latin names of dinosaurs are the most familiar scientific names in the natural world: think of *Triceratops*, *Tyrannosaurus rex*, *Velociraptor*, or even *Pachycephalosaurus*. In the zoological alphabet dinosaurs are more familiar than the aardvark, and more charismatic than the zebra. In my idiosyncratic journey through the history of life I am almost reluctant to reach the dinosaurs. After all, dinosaurs already belong to everybody.

A fascination with monsters is something that long predated the discovery of dinosaurs. Greek myths teem with them. The Medusa even had the geological talent of turning into stone whosoever gazed upon her ghastly visage. There are monsters in Nordic sagas, in Hindu bestiaries, and in Inuit legends. It seems that there is a natural human propensity to cultivate the monstrous. Every small child is convinced at some time of a colossal and malevolent presence behind the curtains. In the dark etching by Francisco coral reefs must surely have broken the surf. There would have been quiet lagoons behind the reef fronts, full of fish that were descendants of Carboniferous forebears. What Garden of Eden might have flourished on the volcanic slopes of such an island? It is a fact familiar to zoologists that oceanic islands are forcing houses for evolution. Remote islands like Hawaii are (or were, until human depredations began) replete with special animals and plants, many of them living on only one or two islands. Very few animals could find their way by chance to these terrestrial outposts, but any successful colonizer could be the founder of a dynasty of curiosities. Who knows what strange Pangaean mutants might have signalled to their mates upon shores of impeccably white coral sand? We can plant the uplands with all manner of gorgeous and improbable plants of our own devising: ferns as big as palm trees, perhaps, for windblown fern spores would surely have reached these mysterious shores. We can imagine a perfumed breeze blowing across translucent lagoons, while some cumbrous crustacean—as improbable in its way as a Pacific land crab—picked a fastidious path on jointed legs through the litter of a silent forest. Maybe, even then, there were giant tortoises, or perhaps some strutting reptilian dodo, as grotesque as anything devised by Hieronymus Bosch. This forgotten oceanic world would have been populated by the vagaries of chance, blossomed through the opportunities that luck created, and had its evidence destroyed by another twist of circumstances. This lost hemisphere was a victim of the chanciness of things, played out on a globe of shifting plates.

Goya a tired scholar sleeps at a table; behind him, the brooding sky is full of horrible imaginings. Goya entitled his unsettling picture *The Sleep of Reason Produces Monsters*. Maybe it is the freewheeling unconscious that conjures up trolls and dragons, monsters which are at best indifferent to humanity, but usually out to get us.

The discovery of dinosaurs dredges the giants of nightmares into the real world. Like things trawled from the deepest layers of the unconscious mind they rise from rocky depths to haunt the imagination. Marine reptiles like ichthyosaurs and plesiosaurs (neither of them true dinosaurs, which were wholly terrestrial) were discovered in Jurassic rocks at about the same time as the dinosaurs themselves. An early artistic vision of Jurassic sea lizards by John Martin (1840) shows them every bit as dark as the visions that haunted Goya's dreamer. When they were first discovered they must have seemed like documentary proof of dragons. Could it be coincidence that Chinese artists had portrayed dragons as distinctly reptilian? One imagines the early investigators half expecting to recognize fire glands in the gullets of their discoveries.

But unlike the creatures lurking behind the curtains, the dinosaurs were securely extinct. Their teeth were rendered impotent by time. They were monsters surely, but safe monsters. This may be why children find them so compelling. They can appreciate a *frisson* of danger as they look at an illustration (lifelike to a fault) of *Tyrannosaurus rex* in pursuit of something less ferocious. But the image is safely insulated by the page; they know that this is unreal flesh propped upon dead bones. They can visit the bones themselves in the Natural History Museum, where they are presented on polished plinths, amazing in their sheer scale, as much monument as monster. When fully fleshed reconstructions, designed to move by hidden mechanics, were first added to the exhibits my small boy bellowed loudly and refused to go into the gallery. Suddenly, his categories had become confused. Dinosaurs were not supposed to bend down suddenly and hiss. They had violated the safety of the page. Yet the realism that we have come to expect from dinosaur reconstructions is a comparatively recent development. It is easy to forget that the exact proportions of these fleshy monsters are still hypotheses, attempts to grasp at a vanished solidity by decking the muscles and tendons, and, ultimately, skin back upon the bones.

Dinosaurs were not discovered until the nineteenth century, and the history of their interpretation has been dogged by revisions and revolutions. The distinction of their first identification is usually attached to the published description of *Iguanodon* in 1825. The find had been made in 1822 by Mary Ann Mantell, the wife of Dr. Gideon Mantell, a country doctor. The first brown, shiny teeth were found by Mrs. Mantell sticking out of pieces of sandstone from a quarry near Cuckfield in the Sussex Weald, then as now a delightful rural area of southern England. The teeth were huge—several centimetres long. They were curiously chisel-shaped, and fluted on one side. Like many other professional men of his time, Mantell was an enthusiastic naturalist, with a particular flair for geology. He knew that the rocks from the quarry were Cretaceous in age, and was reaching an understanding that they represented deposits which had accumuated around a vast freshwater lake. It says much for his independent cast of mind that he insisted on comparing these huge teeth with those of the living lizards of the iguana type, which live in Central America. The name he gave, *Iguanodon*, records his convictions (*don* from the Greek root "tooth"). There was nothing then known to suggest that giant reptiles had once flourished on the Earth, and nothing at all to prepare the world for such a change of scale in cold-blooded, terrestrial animals. Contemporary anatomists did not at first share his opinion. But when the great French anatomist Baron Cuvier accepted Mantell's interpretation, giant reptiles were on their way to becoming a part of popular culture, for their impact was immediate. As so often when a new discovery is made, feverish searching soon followed, with remarkable success. Within a decade much more complete skeletal material of *Iguanodon* and several other animals had been discovered. It was also realized that dinosaur bones already existed in collections, but their significance had not been fully appreciated. Dean Buckland had actually named a Jurassic form, *Megalosaurus*, the year before Mantell named *Iguanodon*, and if we were to follow priority rather than understanding this should be credited as the first identified dinosaur.*

By 1841 nine different kinds of these giant reptiles were known. It was at this time that the term "dinosaur" was coined for them at a meeting of the British Association for the Advancement of Science at Plymouth. The name was invented by Richard Owen, from the Greek meaning "terrible lizard." He distinguished them from crocodiles, ichthyosaurs and flying reptiles. Richard Owen was to become the first Director of the Natural History Museum in London, and thus, I suppose, my ancestral boss. Many of the earliest discovered dinosaur bones are still stored there, ranging in age through the Mesozoic (Triassic to Cretaceous periods). These bones reside in cabinets behind the scenes, away from the scrutiny of the general public, which would, I imagine, find them something of a disappointment. Compared with the set pieces in the galleries they look poor things indeed—small scraps all too

---

*There is an even earlier one. Dr. Robert Plot, a seventeenth-century antiquary, had described the base of a dinosaur limb as *Scrotum humanum*. This name never achieved official recognition!

clearly showing the scrape of the palaeontologist's needle or the stain of some glue boiled from the bones of *living* tetrapods. Their status as the original, or "type material," as it is correctly known, is indicated by a rather modestly coloured spot, red or green. None the less it is these bones which will carry the name given by their discoverer in perpetuity. Such drab fossils seem feeble props on which to hang a changed vision of the world, and now it is difficult to imagine them in the hands of their original collector, trembling with excitement, as he or she carefully removed the vital evidence from its rocky tomb. Yet in these quiet vaults resides the eloquent evidence on which a dozen movies or a thousand children's books are based.

The romantic element in dinosaurian discoveries was appreciated almost immediately. In the first half of the nineteenth century there was not that embarrassed separation between the arts and what would now be labelled science. I find it hard to conceive of a poet celebrating the discovery of Higgs's bosons or black holes with the immediacy with which Thomas Lovell Beddoes (1803–49) responded to the revelation of ancient reptiles:

> The mighty thoughts of an old world
> Fan, like a dragon's wing unfurled,
> The surface of my yearnings deep;
> And solemn shadows then awake,
> Like the fish-lizard in the lake,
> Troubling a planet's morning sleep.
>
> My waking is a Titan's dream,
> Where a strange sun, long set, doth beam
> Through Montezuma's cypress bough:
> Through the fern wilderness forlorn
> Glisten the giant hart's great horn
> And serpents vast with helmed brow.

"The mighty thoughts of an old world" could be a running title for my whole narrative. To Beddoes, at least, the dragon and the serpent were appropriate images to weave into the poem. I suspect, too, that the "fern wilderness" was engendered by too many long afternoons in the cool greenhouse drinking China tea, unless it was a confusion with the Carboniferous.

The gap between dinosaur bones and the fleshly animal has narrowed over the last century and a half. Mantell initially, not unnaturally, reconstructed his animal as an enormous lizard, four-footed, low-slung. On the basis of a few scraps what else had he to go upon but his living comparison, doctored with appropriate changes in scale? These reconstructions were elaborated by Richard Owen, and are immortalized in the gardens of Crystal Palace in south London. As you walk around the landscaped gardens you pass by a rather grand pond, on which there is an island planted with coniferous trees. Out of the trees looms the life-sized model of *Iguanodon*, painted a musty shade of green. Its mouth is slightly agape, and despite its size you might find memories of lizards sneaking into your mind. It certainly stands four-square on very substantial legs. The *Iguanodon* was constructed by Waterhouse Hawkins of a varied mix of materials (some of which are now showing their age and need of restoration). In 1853 a famous dinner party was held at Crystal Palace and a dozen of the more distinguished guests were able to sit inside the half-completed dinosaur. They must have been able to savour the unusual sensation of both being and eating a meal at the same time. Further round the lake you can find Hawkins's idea of *Megalosaurus*, reconstructed as similarly four-footed, as well as a selection of amphibians reconstructed as enormous, stump-tailed frogs. These life-sized models of extinct animals organized around a trail could be a contender for the title of the first theme park. The reconstructions are almost completely wrong in every detail, but they retain a compelling presence, even now: a Titan's dream perhaps.

The models demonstrate too how changeable is our understanding of the past. New material—entire skeletons—of *Iguanodon* was discovered at Bernissart in Belgium in 1877. The pieces that had previously to be sketched in by guesswork, or by comparison with living animals, were now filled in by real and very solid bones. The limb bones could be joined to the hip bones; every vertebra could be fitted in its proper order. The dinosaur's tail got longer and longer, until it was clear that it was a great, fleshy, tapering thing. We are thoroughly used to this appearance now, as displayed by a majority of dinosaurs, but once upon a time it was news. Even more curious, the back legs were obviously and dramatically larger than the forelimbs. It seems that Hawkins's cumbrous animal had to be capable of rearing up upon its back legs to be a truer representation. Far from being large lizards, the extinct ruling reptiles must have been animals which were capable of being bipedal. The teeth, though, remained true informants: it seems that *Iguanodon* was indeed a giant vegetarian. A curious detail was that Gideon Mantell's original reconstruction had placed *Iguanodon*'s thumb upon its nose, thus providing a gesture appropriate to ensure the scientist's proper humility in the face of the evidence.

The bones connected, the next stage was to place back the flesh. This is not such a difficult task as it might appear. While there is always the possibility of some entirely fleshy lobe or crest which would leave no trace upon the skeleton, reptilian bodies generally closely reflect their internal construction. The places where ligaments attach are recognizable; the way the joints work can be inferred quite confidently from our knowledge of living animals; and the size of the muscles can be deduced from the size of the limb bones, and from places where there are scars of attachment. The ribcage defines the chest. As for claws, it has to be remembered that the bones found as fossils were only the core; in life they would have been covered by a horny sheath as sharp as the talons of any roc or dragon that can be imagined. Finally, the skin: it would be dry and leathery, flexible and wrinkled at the joints. It has little chance of preservation, one might think, and yet there are specimens of *Edmontosaurus* from Canada which show impressions even of the skin, finely tessellated like that of a crocodile. This leaves colour. I suppose that there is a prejudice in favour of leaving the large grazing dinosaurs with the matt dun and grey colour that typifies living pachyderms like elephants and rhinos. This is not a showy lifestyle and dull hues seem appropriate. Maybe some dinosaurs were scarlet, or striped like zebras, or dappled with all the hues of the rainbow. After all, many snakes are brilliantly arrayed. This knowledge is something which will probably always elude us, but it would be dangerous to write off anything as impossible. One day, perhaps, traces of decayed pigment will be discovered which will awaken the colours of the Cretaceous.

The story of reconstruction is never complete. Take the dinosaur tail as an example. Older reconstructions of bipedal dinosaurs showed the tail lying limp upon the ground, a kind of burden trailing along behind. This was a cumbersome image, and contributed to a kind of dim impression. Most dinosaurs did not have a very large brain, just measured in terms of its cubic capacity in relation to body size, and—like Winnie-the-Pooh, A. A. Milne's bear of little brain—this carried with it a notion of slowness and ponderousness. The whole image was furthered by a wholly mistaken idea that gigantic dinosaurs, like *Diplodocus*, had "two brains"—so dim was the one in the head, it was alleged, that it had to be supplemented by another at the base of the spine; no matter that growing a second "brain" is no more possible than growing an arm in place of a nose. The supplementary "brain" was nothing other than the posterior ganglion—the nerve centre for the hind end of the animal. Its large size may have suggested "braininess." But when the mechanics of walking and running were considered from a mechanical point of view, it was clear that the tail was more than a posterior appendage. It was a counterbalance for the rest of the animal in an active posture. Those dinosaurs that walked or ran on two legs carried the tail proud and poised; many of those that walked on four used it as a rear counterpart to their long necks. The sluggish image was probably quite wrong for dinosaurs other than the tank-like, armoured species, for which slowness was a virtue. Just to be able to pump blood from the relatively low position of the heart to a head held so much higher than the body required a large and sophisticated heart, probably one of the two-chambered variety found in mammals and birds. The same kind of heart could power lungs suitable for a high level of activity. The emancipation of the tail became a symbol of the release of dinosaurs from their legendary dimness.

One result was that reconstructions had to be altered yet again. At the Natural History Museum in London this did not happen in the main exhibition hall until the 1990s, 170 years after Mrs. Mantell's find. A splendid *Diplodocus* skeleton stands there. This was one of those gigantic but vegetarian dinosaurs whose great barrel of a body was supported on all four massive, elephantine limbs. The length of the spine of this true monster was something like twenty-five metres (82 feet). For a carcase of such size the neck was long and elegant. The head is tiny, so small it looks as if it should belong to some other animal altogether, and equipped with a set of somewhat protruding, peg-like teeth. For many years it has been the first specimen that visitors wandered up to, with that odd mixture of insouciance and respect which is reserved for museum visits. For as long as I could remember the huge tail of this great creature had hung down at the back like that of a depressed greyhound. It trailed along close to the ground supported by a series of little crutches, like something from an oil painting by Salvador Dali. The new interpretation of the wonderful engineering of sauropodomorphs like *Diplodocus* showed rather that the tail was a precise cantilever for the long neck, and the whole spinal structure delicately (for all the massive scale) arched and balanced to sling below it the great load of the body, the whole supported by the columnar legs as if they had been the piers of a suspension bridge of flesh and bone. The tail just *had* to be lifted to be true to life. And so it was. A much more affirmative support now bears the tail aloft, swung outwards purposefully, almost like a whip; perky, you might say. The whole demeanour of the animal has changed: no dim giant this, but a confident, strolling browser. One small, very practical benefit followed. The endmost tail bones of the *Diplodocus* (plaster casts, naturally) had been stolen repeatedly, so much so that there was a box of duplicates hidden away at the back so that they could be immediately replaced the following morning, before the first visitor had come through the door. Now, held above the common herd of *Homo sapiens*, the tail of the Jurassic giant is safe from violation.

Walt Disney's masterpiece *Fantasia* was released in 1940. At that time it was the most expensive animated film ever made. It is still a marvellous thing to see. Disney sought to match music to cartoons perfectly. The animated episodes were linked by rather twee voice-over introductions while the orchestra was seen in dramatic silhouette, until Leopold Stokowski, maestro of maestros, lifted his baton. Then it was magical! One of my earliest memories from the 1950s is being taken to see the film in Oxford Street. I could scarcely breathe for excitement. My hot hands hardly shifted from my bare knees (small boys wore short trousers then). The centrepiece of the film was an animation of *The Rite of Spring* by Igor Stravinsky. This episode was, in its way, another unauthorized biography of life. Like many such histories, as I observed at the beginning of this chapter, it advanced wonderfully fast from creation to the dinosaurs, where it lingered lovingly. The jagged quality of Stravinsky's music must have helped to suggest the design of a primitive world to the animators. I am perfectly sure that Walt Disney took only the best advice on how dinosaurs lived and behaved—scientific advice must have been the least of his expenses—so what we have, coloured and moving before us, is a kind of 1940 conspectus of what many palaeontologists then thought about the Jurassic/Cretaceous world. There was refreshingly little attempt to anthropomorphize the animals, which is often a Disney foible. Huge animals, related to *Diplodocus*, wallowed in endless swamps, where they grazed upon soft and luscious waterweeds. Duck-billed dinosaurs lived in much the same way, scooping water plants into their bills. *Parasaurolophus* had a sail running down its back from its long head-crest. There were actively running, smaller herbivorous dinosaurs, probably *Hypsilophodon*, swiftly scuttling in small herds with tails held aloft. Flying reptiles—obviously *Pteranodon*—launched themselves from high cliffs where they lurked like albatrosses to swoop and dive over the sea to make a catch of fish. All these animals fled in terror from *Tyrannosaurus*, the inside of whose mouth was a terrifying red, the better to contrast with its sharp, white teeth. A battle between a sluggish, plated stegosaur and the carnivorous monster resulted in the inevitable death of the herbivore, whose spiked tail struck the ground in its death throes like the gong of doom, matching Stravinsky's music chord for chord. Throughout this sequence the sky was dominated by a kind of pinkish glow. The heyday of the dinosaurs was drawn as remarkably wet. It seemed like a lush world, richly and damply vegetated, in which most of the dinosaurs slurped lubriciously upon weeds. The end of the sequence showed this bloated world drying up. The poor creatures trudged off across a lurid desert, desperately scooping the last drops of liquid from muddy pools in a useless attempt to survive. Their bones alone remained behind to tell us that they did not.

Rather more than fifty years later Steven Spielberg released the film version of a novel by Michael Crichton, *Jurassic Park*. It was a success almost everywhere in the world. A comparison with *Fantasia* provides an eloquent demonstration of the mutability of our knowledge of the past. Many of the dinosaurs are the same or similar in the two films. Many were Cretaceous in age rather than Jurassic, but to make much of that is too pernickety. But Spielberg resuscitated them into living simulacra that were truly astonishing. So perfect are these dinosaurs that it is difficult to believe that they are still only speculative reconstructions, just as were Gideon Mantell's or Walt Disney's. The reconstructions used by the 1990s director followed several more decades of research by palaeontologists such as Bob Bakker or Jack Horner (who was supposedly the model for the hero of the film). They are in accordance with genuine new discoveries, and doubtless modern prejudices as well. Great herbivorous dinosaurs of the *Diplodocus* type wallowed in swamps in Disney, but gravely walked in herds over the plains in Spielberg, browsing from treetops. In the new orthodoxy, these huge sauropodomorphs were like colossal elephants in their ecology, but weighing perhaps thirty US tons, an order of magnitude more than living pachyderms. The ground shook under their tread. Their legs were colossal pillars to carry their bulk, their toes shortened in support. They could churn up the ground beneath their massive feet. (This kind of disturbance has been recognized in rocks, where it is known by the inelegant name of "dinoturbation"!) Their life habits in the two films could scarcely be more different, yet each seems plausible to the enthralled viewer closeted in the dark of the cinema. The sheer bulk of the body was once regarded as so massive that only water could support it; the long neck conveniently allowed the animals to breathe while immersed. Weak teeth, it was thought, were only capable of eating mushy water plants. Not so, came the revision in turn: careful calculations indicated that water pressure would actually have *prevented* the lungs from operating efficiently, and besides, the pillar-like legs and the spinal design with balanced tail were really only plausible for a terrestrial animal. Furthermore, their tracks had been found, track upon track heading in the same direction, proving beyond doubt that these great animals really did move in herds. Then the elephantine analogy came to mind naturally enough. Now it could be seen that the long neck enabled the huge creatures to attack the crowns of trees, rather in the way that elephants' trunks can attack trees on the Serengeti

Plains. But the dinosaurs were probably delicate croppers for all their size—hence their comparatively delicate teeth. It seems likely that their stomachs were adapted to be microbial "barrels" to help digestion by way of fermentation. Even their droppings were discovered: fossil dung heaps described as coprolites.

In this way truth changes and changes again: even though it is enshrined on celluloid and seems as solid as any documentary footage, a depiction of past life is only as true as the science behind it. Ultimately, it all boils down to bones.

Tyrannosaurus has not changed so much. Spielberg's is the more terrifying because its eyes were allowed to glint with a little sly intelligence as well as implacable hunger. Only a pedant might remark that Disney's had three, rather than two fingers. However, Professor McNeil Alexander, who knows the engineering of animals as enthusiasts know the minutiae of the E-type Jaguar, tells me that Spielberg was guilty of hyperbole when he showed *Tyrannosaurus* giving fair chase to a motor vehicle. Given its size, neither its musculature nor its bones would have supported such speed; maybe the jerky gait of Godzilla might have been more appropriate after all. Ostrich-like dinosaurs (such as *Gallimimus*) were marvellously like their living counterparts: they resembled flightless birds, strutting over a landscape that was much more savannah than swamp. Perhaps the small, bounding dinosaurs were most closely similar in both films; evidently, their close ecological equivalence to gazelles and other vigorous herbivores was appreciated from early in their study. The scariest hunters of all in Spielberg were called by him *Velociraptor*. (More correctly, they were *Deinonychus*, a dinosaur only discovered in 1964 in the early Cretaceous rocks of Montana.) *Velociraptor* were shown hunting in packs, running down their prey. Grisly enthusiasm for the chase and a startling intelligence made them the animals that linger longest in nightmares when the film is over. They were even seen problem-solving—opening a door at one point in the story. At some juncture film-maker's licence crept in, because of course we do *not* know exactly how intelligent these animals were. We cannot run them through mazes as we can rats or chickens. It may be marvellous to see—but is it true? Whatever the truth, Spielberg's film laid to rest for ever the popular notion of stupid giants of reptilian grade lumbering about a dimly appreciated world. These animals were precision-engineered, well-adapted, and with interlocking lives. Only the armoured dinosaurs (*Ankylosaurus*) and horned dinosaurs, like *Stegosaurus*, which were the rhinos of the Cretaceous, continued to plod heavily in search of vegetation as they always had done. But there are still limits to knowledge, where fiction embroiders what palaeontologists have managed to deduce from patient work on bones and rocks alone. These limits matter.

We know that another film which might be made in the future could give us yet another version of the past. We do not grasp the past with the certainty with which we know our own lives. In one sense, the past is always a fiction, an elaboration from the known towards the arcane. There will be no final *Jurassic Park*.

The chase upwards towards the unknown is satisfactorily illustrated by the growth of giants. There is an escalation in size among some of the dinosaur fossils. It is as if, once started, the process was set upon a course. The earliest dinosaurs yet known from the Upper Triassic of Argentina are not much larger than a man. Quite early on there were two types of dinosaurs with different hip structures—"bird-hipped" and "lizard-hipped" respectively—which some authorities believe are not closely related. The latter gave rise to the giants, both among the carnivores and the herbivores. Because of the connected geography in the Mesozoic the dinosaurs spread almost globally. Through the course of their history the largest herbivores grew larger. In a palaeontological parody of the joke about the biggest of everything being in America, every other year brings a new claim about the discovery of the most titanic of the sauropods—always from the Americas, and usually from limb bones. The names are coined to reflect their dimensions: *Supersaurus*, *Seismosaurus* and *Ultrasaurus*. Since *Ultra* is the most superlative of all superlatives it would be difficult to imagine any further advance. One estimate makes *Seismosaurus* weigh fifty tons and extend to eighty metres in length.

The explanation of the size increase is partly based on the notion that each successive species should be able to crane its neck higher than its competitors to get at leaves that would otherwise remain just out of reach. In the warm climates of the time large size also provided a more efficient way of conserving energy. There is also genetic evidence from living animals that increase in size is a comparatively easy modification to introduce: natural selection could readily set to work. That leads to the question: is there an ultimate limit? There is certainly a maximum size to which a tree can aspire, based upon the height to which water can be lifted during transpiration; and even bone and flesh have their mechanical limits. So these, the greatest of land animals, were still limited by the physical laws that stop oaks from growing to the height of the Empire State Building. To build larger animals it would be necessary to release them from gravity, by making them live upon the stars.

The largest carnivores also got larger in the Cretaceous. *Tyrannosaurus* really is the ultimate predator. It might be tempting to turn upside down the aphorism about bigger fleas having smaller fleas upon their back to bite 'em into something like: larger meals have larger diners upon their backs to feed 'em. It is a tempting analogy, this "arms race" between feeder and fed upon, but I somehow doubt its veracity. I once evolved my own theory of larger and larger relating to Italian restaurants. In the Casa Nostra and the like a waiter always comes round to offer you ground black pepper from a pepper-grinder. I noticed that as time and competition went on the grinders got ever larger. Perhaps the proprietors thought that a larger pepper-grinder gave them the edge over their neighbours. I have seen tiny waiters struggling with vast black grinders, all the while trying to smile and keep up the banter. In my scenario, the process continues until the pepper-grinder becomes so large that the waiter staggers around on wobbly legs, unable to lift it. The custom declines, and the restaurant becomes extinct. I do not present this as a serious evolutionary analogy, but something like it has been proposed in the past to account for gigantism, saying essentially that a trend, once started, runs on until it is lethal. In fact, I suppose the pepper-grinder would be abandoned— or the restaurant might hire a bigger waiter.

The changes in the perception of dinosaurs after *Fantasia* were not just a matter of stance and habitat. In the 1970s some of the most energetic arguments that I have ever heard have been about whether dinosaurs were hot-blooded. Hot-bloodedness brackets dinosaurs with mammals, and separates them from cold-blooded living reptiles. Reptiles can tolerate changes in their body temperature in a way that mammals cannot. They have to "heat up" in the sun before they can be fully active, especially in cooler climates. They do not have to sweat, and can survive on very little water. On the other hand, the dinosaurs became so large that their sheer volume would guarantee them a heat retention that living reptiles now lack. Hot-blooded animals require a constant body temperature. Hence mammals need more food than reptiles to keep their internal "fires" burning. They can live satisfactorily in cool climates that few reptiles can tolerate.

Whether or not the dinosaurs were hot-blooded is crucial to understanding their life habits, their energy budgets, and the forces that drove their evolution. In several respects the evidence for hot-bloodedness is quite persuasive. The kind of chambered heart that would have been necessary to oxygenate the giants is of a design that is already known in mammals and birds, hot bloods both. Bone structure of at least some dinosaurs shows a distinctive concentric structure of canals under high magnification—these are vessels for the passage of blood—and resembles mammal bone rather than reptile bone. The active life of hunting and scavenging attributed to carnivorous dinosaurs by the new orthodoxy is ecologically so like that of living African mammal carnivores that it is tempting to extend the similarity to include their metabolisms as well as their habits. Some of the tiny dinosaurs were no bigger than a chicken and many living insectivores of that size are homeothermic (the proper way to describe hot bloods). On the other hand, there really does not seem to be any evidence of structures in dinosaurs that we associate with mammal cooling systems, such as sweat glands (dinosaur skin looks like reptile skin), and there is no reason to assume a thick fatty layer for insulation, let alone fur. Would it really be possible to stuff sufficient food through the small head of *Seismosaurus* or *Diplodocus* to stoke up a furnace sufficient to heat thirty to fifty tons? And how much should a tyrannosaur eat to stay active? Dinosaurs laid eggs, like any bird. There is now considerable fossil evidence that some of them built muddy nests, in communities, like gargantuan gannets. Since 1990 hundreds of dinosaur eggs have come on to the market from localities in China. They have been exported with dubious legality. Very small babies have been found inside the eggs. They have bones which seem to be similar to those of baby birds, and like them might have had warm bodies at this stage—they were chicks sustained by "fast" biochemical reactions at this stage of growth: they were effectively homeotherms while they were young and vulnerable, even if they became more like other cold-blooded reptiles as they got larger. Since the Jurassic and Cretaceous world was generally warm and equable, dinosaurs were ideally adapted to it. They were energy-efficient compared with profligate mammals.

The issue which has stirred even more passion than the matter of hot-bloodedness is whether birds are dinosaurs. At first sight this may seem an extraordinary proposition, almost as odd as asking "Are monkeys fish?" or, "Are weasels frogs?" It was treated as an *outré* idea when it was first introduced more than thirty years ago, but as time passes it seems to have become more and more acceptable. First, it is necessary to unburden ourselves of any prior notions about birds, and look at them afresh. Imagine a farmyard chicken, plucked clean, but still alive and well, strutting round the barn in search of stray ears of wheat. It has a deep breastbone, but its erect head, jerky gait and massive thighs—the drumsticks—might remind you of some small Jurassic animal. The horny talons leave imprints in the dust with more than a passing resemblance to dinosaur footprints. The large beady eyes are thoroughly alert to the world, if not exactly bright with intelligence. Now imagine if the little, naked wings were transformed to grasping, clawed arms, and then supply the bill with jagged teeth, and perhaps the hypothesis does not look quite so improbable. It certainly seems to be more plausible than one which allows crocodiles as the closest relatives of birds, which was for a long time the favoured theory. And the comparison which we *should* make is not with this poor naked chicken, but with the most primitive of fossil birds, the celebrated Jurassic fossil from the Solnhofen lithographic limestone, *Archaeopteryx*. I described this animal in some detail in Chapter 2 (*see* pages 47–9) while I was dismissing Professor Hoyle's claims about it being a fake. At this stage all I have to do is simply assert its bona fides. The early bird takes a very brief bow in *Fantasia*, rather desperately flapping over one of the many lagoons in the film. *Archaeopteryx* has genuine fossil feathers, and feathers define birds; the chance of anything so sophisticated and complex as feathers evolving on several occasions is vanishingly small. The rest of the anatomy of *Archaeopteryx* is extraordinarily similar to that of a small carnivorous (theropod) dinosaur—for example, the Montanan animal *Velociraptor*. The skulls are alike in many of the details that palaeontologists really relish; *Archaeopteryx* retained a full set of teeth, which has been lost in the horny bills of living birds. Talons were present on its forelimbs that otherwise supported the wings, and for all the world these are the same talons as on the forelimbs of *Velociraptor*. Indeed, it is as well that feathers were preserved on *Archaeopteryx*, or it might have been misidentified as simply another small dinosaur. Discoveries of other Jurassic birds in the mid 1990s have confirmed the special place of *Archaeopteryx* in making the link, because other birds were already more "advanced." The answer to the bird/dinosaur conundrum is that some dinosaurs are more closely related to birds than others, and that birds are, indeed, feathered dinosaurs. Up to a point, as craven employees used to say to Lord Copper in Evelyn Waugh's *Scoop*, all birds are dinosaurs, but all dinosaurs are certainly not birds. That does not mean that we have to stop calling birds birds, but a cladist would insist that there should be a name for a group of animals which comprises birds *and* dinosaurs, all of which descended from a common ancestor. Living birds are *all* hot-blooded, and if the earliest bird *Archaeopteryx* was similar, then, by extension, so should be its closest unfeathered and flightless dinosaur relatives. However, the most critical anatomical studies of early birds have suggested that they were more likely still to have been cold-blooded. Hot-bloodedness was a characteristic acquired only by the more advanced birds on their evolutionary tree. So when we watch an ostrich strut its stuff over the African plains we might be seeing something not so far removed from a Jurassic scene.

Not all the flying contemporaries of dinosaurs were birds. Far commoner in the Jurassic and Cretaceous skies were flying reptiles, the pterosaurs, another group of creatures that prove that conquest of the air was won more easily than our prejudices might allow. These animals had membranous wings supported on an enormously extended fourth finger—they were more bat than bird in their construction. Their bodies were gracefully built and carried by hollow bones which were strong, yet light. They evolved into an extraordinary variety of shapes and sizes—from tiny, thrush-like species to flying monsters which had the greatest wingspan of any flying creature. There were pterosaurs that were the flamingos of the Cretaceous, others that were like pelicans, others again like vultures. For once, the modern reconstruction of these bizarre aeronauts is not so different from that shown in a brief episode in *Fantasia*. *Pteranodon* is still considered to have been an aerial angler, which lived by plucking fish from the sea with its long "bill." But exactly how they flew has been the subject of much debate. At one time the larger species were thought to be gliders. Aeronautical engineers were called upon to calculate their stalling speed, and there was learned discussion about whether or not they could take off from the ground. For a while, they were pictured perched on the tops of cliffs, from which they could launch themselves into the air. However, the mechanical engineers have thought again, and the anatomists have reconsidered the musculature, and as a result the pterosaurs are generally claimed as flappers once more. But the pterosaurs were also prone to gigantism in the Cretaceous, and the greatest of them all, *Quetzalcoatlus*, named for the appropriately awful Aztec god, is reputed to have had a wingspan of no less than fifteen metres. There are aircraft which are smaller! This is a puzzle to the mathematicians and palaeontologists who have attempted to reconstruct its everyday life. Was it some kind of reptilian vulture that cruised the Cretaceous skies over what is now Texas in search, perhaps, of the bloated carcase of a brontosaur? The sky would have darkened as the scavengers spotted their grisly, putrifying feast. . . . The tableau that suggests itself is darker than anything that the poet Beddoes could have desired, or that might have haunted Goya's sleeping philosopher. More prosaically, *Quetzalcoatlus* would itself have been easy prey to one of the small theropods that abounded at the time. And how could such an ungainly creature have taken to the air again once grounded?

There is an entertaining footnote to the dinosaur/bird story, which illustrates, again, how transient are our certainties about the past. Dinosaurs laid eggs, and some of them probably nested communally. *Maiasaura*, a duck-billed dinosaur, was considered to be a nest-builder and, by implication, a good mother. For once, a dinosaur genus name was assigned the feminine gender (suffix *saura* rather than *saurus*) in a slightly sentimental and grammatically dubious acknowledgement of its alleged caring qualities. Another smallish Cretaceous dinosaur associated with eggs was called *Oviraptor*—literally "egg-hunter"—and was considered the villain of the times, sneaking in like any magpie to rustle the eggs from innocent herbivores. These eggs were frequently discovered in the Flaming Cliffs area of the Gobi Desert, a famous hunting ground for palaeontologists, where new bones can still be discovered scattered about the dry slopes. Some doubt attended the carnivore interpretation when a particularly perfect egg was discovered containing a hatchling dinosaur. But the small body seemed to belong to *Oviraptor* itself, rather than any other dinosaur, and hence the alleged egg thief began to appear more guardian than invader. Had it been judged too hastily? An amazing proof that it was innocent of all charges was published in the journal *Nature* on 21 December 1995. An *Oviraptor* was found, fossilized, still squatting over its nest of eggs, in the same sitting pose as that of a present-day ostrich. It had been overwhelmed by a violent storm, but loyally stayed by its twenty-two eggs—the very model of maternal devotion. Sadly, its name will for ever testify to its former bad character, even though it is now, perhaps, the original poacher turned gamekeeper.

Oviraptor is one of the dinosaurs most closely related to birds. Could it be that feathers first evolved not for flight, but for incubating eggs? This seems a plausible idea because even a small patch of feathers would serve some purpose. Their subsequent use for insulation and eventually flight would be less remarkable than if the whole flight "package" had had to appear together.

It must have been at this time that Earth began to sing. Many birds are songsters, and even those that do not sing are prone to squawk. Sound serves several purposes: to define territory; to attract a mate; even to repel enemies. What a transformation the appearance of animal sound must have made to the world. The silent forests of the Palaeozoic were replaced by honking, snorting, singing environments. The "crested" dinosaurs added their own reverberating cacophony: the crests are now thought to have functioned as resonating chambers, which could have blasted fanfares like a band of alpenhorns over the wide Cretaceous plains. This is yet another transformation in interpretation—Disney had the crest of *Parasaurolophus* doing little more than supporting a sail running down its back.

I have frequently speculated that just as dogs and their owners are supposed to look increasingly alike over time, scientists come to resemble the organisms that they work upon. Occasionally, they even have a name appropriate to their favoured calling. At the Natural History Museum in London we had a Mr. Wrigley who worked upon worms. It is surprising how many botanists are called Green and Bush. There is a well-known fish specialist called Herring. Another scholar in the Natural History Museum was an expert on bees. He was rather stout, and given to wearing bold, stripy jumpers. He could frequently be found buzzing about the corridors, humming to himself. Logically, therefore, dinosaur workers should be large, or at least larger than life. Some evidence for this is found in the battles that have scarred the history of their study.

In the latter half of the nineteenth century dinosaur bones were so abundantly strewn over the ground near Medicine Bow, Wyoming, that a shepherd is reputed to have used them to construct a makeshift hut. Two pioneer palaeontologists, Professors Edward Cope and Othniel Marsh, fought tooth and nail to be the one to name the greater number of these hitherto undescribed animals from Colorado, Wyoming and Montana. They raced to be first to each new site as it was discovered; they bribed and schemed and indulged in all manner of shenanigans. Between them, they named no fewer than 136 dinosaurs. The most splendid specimens known of many species, and now resident in pride of place in major museums, were the product of a rivalry stoked by fierce hatred, which proves that fine ends may very occasionally result from ignoble means. Serious battles have been fought since, but few with the venom of the early days. The debate over hot-bloodedness was a case in point: there was plenty of huff and puff, but little bile. One of the leading figures in transforming our view of dinosaurs as slow dullards, Bob Bakker, is certainly colourful enough to qualify for the outsize tag. He has a wild look and long, lank locks, topped by a battered cowboy hat. He is a wonderful artist, and it was, to a great extent, his drawings of bounding and energetic dinosaurs which served to alter the way many other artists portrayed the world of the vanished giants. Spielberg merely animated them. John ("Jack") Horner is the Buffalo Bill of the palaeontological world, with the same smack of the wilderness, and something of the mystique of the tracker—but of bones rather than hostile tribesmen. There is a species of hero which combines learning and derring-do, a kind of muscular scholarship that extends back at least to the popular romances of H. Rider Haggard. Archaeologists are particularly suitable for the role, as the movies starring Indiana Jones will affirm. Jack Horner fills the same niche among palaeontologists. As for the younger generation of vertebrate specialists, Drs. Mark Norell and Michael Novacek from the American Museum of Natural History have dark beards and long hair and wear dark glasses with circular rims. At symposia they sit apart. They are as "cool" as it is possible for any scientist to be, and have no resemblance at all to the common notion of the boffin as a style-less monomaniac. Qualified support, then, for the theory that bone people are outsize and colourful characters.

THE MARINE REPTILES WERE contemporaries of the dinosaurs. They had descended from terrestrial ancestors, and hence were early examples of animals that had returned to the sea—the same sea which they had deserted more than 100 million years before. These were Thomas Beddoes's "fish lizards" or John Martin's "sea dragons," although they are now probably at least as widely known by their scientific names of ichthyosaurs and plesiosaurs. The plesiosaurs had long necks and comparatively small heads, and four, often lozenge-shaped, flippers. The ichthyosaurs are thoroughly fish-shaped, complete with dorsal "fin" and a forked tail. However, their skull anatomy at once reveals their true, reptilian nature. Unusually among vertebrate remains, it is not uncommon to find preserved some outline or impression of the smooth skin of these remarkable animals. Their bodies are preserved in shales—originally dark marine muds—that can preserve exceptional detail. Their skeletons are surrounded by a shadowy ghost of the flesh, as if they had been wrapped in a black shroud upon their death. There is no doubt at all about what they ate, either, for their stomach contents reveal to the pathologist the dead animal's last meal—often traces of ammonites, or other squid-like animals. Ammonites have even been found with a neat set of plesiosaur toothprints across their midriff; as evidence of guilt, this is the palaeontological equivalent of a smoking gun. No doubt they ate fish, too, like the living marine mammals, dolphins and porpoises, with which they have often been compared in their supposed ecological adaptations. This comparison is justified to some degree, but not in the simple way that was once claimed. Perhaps the analogy works to the extent that one can recognize soldiers by virtue of their bearing arms, even though this tells you little about their way of fighting: same trade, different battles.

Consider the virtues and limitations of the arguments from analogy. Plesiosaurs may have used their flippers to beat up and down in the same way that turtles propel themselves through the Pacific Ocean today—rather than for rapid sculling, as was once thought. The flippers have a curious hydrodynamic structure that would only work satisfactorily in this way. On the other hand, living turtles use only the forelimbs for swimming and the hind as brakes and for swerving, so the gait may have been quite different from any animal still swimming in the sea. Debate still continues about how the flippers beat, and in what order. Some workers maintain that the forward pair were "upbeat" while the hind pair trailed. Ichthyosaurs are obviously streamlined in a fashion which invites comparison with a dolphin, but there are several differences—for example, in the vertical orientation of the tail of the extinct reptiles. Some ichthyosaurs were very large—more than ten metres long. A fine specimen of one of the largest species was collected in 1985 near Charmouth in Dorset, and is now displayed in the Bristol City Museum. One name acknowledges their former monstrous reputation: *Grendelius* is named after Grendel, the first literary monster in the English language, who carried off thirty of Hrothgar's thanes, as described with full gore in the poem *Beowulf*. Other species were comparatively diminutive, and may have been the pursuers of small prey species. Some had narrow, streamlined bodies, built for speed; others with wider, deeper bodies may have cruised in a more leisurely fashion. Their hydrodynamics can be plausibly reproduced by scale models. Several kinds had enormous eyes, fringed with a circlet of plates. Their vision was evidently remarkably acute. It has been claimed that these ichthyosaurs may have been able to dive to great depths in pursuit of prey. There is nothing similar in the living fauna with which these eyes can be directly compared. However, it is certain that, like living marine mammals, Jurassic and Cretaceous marine reptiles would have been compelled to return to the surface to catch their breath. A veritable school of ichthyosaurs (and plesiosaurs) lines one of the galleries in the Natural History Museum in London. The slabs containing the fossils are mounted from floor to ceiling, still in their 1920s display cases. Most were collected from the Blue Lias outcrops along the Dorset Coast, and some at least by Mary Anning, whose demure portrait can be seen alongside them. If you pause for a while to allow the giggles of visiting schoolchildren to fade you can imagine yourself floating beneath the Jurassic sea while groups of marine reptiles flash past you, sculling towards the depths. The dark stone on which the fossils lie reinforces the illusion. You scarcely require artificial animation in this setting to re-awaken these dead bones: the sea monsters oblige by opening their mouths in dumb greeting, as if freshly turned upon the slab.

Along one end of this display there is an ichthyosaur that contains inside it several smaller examples of the same animal: little skeletons inside the bigger one. Could they have been cannibals, these elegant fish-lizards? Look closer, and it will be plain to see that the tiny creatures are unharmed, or at least not crunched up and half-digested. Many of them are not even in the right part of the body for digestion; they are behind where the stomach was likely to have been. A few seem to be in the right position to be sitting in the birth canal. It is more than surmise that these ichthyosaurs gave birth to their young alive—ready-formed, in a state to take to the sea at birth. Some unfortunate mothers died in the very act of giving birth. Once again, there is an analogy with members of the whale family, such as dolphins, which also bear their young at sea. Ocean-going habits evidently fostered such early independence on more than one occasion. Live-bearing reptiles are not unknown: some of the more northerly snakes do the same, and the process has been recorded on film several times.

These extraordinary specimens were nearly walled in! Those in charge of modern museum displays have reacted in recent years against what has been perceived as old-fashioned parades of objects. There is something intimidating—or so it is alleged—about ranks of *things* in cases, stretching through gallery after gallery. Rather, a museum is supposed to be there to inform, explain and to tell a tale. The great ichthyosaur gallery did not conform to modern exhibition practice—and might well have been dismantled, were it not for the sheer size of the plesiosaur specimens comprising the centrepiece of the exhibits. They were too heavy to be removed. Eventually, in the early 1990s, it was decided to make a virtue of necessity and to clean and restore the whole Jurassic marine wall. So it is still possible to see the direct evidence of a small tragedy that happened in the sea during a live birth 150 million years before the museum opened its doors to the public for the first time. It is a much more eloquent testimony to the drama of the past than some meretricious piece of talking plastic.

It is sad that the ammonite crushed by the bite of a plesiosaur is not on display. But there are many other ammonites, and they serve as a reminder that not all life in the Mesozoic tended to the monstrous. These fossil capricorns are to be found in every amateur collection of fossils; they may be curled tightly, ribbed or scalloped, knobbly or smooth. They are often the size of a coin or medallion, but sometimes approach the diameter of a shield. They may have been comparatively modest animals, but they teemed in ancient seas. Occasionally, they comprise whole beds of rock. Appropriately enough, the Ammonian Horn was the original cornucopia, endowed by Zeus with the property of granting wishes without end. He presented it to the nymph Amalthea. With only a deft change of gender, one of the best-known Jurassic ammonites was named *Amaltheus*. Ammonites evolved so rapidly during their long history that they are ideal fossils for use in subdividing geological time: thousands of different kinds of these fossils have been recognized. It requires expert eyes to identify them. I have been known to tease my ammonitical acquaintances with the claim that there are really only about ten different ammonites, which just pop up again in different rocks from time to time. It is certainly true that broadly similar designs of ammonite shells re-appear on many occasions through their long geological history (these are known as homeomorphs), but they really do offer an almost limitless set of variations upon a few simple themes. It is amazing how a coiled shell carrying external ornament can be so exquisitely variable. There are forms as smooth as a discus, and as thin; others are fat and tumid, almost spherical; they can be spotted with lumps, or covered with ribs as delicate as weaving. Some have keels along their backs. Several years ago it was discovered that ammonites came in pairs. There were small-shelled individuals and large-shelled indi-viduals of the same species. These dimorphs are now regarded as being the two sexes; it is hardly necessary to add that the male is usually considered the weedier of the two. The shell of the living ammonite housed an animal some-what like an octopus, being equipped with tentacles and having a complicated jaw which could shred shrimps and other small prey. The inner part of the shell was divided into chambers, which show to advantage in those specimens which are polished and cut through; such cabinet specimens are a staple of rock shops. The growing animal built itself a new chamber from time to time and sealed off its old living quarters with a wall. It maintained contact with its former dwellings by means of a hollow tube, through which it could control its buoyancy by controlling the amount of light gas retained in the chambers in the early whorls. Clearly, this animal was sophisticated enough to flee from the jaws of a sea lizard when the occasion demanded.

One of my earliest fossil finds was in an old quarry in the Cotswold Hills. The richly ochreous Jurassic limestones from this area are famous for yield-ing the building stones that have made many villages of almost overwhelming picturesqueness. Occasionally, fossils of sea urchins are to be seen in the stone walls that crisscross the area. I was fortunate enough to discover one such fos-sil in the quarry. It was about the size of a Danish pastry, but more the shape of a bun, a little tumid in the centre. It was a pleasure to hold, and it became one of the prized possessions of my schoolboy collection. I labelled it in my best writing "*Clypeus ploti*" and stuck the label on with glue. No dinosaur bone could have been charged with more significance for me. Sea urchins are comparatively common discoveries in rocks of Jurassic and Cretaceous age, but they are always exciting. Their bodies are plated with calcium carbonate, like a three-dimensional mosaic. Urchins are still very numerous in the sea today, and anyone who has visited the Mediterranean will surely recall seeing them clustered like oddly sinister, black hedgehogs around sewage outlets, where they doubtless perform some useful and hygienic function. To have trodden on one of these animals is another, but also altogether memorable experience. It is not surprising that the stout spines of species like these are frequently found as fossils even in the absence of the urchins that once carried them. In the white Chalk of the late Cretaceous, sea urchins are comparatively glamorous fossils alongside more everyday molluscs and brachiopods. The heart-shaped urchin *Micraster* had tiny, felty spines, but these are lost in the fossils. *Echinocorys* is shaped something like a bishop's mitre. They are sometimes cast in durable flint, which outlasts the soft chalk. These urchin fossils can then be discovered where they "weathered out" just lying on the ground, and they are one of the few fossils to have acquired a common name—"shepherd's crowns."

Chalk is a remarkable deposit. It is probably the most widespread rock formation. It is quintessentially English, as in the White Cliffs of Dover or the Seven Sisters. But the same chalk (and of the same age) crops out along streams that run through the Texas city of Austin, where you must walk along the ends of back lots to examine it. It is present too, in arid ranges of Kazakhstan and in the Midi of France. At exactly the same time as the most recent, largest and most fantastical dinosaurs walked over the plains of the Midwest, the warm and equable climate filled the seas with soft, calcareous ooze that was destined to become chalk.

Chalk is worth examining closely. If you first scrub a piece gently with an old toothbrush, and then peer at it under a strong lens or a weak microscope, you may see tiny, chambered shells—no more than a millimetre or two across—etched out upon the surface. These are the tiny tests (hard shells) of single-celled foraminiferans, protists with a record extending back to the early chapters of this book. But these are mostly novel planktonic species, built as a series of bubble-like chambers, and they must have drifted in uncountable billions in the sunlit waters of the Chalk sea. Their dead tests contributed to the pure, white ooze on the Cretaceous sea floor. They may be admixed with fine molluscan fragments and other shell debris, too. But focus down finer. The dust that you brush away to reveal the foraminiferans can be examined under a very powerful microscope—or, preferably, the scanning electron microscope—to reveal a further scale of life which might otherwise have been unsuspected. The dust itself is almost entirely made of fossils—minute rosettes of calcium carbonate plates, regularly arranged, like wheels with delicate spokes, or minuscule stony flowers, of which a hundred or so might fit on a pinhead. They are called coccoliths. These are the debris of another single-celled organism, a planktonic "alga." During their life, the coccoliths were secreted by the living cells, many coccoliths to a cell. No tinier fossils exist, yet each is, in its way, as organized as *Supersaurus*. These most modest remains are in proportion to the largest dinosaurs as the earth is to the sun. The Chalk, including all the cliffs from Texas to Dover, the downs that floor much of southern England, the bowl that holds the Paris Basin, all this is made of fossils, the sheer numbers of which boggle arithmetic. Chalk may be a commonplace rock, but it is surely one of the most curious.

In medieval times an ability to tell chalk from cheese was one of a series of questions posed to questionable individuals as a test of their sanity. If we had attested then to the composition of chalk as a pot-pourri of tiny animals and smaller plants, to which the occasional sea urchin was added, like a plum in a pudding, we would surely have been confined to a straitjacket without a chance of a second question. It is easy to take the limits on our vision for granted. What we know is partly what we see. Louis Pasteur fought battles over the existence of microbes which lay close to the limits of visibility in the nineteenth century. Now, we are slightly disappointed by portrayals of viruses that are slightly out of focus, as if we had some kind of *right* to see everything. Images sent back from deep-space probes serve to solidify astronomical speculations into facts, albeit facts that are ever further and more remote. Microscopy has had a similar effect on the reverse scale: an exploration of small and even smaller. Modern microscopy has shown chalk for what it is, and it has been turned upon cheese to similar effect, revealing both to be seething masses of life. The way in which our view of past life has been changed at this small scale is probably even more profound than the several changes in our perception of the past lives of the dinosaurs, which was a matter for new interpretations as much as fresh observations. To be sure, the electron microscope has revealed details of dinosaur eggshells which can be gathered in no other way. But the techniques that have enabled small fossils to be studied in as minute detail as large ones have had extraordinary implications, not least because small fossils can be obtained in very large numbers, which allow statisticians to work their mathematical skills upon changes in whole populations.

There is sight, and there is insight. One of the most remarkable scientists I have met is totally blind, yet "sees" better than most. Professor G. J. Vermeij knows about shells. He sees them with his fingers, appraising their most subtle details by palpation alone. He can identify hundreds of species by touch, being preternaturally sensitive to ornament and shape, the kinds of features that sight cannot substantiate as effectively as touch. He uses the eyes of amanuenses to read scientific papers to him aloud, so even where his fingers cannot travel his mind certainly can. He is adept at explaining not just how one shell differs from another, but why. He noticed an important change in the marine life of the Mesozoic, and especially during the Cretaceous. During this time the scale and extent of predation increased. New kinds of hunting snails appeared, equipped with highly efficient means of boring though shells with their toothy radulae; crabs and lobsters with massive claws that could crush shells bodily—or peel back the "lids" on snail shells to get at the succulent and vulnerable meat within; starfish of modern kinds which could grapple with and prise open quite substantial clams, before digesting them in grisly fashion with their everted stomachs. Then "bony fish" also diversified, with their myriad ways of nipping and biting and probing.

In sum, the undersea world became a more dangerous place to live than it had been before. Vermeij remarked that prey animals responded to increased variety of attack by increased variety of defence. It was almost like an arms race, in which every ingenious thrust was matched by a matching skilful parry. The result was an increase in diversity of life, as predators and prey sought to outwit one another. Accordingly, some prey species developed very thick shells in order to place them beyond the competence of all but the most persistent borers. Others, clams especially, burrowed ever deeper into soft sediments, while wood- and rock-boring clams effectively escaped from any kind of attack at all (they are roundly cursed by mariners and structural engineers today for their particular talents). Scallops learned to swim more efficiently by "clapping" their valves together like animated castanets. Some snails developed an external armoury of long spines. Limpets and barnacles colonized shallow environments on the foreshore where only the most tenacious hunters could venture, and where they now cause bathers with sensitive feet to fret and hobble. There were even subtle beneficiaries who prospered in novel ways from increased confrontations elsewhere, much as arms merchants and black marketeers benefit at the fringes of human conflicts. Thus hermit crabs evolved which could occupy cast-off shells (the thicker the better) of several species of snails. These crabs themselves then sought to disguise and protect themselves by planting sea anemones on their backs, when many other animals, especially bryozoans (sea mats) came along as uninvited guests. All parties seem to have benefited by the association. In some rock pools it seems as if every snail shell is occupied by a crustacean tenant, and there is so much competition for space that I have seen a bottle top stranded in such a pool tottering around on jointed legs, temporarily occupied by a desperate hermit. If such descriptions apply human motivation to what is, in evolutionary terms, a struggle for reproductive superiority, then for once I feel inclined to let the military metaphor have its head. After all, Professor Vermeij has justified these assertions with all manner of experiments and graphs, but somehow the image of escalation of attack and defence during the Cretaceous is most readily understandable in terms of the Cold War arms policy that many of us have seen in our own lifetimes: to maintain equilibrium, the constant invention of new methods of attack and defence.

I draw no moral conclusion here: this is merely a descriptive device. It could, with a little tinkering, equally well be applied to the progressive increase in size of dinosaur prey and predators. Quite different kinds of interactions are possible—and, indeed, happened at the same time.

THE CRETACEOUS WAS WHEN flowering plants (angiosperms) became an important part of the flora. The relationships between insects and flowering plants afford a more persuasive and complete comparison with the rich complexity of human interactions. There is a mixture of collaboration and aggression, of rivalry and reward. Plants continually fight with insect foes, yet woo insect pollinators. In the case of butterflies the two roles are combined within a single species: the caterpillar that destroys becomes the beautiful imago that helps vegetable reproduction. The love–hate relationship that powers so much stage drama is as good a metaphor as any for the way in which flowering plants have evolved together with insects, the two linked by a cord of interdependence as binding and ambiguous as any relationship devised by Chekhov. I was once responsible for a research student who spent three years studying the changes in nibbling and boring inflicted by insects upon leaves, signs which can be preserved in their fossils. He was able to show how the variety of such attacks upon leaves increased progressively through the Cretaceous rocks and into the Tertiary. You could not *see* many of the plant's responses—but it was easy to infer that this included the inception of chemical warfare. Plants became poisonous in order to counteract the inroads of nibblers—but then specialized insects in turn developed tolerance to the poisons. Some insects even recruited the poisons into their own bodies to become poisonous to their own predators, a kind of chemical double bluff. Those caterpillars with yellow-and-black warning coloration are likely to have come along this evolutionary route. Other plants developed finely hairy leaves upon which insect legs could not gain a purchase. Aromatic oils have their origin in this conflict. Whenever we use basil, fennel or fenugreek to enliven our cooking, or lavender to sweeten our pillowcases, we should remember this arms race between leaf and six-legged pests. The gorgeousness of flowers is just the opposite: the outcome of attempts to *attract* insect pollinators; these are honeyed rewards, nectar most seductive, sugared recesses held in floral purses, and the thousand fragrances which have kindled erotic thoughts and made the fortunes of *parfumiers*.

In the Cretaceous there are several kinds of fossil pollen—which is recovered from digesting rocks in acid—that can only have been produced by flowering plants. Actual fossils of Cretaceous flowers are understandably rare, because they are of their nature ephemeral, but fossils which are unequivocally related to living magnolia trees and to shrubby relatives of poppies are widely known. These are simple flowers—little more than coloured leaves surrounding the seed capsule. From this simple start, all manner of exotic blooms derived, from the waxy orchid, or the nocturnal cactus known as Queen of the Night, to many-headed umbellifers and daisies. They have each seduced different insects to do the work of pollination. Some have attracted bees; others, night-flying moths; others, butterflies, or even plain, unglamorous flies. Since the Cretaceous, insects and flowers have evolved together in a kind of mutual dance of ever-increasing complexity to which the term "co-evolution" has been applied. In its own way this is another kind of escalation, but it is an escalation of inter-dependence rather than antagonism, since the end is a flower which depends on one species of insect and an insect that can hardly exist without its particular flower. The elaborate devices by which orchids achieved pollination were described by Charles Darwin in *The Fertilisation of Orchids* (1862). The ruses adopted by these plants even extend to their flowers imitating particular species of bees or wasps—so that their own fertilization can take place when the fertilizing insect makes a doomed attempt to copulate with its floral surrogate partner. In other flowers, the length of flower tubes (corollas) can be precisely tailored to one or another species of butterfly. The feeding organ of the animal changes in harmony with the plant. Or maybe there is a secret route to nectar which can be cracked by only one species of pollinator. But in all these cases the effect is the same—and similar to that induced by escalating interactions between prey and predator: an increase in the glorious complexity of the biological world. We know that many of the relevant insect groups for co-evolution had a fossil record extending to the Cretaceous, or even Jurassic. To use a word which has now been employed so often and so variously that it has become diminished by its very usefulness—there was an enhancement of "biodiversity" since that time: the diversification of life after the trauma of the Permian–Triassic extinctions. I imagine that insect–plant interactions effecting proliferation of the richness of species were taking place in some quiet corner of the world even as *Tyrannosaurus* paraded with such flamboyance. Quite probably, these quiet botanical and entomological enhancements were the more durable.

I like to think of Charles Darwin reflecting upon such wonders of pollination as he took the Sand Walk every evening at his home, Down House, in Kent, not so many miles to the south of Piccadilly, in the centre of London. Down House is set in the midst of Cretaceous Chalk countryside, and so has an oblique chronological link with the onset of the pollination *pas de deux*. It is still possible to retrace Darwin's steps at Down. It is not a grand place, no stately home, but a country squire's house, and it breathes comfort. The furniture is just a little worn. You feel that Charles Darwin had something more important on his mind than a slightly frayed carpet. There are notebooks and writing implements on display, but they do not have quite the formality of many of those laid out in other studies of great men. I found myself staring intently at these everyday objects as if they might somehow hold the key to his extraordinary insights. They offer no clue, of course. But you can readily imagine Down as a family home of one of the few intellectual giants who was also beloved by his family and intimates. He was far enough from Piccadilly to have a perfect excuse *not* to attend the meetings of the Linnean Society at Burlington House if he did not have a mind to—and he frequently did not (there has been much debate whether the ailments which plagued him were psychosomatic, or a legacy of his time on HMS *Beagle*—perhaps some parasite picked up in South America). He had plenty of time to think about pollination, or barnacles, or coral reefs, and to conduct his extensive correspondence. The view from the French windows over the lawn at Down has not changed much since Darwin lived there. The yew tree must have grown a little, and I suspect that the herbaceous border was once in better order. The mossy lawn erupts with interesting fungi late in the year. It is easy to understand how the calm of the countryside beyond might unstick a reluctant pen. The Sand Walk was made around the perimeter of a copse at the end of the garden, a place for a modest constitutional, where ideas could be mulled over and phrases polished for the press. The domestic scale of some of his researches is something the visitor might not be prepared for. In the hedgerows in the lanes round about there still twine white bryony and honeysuckle, the raw material for his essay upon "The Movements and Habits of Climbing Plants." Orchids still grow on a nearby chalky slope, much as they did in Darwin's time, and there he could observe the bee orchid for himself. His investigation on the importance of worms was also conducted on the premises. Some of his greatest conclusions were drawn from comparatively humble materials—finches rather than dinosaurs. His reflections upon the origin of species changed the way in which we understand the natural world, and inform my narrative as pervasively as grammar does a novel. The monstrous and the modest were embraced with equal favour by his great unifying theory.

Chapter 10

Theories of the End

T HE DINOSAURS DID NOT SURVIVE beyond the Cretaceous—
save those that were transmuted into birds. Their end was apparently sudden. Nor did they die alone.

There must have been a catastrophe between the Cretaceous and Tertiary,
some 65 million years ago. There had been previous mass extinctions, but
this is the one that has exerted a grip on the imagination like no other. The
death of reptilian giants is *the* definitive death, and has somehow become an
emblem of the precariousness of our own place in the firmament. We must
change and adapt; otherwise, we might go the way of the dinosaurs—or so
the cliché would have us believe.

Jean-Anthelme Brillat-Savarin was the first great French writer on
gastronomy. His *Physiologie du Goût* (1825), published one year before his
death, is a delightful soufflé of anecdote and observation centring—but
eccentrically—on the glories of gourmandism. His asides on the vagaries of
human nature are every bit as entertaining, and probably rather more accurate, than his observations upon the virtues of the table. His faith in the
omniscience of food is absolute. "Tell me what you eat," he remarks as an
opening aphorism, "I will tell you who you are." It comes as a surprise to find
in the midst of his philosophical reflections a section on "The End of the
World."

Incontrovertible evidence tells us [he writes] that our globe has already suffered several absolute changes, each one nothing less than an *end of the world*; and an indefinable instinct warns us that more revolutions are to come. Many times already it has been thought that such revolutions were close at hand, and there are still people alive whom the watery comet foretold by the worthy Jérôme Lalande sent hurrying to the confessional. Judging by what has been said on the subject, men are prone to invest this catastrophe with vengeance, and destroying angels, trumpets and other no less terrifying accessories. Alas, there is need of no such fuss for our destruction; we are not worthy of so much pomp; and if the Lord so wills He can change the face of the globe without the help of ceremonial apparatus . . .

Let us suppose for example that one of those wandering stars whose course and mission are alike unknown and whose appearance has always been accompanied by a display of terror; let us suppose, I say, that some comet passes close enough to the sun to be charged with superabundant heat, and comes close enough to earth to cause for the space of six months a temperature of 168 degrees . . . At the end of that fatal season, all living things, both animal and vegetable, will have perished; every sound will have died away; the earth will revolve in silence, until new circumstances develop new germs.

This exhilarating litany of doom shows that the notion of catastrophic revolution mediated from "outer space" has a long pedigree. Brillat-Savarin thought of himself as a man of rational method, even of science, and the ideas he articulated were tossed through the salons of a fashionable and comparatively free-thinking society. He believed in "incontrovertible evidence." It is necessary to be wary before taking too literal a view of his prescience. He might well have had revolution on his mind because he was forced to flee France during the French Revolution, and we may be certain that his several ends to the world were not the same as those mass extinctions revealed by the fossil record, which were scarcely catalogued at the time. Rather, they may have been the cycles recognized by the pioneer palaeontologist Alcide d'Orbigny in the rocks around Paris—separate and successive fossil faunas which subsequently provided the basis for correlating the formations in much of France; or they may have been the series of successive "creations" described by the great anatomist Baron Cuvier to account for the sequence of fossil mammal faunas he was studying. Brillat-Savarin's ideas of catastrophic extinction were not factually watertight, but they remained, as it were, an ingredient in the intellectual cuisine. The time would eventually come when a new recipe would incorporate this extraterrestrial ingredient in a new dish of scientific cookery which would have delighted the palate of Brillat-Savarin.

There was a long intervening period in intellectual history, however, during which catastrophic explanations of events in the Earth's narrative were unfashionable. The triumph of Charles Lyell's Principle of Uniformitarianism in the 1830s gave geology a method for interpreting the past from observations made at the present day; it reduced the past to chemistry and physics. The simplest formulation of the principle stated that the physical processes operating in the past were the same as those operating today, time and process uniformly stretching in unbroken continuity to the beginnings of the world. Thus, the workings of a Devonian stream could be deduced by looking for the appropriate river today; ancient volcanoes had their rumbling counterparts in Etna or the Azores. The catastrophic interpretation of fossils as the literal evidence for the Flood was banished to the realm of myth and fancy. As Lyell succinctly expressed it: ". . . we are not authorized in the infancy of our science to recur to extraordinary agents." Uniformitarianism held the field through much of the nineteenth and early twentieth centuries; after all, why invoke the extraordinary when readily observable phenomena explain so much so satisfactorily?

Lyell was equally seminal to Charles Darwin, donating to his development as a scientist not only the leisurely timescale needed for species to proliferate, but also the idea that the *process* of formation of species may be observed by studying the world as it is. Darwin was hesitant at first about rigorously applying Lyellian principles, but it is unlikely that he would have so confidently extrapolated from his experiences of finches on the Galapagos Islands to theories about the origin of *all* species had he not by then been thoroughly imbued with this fashion of reasoning; indeed, it infuses all his writings. So the world and its long history were explicable by the ebb and flow of processes still at work, now quickening, now slowing down, to be sure, but never changing in kind. Evolution itself was an unfolding process prompted by natural selection, in which competition and heredity joined hand in hand to effect what were nearly always termed "improvements." This idea is with us still in various kinds of social Darwinism, although recent, fascist misinterpretations of it have given birth to deformed progeny with terrible racist consequences. But despite these perversions, the method of reasoning still has a tremendous appeal to those who like their deductions to be rigorous and clear. For catastrophes are the stuff of panic and confusion, where rules get broken, and survival may depend on cryptic luck or obscure virtue. Each one may be unique. It is hardly possible to run a field trip to study a contemporary catastrophe in order to compare it with one supposed to have happened long ago. Worse, once admitted, catastrophes can be claimed whenever some unexplained change occurs in the record of the rocks. Any problem? Bring in a catastrophe! This, the *ad hoc* explanation, is anathema to all those brought up with the scientific and philosophical rigour of Karl Popper and Ernst Nagel. Scientists do not trot out *ad hocs* the way a magician pulls flowers out of a top hat; it is not considered proper behaviour.

A tradition of catastrophic, extraterrestrial explanations none the less continued to propagate into the twentieth century, even if it was mostly on the wrong side of the intellectual blanket to achieve respectability. The comets we have met before. They are the most obvious and regular of extraterrestrial visitors. They have a dubious ancestry in geological explanation. In the 1950s Immanuel Velikovsky used their disastrous interference for an explanation of almost everything: to raise mountains and rend continents, to induce flood, to spawn myths. The Flood is a tribal memory of such a visit, for example. His book titles—*Worlds in Collision, Earth in Upheaval*—are juicy vade-mecums into a world of twisting and slithering rock masses and quivering tribes fleeing before flaming heavenly bodies; seas of blood, plagues of frogs are all proof of the astronomical theory. It was great fun, not least as a demonstration of how wayward learning can be recruited to support grand theories. As I described earlier, twenty years later Fred Hoyle was using comets to visit viral plagues upon the Earth—not least, to extirpate the dinosaurs. Under this scenario they died a grim, painful death. There were several professors who held university positions who could continue to argue for extraterrestrial influences: Dr. Trechmann claimed that the tidal pull of the moon could raise mountains. Professor Otto Schindewolf of the University of Tübingen noticed the contrast between mass extinctions and a normal rate of turnover of species. In 1962 he published a paper with the challenging title *Neokatastrophismus?*; he stated there that the great Permian extinction may have had an extraterrestrial cause, partly because he could identify no Earth-bound agent that could possibly do such damage. He was ignored. Curiously, he is practically alone in being given credit for the paternity of the new catastrophism, in which the fate of the Earth is once again written in the stars. But then he was, after all, a real scientist, and thus his antecedence is considered respectable. To acknowledge that any of the other renegades might have been vaguely along the right lines is to admit oddballs as associates, and may taint the alleged novelty of any new observations. That would never do.

Evidence—that is, hard facts, field evidence—must be collected from the rocks, one rock bed after another, in order to find out what happened at the end of the Cretaceous. There are problems. In many rock sections there is a break in the rock record at exactly the time of extinction: you can see the world before and the world after—but not the critical moments. It is like losing a few frames from the print of a whodunnit; one moment the victim is alive, the next he is dead, but the critical frames featuring the criminal and the smoking gun have vanished. Common sense suggests that one is most likely to discover full and complete rock profiles in places once covered by the deep sea, and well away from whatever dramas were being played out on land. Here, naturally enough, there will be no direct evidence of the demise of the giant reptiles, and instead the chronology must be read in the changes that happened to marine animals and plants through the same interval. Whatever explanations are invoked must account for the changes in the sea *as well as* those on the land. Explanations for the demise of the great reptiles—and my former colleague Alan Charig has listed several dozen—have included such odd candidates as death by indigestion (supposedly related to floral changes). This could hardly apply to tiny foraminifera in the open ocean, which were also drastically reduced at the same time. Nor could the total demise of the ammonites—which brought to an end an illustrious history of several hundred million years—be attributed to a drought on land, as depicted in Walt Disney's *Fantasia*. It has become a convention to refer to the time of extinction as the K–T boundary. T stands for Tertiary, and K for a Greek (*kreta*) alternative to C for Cretaceous. One might describe the extinction problem as *What K–T Did*.

Let us travel to Gubbio in central Italy in search of an ideal marine section. The rocks that are now elevated as part of the spine of Italy—the Apennines—were originally laid down under a comparatively deep sea, where they solidified. The construction of the present Mediterranean region brought the depths to the surface, and threw up these hills in one of several convulsions of tectonic plates, the last legacy of which is the hot springs that bubble up along the range. The rocks have been upturned steeply in places. You leave the ancient town of Gubbio by the medieval Metauro Gate. The road winds northwards up into the hills, and as you look back the red pantiles, and the cypresses that rise like dark flames, gradually retreat, and the gorge which the road follows soon shows much bare rock, and is relieved only by the sparse vegetation typical of limestone country. This is the Bottaccione Valley. Beds of limestone form bluffs which line the winding road; in places the steeper slopes have been held back by nets (which the geologist curses because they make the rock so hard to get at). The rocks dip to the north regularly in such a way that they get progressively younger along the road, marvellously displaying the story of the late Cretaceous and what followed. From time to time one sees wreaths by the side of the road, reminding the visitor of those who took a bend too fast. Roadside geologizing must be a cautious business. A kilometre from town one encounters a series of warm-pinkish limestones dipping away quite steeply. This is the Scaglia Rossa Formation. It is the part of the section that has to be examined for evidence of mass extinction, for within it there are tell-tale fossils which allow for recognition of the very latest Cretaceous and the very earliest Tertiary (Palaeocene). The limestones form rather regular layers, a foot thick or less, sometimes softer and marly, and with periodic beds of hard, siliceous rocks—chert—which contain the fossils of radiolarians. There are several hundred metres of such rocks, and it has to be admitted that they are rather boring. This is good news! A touch of tediousness is just the thing that is needed to indicate a continuous, uninterrupted geological record through a few million years, spanning the death of mighty and modest organisms alike, for it shows that, whatever was happening elsewhere in the world, sediment continued to rain down regularly in this site at least. The collecting has to be utterly methodical, bashing away at bed after bed. Large numbers of the tiny fossils can tell tales about what happened to whole ecologies. The routine of the labour can be relieved by a visit to the Ristorante Bottaccione; sadly, even the most tolerant lover of Italian cuisine would have to admit that it is disappointing. But the red wine is good, and matches the tint of the Scaglia Rossa rather well. On occasion, it is tempting to spend more time speculating about what might lie out there than actually thumping the hard rocks in the hot afternoon sun.

What lies concealed in the rocks proves to be extraordinarily interesting. The pattern found at Gubbio has been confirmed in several other rock sections in places as far away as Mexico, Antarctica and Tunisia: one may conclude that it was global. The first striking fact is the sharpness of the change between Cretaceous and Tertiary. Few species of fossils, large or small, pass straight through the K–T boundary. Ammonites disappear, never to return, and there are extinctions in other marine animals, too, even if their families did not become extinct as a whole. So sharp is the change that it is possible to put your finger on the exact boundary; at Gubbio it is a thin clay, no more than one centimetre in thickness—nothing remarkable to look at. A layer of clay has been found in exactly the same position in the other sections also. It is likely that, whatever caused the change, it operated fast; exactly how fast has stimulated considerable argument, as we shall see. After the clay layer there is a thickness of strata in which there is an impoverished fauna predominantly composed of microfossils. Some of the few species are "hang-overs" from the Cretaceous world; some are strange forms that seem to occur nowhere else; one or two herald the world to come. But it was clearly an odd and diminished sea that followed the K–T event, whatever the nature of that event. A few metres up in the rock section and some of the animals that were to typify the seas of the Tertiary had established themselves; the crisis was over, and the world had changed irrevocably. What brought extraterrestrial interference back into the explanation, for the first time as a serious (and now dominant) factor in the K–T controversy, was something so subtle it could not be seen in the field. It was hidden in that thin and unexceptional clay.

The discovery was a curious marriage of technique and imagination. Technique first: there are precise methods for measuring the quantities of chemical elements in rocks. They work by discriminating the differences in the weights of the atoms themselves. Until comparatively recently it was not possible to measure accurately the presence of very rare elements. Now it is routine to measure in parts-per-billion. When Walter Alvarez investigated the section at Gubbio in the late 1970s for the trace element iridium he was in pursuit of a different problem from the cause of mass extinction. He, too, logged the section bed by bed, but he was looking for something other than fossils. The surprise came when the thin clay at the K–T boundary was analysed. The concentration of iridium was enormously enhanced in this one thin layer by comparison with the rest of the rock section—indeed, at the time, the concentration was unique in a sedimentary rock, nearly ten times the amount in the rocks below. Walter Alvarez looked for advice to his father Luis, a Nobel Prize–winning astronomer in the University of California. Iridium is a relatively rare element on the surface of the Earth. However, it is comparatively abundant in meteorites. Indeed, it was established previously that most of the iridium which reaches the ocean floor comes via meteoritic dust. The imaginative inference seemed at the same time astonishing and obvious, at least to its proponents: there must have been a massive influx of meteoritic material at exactly the same time as the K–T extinction. It was then one small step for logic, if a great leap for its implications, to link one directly with the other. The K–T extinction, affecting little and large animals, over land and sea alike, was a catastrophe caused by the impact of a massive meteorite: this was death, global death, by sudden, inflicted disaster.

Possibly because Luis Alvarez was neither geologist nor palaeontologist, this must have seemed to him both a plausible and a "natural" explanation. He would not have been prepared for the furore it might cause for its infringement of the tacit acceptance within the geological community that explanations should not invoke catastrophes. The scientific paper that broke the news was published in the leading American journal *Science* in 1980. It was co-authored by the two Alvarezes—Luis first author, with F. Asaro and H. V. Michel as technical collaborators—baldly entitled *Extraterrestrial Cause for the Cretaceous–Tertiary Extinction*. The paper had had a slightly rocky road to publication. The distinguished Chicago palaeobiologist David Raup was one of its referees—readers who have the job of assessing the worth of a paper before it is approved for publication. Many papers have plunged helplessly into the wastebasket on the instruction of negative referees (I regret to admit that it has happened to me). Raup relates in his book *The Nemesis Affair* how he had major reservations about the work at the time, and I dare say that if similar doubts had been expressed about one of mine (rather than one whose senior author had won a Nobel Prize) it would have plummeted bin-wards like some literary meteorite without much regret on the part of the editors. As it was, it was published with a few modifications, and soon attracted more informed attention than Otto Schindewolf had ever enjoyed, let alone any of his more disreputable catastrophist bedfellows. Not all the doubters have been satisfied, but regardless of their caveats the fate of the Earth was once more joined to the stars from which the planet had been born: relics from the Creation with which my story started visited belated havoc upon an older and infinitely more fertile Earth.

Meteorites fall continuously upon the Earth, and most of them burn up in the atmosphere as shooting stars. Some survive to reach the ground, and, very rarely, a truly huge one may push onwards to wreak great destruction. They leave craters even when their substance explodes to nothing. But preserved in museums there are famous examples of large meteorite fragments weighing several tonnes. In northern Greenland, the Inuit people knew of pure nickel/iron meteorites, from which they fashioned knives—directly, by cold working. These knives became valuable items of trade. There were several meteorites of this kind, which were probably derived from a single giant one that fell several thousand years ago, because they could be directly related to break-up along one path, or trajectory. If they had not landed in the high Arctic they would have caused major disasters. A huge fragment was discovered in 1936, which was laboriously transported, by way of sledge and boat, to Copenhagen, where it can still be studied. This story inspired the writer Peter Høeg to devise a widely admired novel about Greenland, *Miss Smilla's Feeling for Snow*. Rather than Brillat-Savarin's "watery comet," the dinosaurs and a myriad other species were felled by one of these very, very rare giant bolides; Alvarez estimated it would have been something like nine kilometres across. How appropriate that the largest zoological giants should have been felled by another, but mineral giant. Iridium would have spread around the world as the great mass impacted; its signature should be found in other, complete rock sections spanning the critical interval, whether they were originally on land or sea in Cretaceous times. As for the cause of death: great clouds of dust thrown up by the impact would have blotted out the Sun. The leaf would wither on the vine. The phytoplankton of the sea would perish, destroying the food chain. Gross darkness would cover the Earth. As W. B. Yeats wrote in a different context:

> A brand, or flaming breath,
> Comes to destroy
> All those antinomies
> Of day and night . . .

But this would be night perpetual. Deprived of leafy food, giant herbivorous dinosaurs starved and perished, and shortly afterwards their colossal predators followed suit. *Tyrannosaurus* would be *rex* no more, reduced below the status of its humblest subject by lack of food. Mere temperature change alone would account for millions of species fatalities. Some authors add torrential rains of sulphuric acid to the horror of those times. Others find evidence of great tidal waves—tsunamis—set off by the greatest pebble ever to fall into the pond of the sea; or great wildfires raging through the unrelenting night, contributing yet more smoky obscurity. It was a dark nightmare, through which blinded, cold, confused dinosaurs stumbled and then were still. The world was changed, in geological terms, in the twinkling of an eye. Could any scenario sanctioned by Cecil B. De Mille be more dramatic? Would not the pencils of Walt Disney's top draughtsmen snap in frustration at attempts to animate such destructive mayhem?

Science does not deal in such romantic hyperbole. The convention of cold portrayal is designed to bleed drama from even the most momentous events. In the case of the K–T question, most of the emotion has been expended in the back rooms of the conference hall, but only the smallest inkling of this seeps into the sterilized pages of the scientific literature. Many were truly incensed by the extinction scenario I have just described.

When a debate reaches a crucial stage, and if there is somebody involved of sufficient influence to conjure up funds, the usual response is: convene a conference. So it was that in October 1981 a conference was convened in Snowbird, Utah, under the title "Geological Implications of Impacts of Large Asteroids and Comets upon the Earth."

*   *   *

THERE IS A POPULAR VIEW that scientific conferences are forums for intellectual exchange, where like-minded colleagues freely swap information, motivated only by a disinterested love of truth. The search for advance in knowledge is lubricated by enthusiasm and buoyed by optimistic fervour for new ideas. How such a platonic ideal became established is curious. Perhaps accounts of such affairs tend to be written only by the top scientists, the winners of Nobel Prizes and the like, who recall their past triumphs either disingenuously, or in the rose-tinted revisionism of old age. For most workaday scientists the conference is fraught with danger and frustration, and is as aggressive an environment as any sales convention. Advancement is at stake. The long, long ladder of academe has few promotions. Any wrong-footedness is seized upon with glee by sharp-eyed rivals alert to the possibility that old so-and-so has peaked, and what a pity that he is no longer up to the ground-breaking work he did in 1976. The rule is to acknowledge the seminal work of one of the handful of scientists sitting securely at the top of whatever tree it happens to be, who control the research grants, write the job references, and thus wield much power. The ideal research paper demonstrates that an idea generated by one of these people can be applied in some new situation—to the Precambrian or to the high Himalaya, perhaps—thus demonstrating that the speaker is right up-to-the-minute while at the same time confirming the brilliance of the idea of his chosen mentor. When a new idea appears which is thought to have grants and advancement attached to its tail, the hunting instinct to pursue it around the world and into new situations is astonishing. So, a short time after the iridium anomaly was discovered at the Cretaceous–Tertiary boundary in Gubbio, there were papers published confirming its existence in nearly a dozen other rock sections; the important thing is to get your name attached to an idea while it is still "hot." Even the conference cocktail party is a kind of desperate bazaar where the ambitious mill around trying to catch up with the latest thoughts. Links are forged, troths given.

The philosopher Feyerabend has shown that a ruthless mix of ambition, competition and true intellectual gymnastics is what drives scientific advancement. The desire to be first, to beat the other team, to profit from association with a famous name, all contribute to the advance of knowledge. When the meteorite theory of dinosaur extinction was first proposed it came ready wrapped with the glamour of a famous name. Luis Alvarez was a Nobel Prize winner, full of urbane confidence. Such famous and clever people do not often dabble with anything as mundane as history. Some speakers at the Snowbird Conference showed that there were several rock sections for which evidence of meteorite impact might be adduced, while others attempted to demonstrate that the lethal effects proposed were scientifically plausible. After this conference the impact hypothesis was on its way, but David Raup maintains that adherents were still a minority. Palaeontologists are accustomed to being poor relations in the scientific family compared with physicists and mathematicians, like obscure cousins only asked to the big house for Christmas and Easter. But suddenly, they could be at the head of the table! There was a rush to assess the new idea. Those who knew of sections with rock deposition continuously spanning the junction between Cretaceous and Tertiary re-studied them in the light of impacts (not always in favour, it should be said). What before might have seemed like a small change in a uniform chalk cliff marking the boundary between the two eras was suddenly seen for what it may have been: the flash of a mighty darkening event which changed the world for ever. The excitement was extraordinary. All around the world evidence could be drafted in to support the meteorite theory, as section after section revealed an iridium anomaly. The words "bolide impact" became a commonplace in conversations around the fringes of lecture halls, uttered with an easy familiarity between slurps of cappuccino during intervals between the presentation of papers. The possibilities of jobs and grants seemed to bob to the top of this ferment like dumplings in a rich stew. Tiny foraminiferans could be recovered in considerable numbers from rock sections deposited under the sea and spanning the critical event; changes in their abundance and their patterns of extinction might provide a chronometer for the catastrophe, a history much more precise than could be obtained from studying rare dinosaur bones alone. Oxygen isotopes measured from their shells revealed how poorly productive the ocean was after the impact. Suddenly, those who knew about these minute, single-celled animals were projected from comparative obscurity to international prominence. Their papers could be published in *Science* and *Nature*, their opinions listened to with respect by astrophysicists and geochemists. It was a heady experience, and what palaeontologist could fail to be enthused by the prospect of so overturning the comprehension of such a pivotal event in Earth's history?

Better still, there arose almost immediately a sceptical school who thought that the effects seen in the rocks could be explained in ways other than by the impact of bolides, or even that the rock record actually went against impact. This focused the protagonists of the new ideas—against their enemies! The ingredients for a juicy controversy were in place: a radical new interpretation versus an onslaught from what could be labelled as a phalanx of fuddy-duddies and traditionalists. Wherever two palaeontologists gathered together the conversation would work around to a tentative enquiry: "Do you or don't you believe . . . ?" Friendships were sundered on the reply. If there *were* a bolide 65 million years ago, the extermination of affection today was the last of its fallout.

The two schools recruited their supporters. But where would the famous figures ally? Ambitious young postgraduates would be on the lookout for PhD theses safely located on the winning side. David Raup, doyen of the Chicago school of mathematical palaeontologists, let it be known that he found the evidence compelling. Dr. Raup's home territory is probability theory, the distinction of chance from design. When someone as prodigiously clever as Raup becomes a convert, having weighed the probabilities in his vast calculating machine of a mind, more feeble intellects are bound to waver. However, there were dissenting views on the western side of the Atlantic Ocean from Drs. Officer and Drake. In England, Tony Hallam, the ebullient Lapworth Professor at Birmingham, joined the ranks of the sceptics. The scientific world waited with its collective breath badly bated to see how the head honcho, Stephen Jay Gould of Harvard University, would vote. At last, in an article in *Natural History*, he added his approval to the explanation of the K–T extinction as an extraterrestrial catastrophe. The sociology of conference and influence had run its course. Now was the time for all ambitious students to scuttle to the best rock sections they could find to add their own embellishments to this extraordinary story. In campuses around the United States lectures were rewritten for the spring semester. In 1994 I attended a meeting of the Geological Society of America at which there were twenty papers predicated on the K–T disaster, and not one dissenter.

Dozens of K–T rock sections were studied in the years from 1981 onwards: it would be tedious to list them all. But the global span of the sites shows how widespread was the attention paid to the K–T question. Here are just a few: El Kef, in Tunisia; Zumaya in Northern Spain; rock sections near the famous resort of Biarritz in France. Denmark had several chalky sections which became standards—iridium and all—such as Stevns Klint, a sea cliff resembling many in southern England, but spanning the crucial interval. New rock sections were discovered in Texas, in Montana, and many different sections in Alaska. On the opposite side of the world, Dr. Zinsmeister was struggling over the cold desert of Seymour Island in the Antarctic looking for the last of the ammonites.

At the same time there were striking additions to the evidence of impact which could be used to supplement iridium. From several of the K–T rock sections tiny pieces of "shocked quartz" were recovered. Quartz, silicon dioxide, is a very common mineral indeed—most sand is quartz, for example. But it develops very curious properties when impacted at high pressures, as had previously been recognized around several well-known meteorite impact craters. A team at the United States Geological Survey discovered unmistakable examples of this special form of quartz from the same layers that yielded iridium anomalies. There really was no simple alternative explanation for the presence of such material, and I can remember my own doubts about the reality of a major impact withering away when, in the mid 1980s, I first saw colour slides of tiny, stressed pieces of quartz glowing iridescently in cerulean blues beneath polarized light. An example was even recovered from a deep ocean core drilled through rock spanning the fateful boundary. In a few sections tiny, glassy meteorites (known as microtectites) were discovered at the critical level, allegedly companions in the fall of the great doomster.

THE ROCK SECTION AT Gubbio was developed wholly in marine strata, and hence without terrestrial vertebrate fossils. Before my story gets too far ahead of itself I am obliged to return to where the dinosaurs lived, in former terrestrial sites, to seek out the evidence there at the end of the Cretaceous. There is always the risk of more drastic breaks in rocks that accumulated in lakes or rivers, for terrestrial deposition is often wayward and difficult to interpret. But in the Badlands of Montana there are sections with dinosaur fossils which seemed to be complete. The rock section at Hell Creek was one of the most fully studied in spite of its diabolic description—for during the summer it can get as hot as the fires of Hades in this ruthlessly eroded countryside. The youngest dinosaur skeletons were discovered in strata about two metres below the iridium layer, which was recognized in 1983. "Aha!" cried the critics. "Are we to believe that dinosaurs had precognition? They died out *before* the event. The meteorite (if there was one) was more of a wake than an assassination. . . ." In fact, finding the *very* last dinosaur is an extraordinarily difficult thing, worse than discovering the whereabouts of the last guest at a party, who may have gone wandering off to snooze in the airing cupboard. Dinosaurs are often uncommon at the best of times, and a little thought will show that mere sampling defects might account for an alleged absence. Where commoner fossils like ammonites have been studied, really intensive collections have shown how closely they approach the K–T boundary before they disappear. One or two claims have been made that they actually outlast the event—in Seymour Island, for example—but I gather that these have been shown to be the result of errors in the location of specimens.

So, in Montana and elsewhere on what was Cretaceous land, dinosaur remains get very close to the critical iridium layer, even if a felled giant cannot actually be found astride it. Terrestrial rock sections also include the remains of plants, naturally enough. The reader will now be familiar with the prospect of recovering tiny fossils with resistant cell walls, like pollen grains, from such rocks by their dissolution in strong acid. Several land-based sections showed an extraordinary feature when this was systematically carried out through the K–T interval. In Alaska, for example, a varied array of pollens below the K–T boundary was replaced at the boundary (marked by the iridium anomaly which had ceased to surprise) by a sample hugely dominated by the spores of ferns. Upwards in the same sections the pollen would return and the fern spores recede. The abundance could be plotted from one rock bed to the next like a graph, when the extraordinary surge in ferns appeared as a sudden peak . . . it became known as the "fern spike."

But why ferns?

When volcanic eruptions create a new island today, at least in a climate with regular rainfall, then some of the first colonizers to appear are ferns. Their spores are able to spread—invisibly but potently—because they are very small and can be carried by wind. After Mount St. Helens erupted so dramatically, spreading ash and destruction in equal measure, some of the first plants to appear in crannies, or where rains cut through the barren wastes, were ferns. Their fronds waved upwards as if to greet the resilience of life in the face of disaster. Some of the Mount St. Helens ferns were regenerated from black, creeping roots called rhizomes, which are extraordinarily tough, and far more indestructible than the roots of trees. It seems that once again what was observed in the rocks was consistent with the great global disaster. In other rock sections thin carbonaceous layers were claimed to be the graphic—and graphite—legacy of wildfire, the bonfire that burned away what dust had not blacked out, or acid poisoned. What a strange sight this world must have been just after the disaster, some kind of grey waste, which lasted maybe for a decade (the shortest time period I have mentioned in this book); then fern spores spread, originating from some small refuge. They germinated freely, released as they were from all competition, until very quickly there were groves of delicate fronds, taking us, briefly, back to a scene which would not have been out of place in the Devonian. This is not merely turning back a page of history, but several chapters.

The dinosaurs perished not because they were badly adapted and refused to change, as in the cliché, but because they were laid low, quickly and irrevocably, just as surely and inevitably as those poor inhabitants of Pompeii, who did nothing to deserve their grisly fate other than being unable to flee. What we still lack as evidence is a mummified body, twisted in final agony, vainly trying to ward off oblivion.

This story seems to be so consistent in its details, with each discovery following logically on from the last, that it might seem difficult to understand why it should have met with resistance—indeed, there are still many sceptics, most of them reasonable people entertaining reasonable doubts. The prometeorite evidence seems overwhelming to me, although it is undoubtedly true that there were also major volcanic eruptions at more or less the same time. These were connected with "cracks" opening in the crust, timed with the rapid splitting-apart of Pangaea, as the continents moved inexorably towards their present positions. Since volcanoes also have a high iridium level, this might, possibly, be the source of the exceptional boundary levels, especially if eruptions were combined with changes in sea level producing unusual concentrations—and there is some evidence for that, too. Also, there may be time missing, or concentrated into a single layer, at exactly the level of the iridium anomaly at Gubbio: distillation rather than input. There are certainly proper questions to ask about *how long* the event would have taken: the meteorite theory demands a short timescale, the kind of global devastation envisaged by many science fiction writers, not to mention nuclear disarmers, as following total nuclear war: months or a few years. There are various ways of calculating the time represented at the K–T boundary clay, and not all of them came up with the short timescale needed.

All this might seem to be the huff and puff of scientific debate, but what has surprised me is the vituperation with which the arguments have been prosecuted. This is surely more emotional than can be accounted for by a perceived threat to traditional arguments, even if they did hold sway for so many years. After all, the extinction did happen a *very* long time ago, but the violence of some of the reactions carried more of the flavour of a personal insult. It is a curious parallel that the supposed violence of the Cretaceous end is matched by the violence of the twentieth-century exchanges. Here is a 1985 *New York Times* comment from R. T. Bakker, who will be recalled as one of the originators of the "hot-blooded dinosaur" theories. He writes of the physical scientists who originated the impact hypothesis: "The arrogance of those people is simply unbelievable. They know next to nothing about how real animals evolve, live and become extinct. But despite their ignorance, the geochemists feel that all you have to do is crank up some fancy machine and you've revolutionized science." There is a degree of outrage here which is out of proportion to the stimulus. After all, both sides of a controversy pay lip service to what Brillat-Savarin called incontrovertible evidence. Bakker's fury is rooted in the invasion of what might have seemed the last inviolate territory for the palaeontologist—the past itself—by the big boys from the labs. Those who had rejoiced in the complexity of the past, perhaps celebrated the undoubted truth that it can never totally be unravelled—or at the very least that an infinite variety of unravellings were possible—suddenly had to confront an explanation of great simplicity asserted by its protagonists with a certainty that belied a century of other speculations and brooked no doubt. How dare they! It rankled.

I can suggest something darker and more subconscious as a reason for the anger: this is the apocalypse. Even the driest of scientific prose cannot resist an apocalyptic flavour when describing the meteorite event and its aftermath. There is fire and there is brimstone—sulphurous acid rain—and there is darkness obscuring the Earth. Velikovsky's prose and Brillat-Savarin's philosophical diversions could scarcely be more dramatic. The great battles which palaeontology fought in the nineteenth century (and sadly continues to fight in some quarters) have been against an over-literal reading of biblical history. Recall that Dean Buckland's *Relics of the Flood* (*Reliquae diluvianae*) was offered both as proof of the Flood and as an explanation of the succession of fossils. Just recently I was shown a fundamentalist text (written, I believe, in the 1970s) that sought to explain the history I have outlined in this book in the following fashion: the oldest and most primitive animals had the lowest intelligence and thus were drowned first in the Flood, while the more intelligent fled uphill to comprise the higher strata. Animals like horses were bright enough to finish up near the top of the mountain! Thus the succession of remains paralleled the ladder of life. Astoundingly naive though this story might seem, it was being purveyed as a serious alternative to the geological narrative. Perhaps the violence of the opposition to any theory which smelled of brimstone was the expression of a deep fear of returning to a time when the hard-won accounts of history were still vulnerable, a time when flood and fire arbitrarily meted out punishment during eras of judgement.

Regardless of the outcome of the *brouhaha*, there are some puzzling features about the impact theory. For example, there are many animals and plants that *did* survive, and somehow it does not seem satisfying just to call them "lucky ones" and leave it at that. Their survival should chime in with the fatal scenario. Birds and mammals, of course, survived and prospered. We know that they were hot-blooded, and thus it can be suggested reasonably enough that they could have outlived a cooling event that would have finished off a cold-blooded reptile (assuming dinosaurs were cold-blooded, of course); their small size might also have helped them to shelter from the worst effects. Flowering plants survived as seeds: after all, seeds are built to survive hard times. I have a problem with the insects. We know that very many insects rely intimately on living plants for food and shelter, and most have short, or at least annual life cycles, so that they cannot just shut down for a while until things improve. There are few insects which can survive for years on dead wood, but this seems to be a special talent reserved for the beetles. Although there were doubtless some extinctions, the major kinds of insects that originated in the Mesozoic survived the catastrophe. I was particularly impressed by the fossil of a Cretaceous moth, one of the earliest of its kind, which belongs to a family of moths that earns a living today solely by eating pollen: no pollen, no moth. It could surely not have survived even one season of withered flowers—the shortest possible catastrophe. Therefore, we have to assume that there were parts of the world sheltered from the worst, where flowers could still open up to a watery sun. Leaves there could still sustain ants and leaf-borers and all the other legions of nibblers. Of course, insects breed, it is hardly necessary to say, like flies, so that they could recover rapidly from a crash in population size, even had they been reduced to just a few individuals. But even so you cannot divorce the biological facts from the narrative of extermination. It is easier to understand how large crocodiles could continue through the whole crisis period unaffected, because they can feed on carrion or a variety of small animals, and are able to survive for a long time without being fed at all. Even if the evidence for a collision of a major astral body into the Earth at the level of the K–T boundary is unequivocal, it seems to have taken different animals in different ways.

In the sea, crabs survived, but then they really are the toughest of all creatures and can live on almost anything. But the "bony fish" also braved the events, which is curious, since so many of them are fastidious. If the oceans suffered as extensively as has been claimed, how did the fish pass through while the ammonites perished? Ken Hsu had dubbed the conditions following the K–T event the "Strangelove Ocean," a rather knowing reference to Stanley Kubrick's atomic age *film noir, Dr. Strangelove;* it was an ocean in which the normal circulation of nutrients had broken down (Dr. Strangelove was no friend to the survival of organisms). The impoverished variety of small planktonic animals that could tolerate these conditions have been claimed to be unusually small—stunted—like orphans deprived of enough broth. Since fish ultimately depend on the plankton at the bottom of the food chain, why were they able to survive on such thin rations? Then there are the corals. As far as I know, the closest relatives of Tertiary corals are to be found in Cretaceous strata, so that their chain of descent was not broken. The massive corals that form reefs today keep algae in their tissues as symbionts—and without them they cannot prosper. Yet, since algae require light to thrive (to say nothing of a minimum temperature), how could these sensitive animals have negotiated such a crisis, particularly in league with such choosy plants? All this is negative evidence, and is the kind of stuff that spoilsports trot out to dampen spirits at the party. It is always possible to devise little islands where one or another of the survivors might linger on, and maybe that will eventually prove to be the case. But the image of a tiny moth seeking a flower where none can open still sticks obstinately in my mind's eye.

Another criticism—and the most obvious one—was: where is the hole? Craters left by the impact of meteorites can be found in various sites around the world. They can now be dated by the radioactive "clocks" that they set off when they impacted, and in some cases by fossils included in sediments that filled in their floors once quiet conditions returned. The K–T crater should have been one of the largest, but if so, it was also the most elusive. Ancient scenes were reconstructed in which the bolide entered a deep part of the sea, only to vanish; or it was suggested that the agent was a small, interstellar black hole, invisible but infinitely destructive. But then an Earth-bound answer was supplied: the night of destruction was preserved at Chicxulub. The reason that it was not seen earlier is perverse but rational: the crater had been buried. Subsequent sediments and a rich growth of tropical forest had concealed it. For Chicxulub was on the Yucatán Peninsula, virtually on the equator, on that part of Mexico that sticks out like an appendage from the central Americas, looking on the map like a Florida in reverse. The vast crater is supposed to be some 200 kilometres across, but has left only the most subtle physical expression on the surface topography, for it has been filled in with sediment, as perhaps might have been expected after the passage of 65 million years. None the less its dimensions can be plotted with sensitive instruments that can measure minute changes in gravity which result from the concealed geology. Boreholes revealed that the impact must have vaporized sulphur-bearing sediments—hence the clouds of acid rain that added further grisly misery to fatality. In the sediments near the crater there were discovered tiny, glassy balls that were alleged to be a legacy of the spatter from the impact itself. Further away, there were deposits at the K–T boundary claimed to have resulted from a great tidal wave—a tsunami—set in train by the momentous events attending the impact. Thus it was that in 1991 A. R. Hildebrand reported, as one criminologically minded observer put it, the "smoking gun."

Yucatán was the home of the Mayan civilization, whose temples survive among the undergrowth, their ceremonial buildings rich in perfectly carved, but oddly alien bas-reliefs. It is strangely appropriate that these people had such regard for astronomy; at the head of their extensive pantheon was Kinebahan, the mouth and eyes of the Sun. They were obsessed with time, to the extent that their lives were ruled by a calendar drawn from the heavens. They could not know that under the soil on which they erected their pyramids to the gods lay evidence from another time, when the celestial visitation they most feared had wrought damage more profound than they could have guessed at. Or did some shaman doped with *peyote* see a vision of a flaming ball, plummeting towards the Earth and carrying destruction in its wake?

I should add that there are trenchant critics of Chicxulub as the site of the great impact. I have before me several recent issues of the periodical *Geology*. Three articles published in 1994 offer different versions of the authenticity of the Chicxulub crater as the site of the K–T impact. One asserts that the "tsunami deposits" are nothing of the kind, and are instead a much more routine kind of sediment produced by currents. Having read the evidence, this seems quite possible to me. Another article states that there is something wrong with the dating of the K–T interval—that it might well embrace much more time than the fatal instant. The crater, it is alleged, might be too old to be the "smoking gun"—making it more of a damp squib. However, another series of letters defends the evidence as vigorously, invoking chemical arguments comparing glassy objects found near Chicxulub with those from other known impact sites. "The reality of a large impact event marking the K–T boundary cannot be denied," one comment roundly asserts. Well, it obviously *can* be denied, since that is exactly what the critics are doing. Even if an impact is accepted, a direct link to extinctions is a separate step. This discussion of the K–T crater is at that critical stage in a historical argument where evidence and counter-evidence are paraded through the pages of the journals. Things have changed since the early days, because it now seems that it is the *critics* who are in the more defensive position. A slightly righteous whiff now pervades the meteorite camp. "Surely these people can see what is obvious?" is the tone of the replies. The outcome of the debate may eventually result in a bald new fact in popular encyclopaedias, but for those bloodied in the controversy I doubt whether the passion of the debate will ever totally fade away. I suspect that some of the enmities are for life.

*   *   *

42. The strange Carboniferous shark *Stethacanthus* with its "toothy" anvil. This specimen still shows the last two meals that it ate. It even preserves its sexual organs; it is a male.

43. Fossil shark teeth—the most enduring parts of the most enduring predators in the sea. These examples are of *Otodus* (large, natural size) and *Striatolamia* (x2) from the Eocene London Clay.

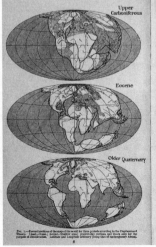

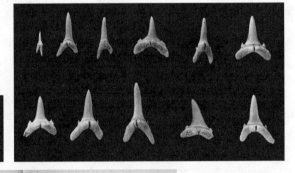

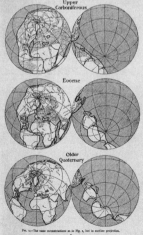

44. Alfred Wegener's reconstruction of the continents at the time of Pangaea; the great seaway Tethys lies to the north of Africa and extends to the Orient.

45. Giant single-celled foraminiferans—fusulines—in thin section to show their internal chambers. These inhabitants of the Permian Tethys ranged from the size of a grain of wheat to several centimetres long, and made extensive rock beds.

46. *Glossopteris:* large leaves of a distinctive deciduous fossil tree which served as evidence both of the united southern continent and of ancient seasonal climates. This "leaf fall" is from the Permian of India. The large leaves are the same size as living laurel leaves.

47. A fern frond unrolling. Ferns have survived from the time of Pangaea to the present day.

48. John Martin's illustration for Hawkins's *Book of the Great Sea Dragons* (1840): dark and terrible imaginings

49. The old, droopy-tailed *Diplodocus* in the magnificent main hall of the Natural History Museum, London

50. *Diplodocus* with its tail lifted into an active, perky position in accordance with current views of its life habits

51. A reconstruction of *Brachiosaurus* in aquatic habitat by Zdenek Burian (1960) "under the direction of Dr. Joseph Augusta, Professor of Palaeontology." This is little different from Walt Disney's semi-aquatic conception of large sauropod habits in *Fantasia* (1940) (for copyright reasons the Disney Corporation does not permit reproduction of stills from the cartoon).

52. The modern version of sauropod habits exemplified by Spielberg's *Jurassic Park* (1995)—but will this view change again?

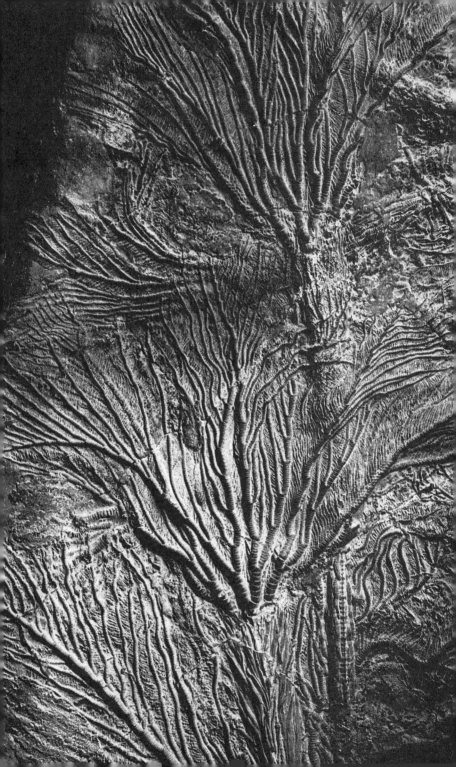

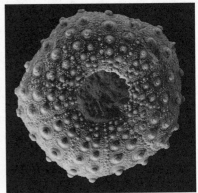

53. OPPOSITE: A beautiful Jurassic sea lily (crinoid) from the Lias Formation. The feathery arms resemble petals, and the animal was moored to the sea floor by a long stem.

54. ABOVE: Two Cretaceous fossil sea urchins, *Micraster* (left) and *Phymosoma* (right)

55. RIGHT: A Jurassic ammonite, *Dactylioceras*, one of thousands of variations on the spiral form

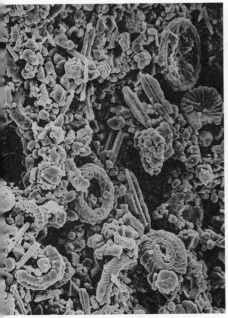

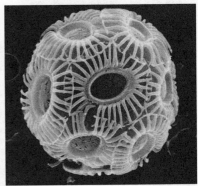

56. ABOVE: A living coccolithophorid *Emiliania huxleyi*, carrying the calcaeoous plates—coccoliths—on its exterior, enlarged 10,000 times on the electron microscope

57. LEFT: An electron microscope examination of Cretaceous chalk shows the broken cartwheels of coccoliths forming much of its substance

58. A fossil insect (*Simulium*) preserved in Baltic amber of Tertiary age. The oldest amber is Mesozoic.

59. BELOW: Charles Darwin in middle age

60. BOTTOM: Down House, Kent—Charles Darwin's house—in winter

61. The passage of a meteorite through the atmosphere, recorded by Dr. Ian Halliday (Herzberg Institute of Astrophysics, National Research Council of Canada)

62. BELOW: The point in the rock section at Gubbio where the Cretaceous ends and the Tertiary begins. The coin rests upon the boundary clay.

63. BOTTOM: Single-celled Cretaceous planktonic foraminifera—each one a millimetre or so across. Changes in these tiny organisms have figured prominently in the debates about extinctions across the K–T boundary. This example, late Cretaceopus Mendez Shale of Mexico, courtesy Norman MacLeod

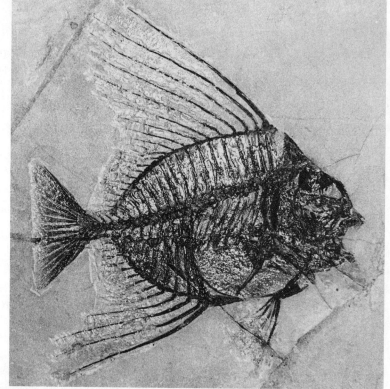

64. The bony fish passed through the K–T event. This is *Acanthonemus subaureus*, from the famous Monte Bolca locality (Tertiary, Eocene), Italy.

65. LEFT: The skull of a whale that could walk, *Pakicetus* (Eocene, Pakistan)

66. The skull of a carnivorous roo, *Ekaltadeta*, from the Riversleigh fauna—an Australian endemic marsupial, and a surprise. This was collected from the picturesquely named Camel Sputum Limestone.

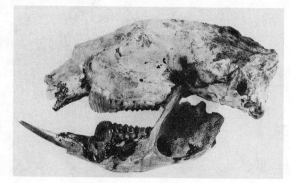

THERE REMAINS A further surmise. If a giant bolide can wreak havoc once, why should it not have happened more than once? Could this be how the Universe intercedes in the narrative of life? Could the repeatedly disastrous dramas described so vividly by the gourmet Brillat-Savarin more than 150 years ago be on the menu once again?

Several important extinctions have been mentioned during the course of my story. I have especially emphasized one at the end of the Ordovician and another at the end of the Permian, because every scientist I know who has looked at the evidence recognizes these as major interruptions in the narrative of life. There are more extinction events that could be cited—at the end of the Devonian, for example, or between the Triassic and Jurassic. What is regarded as a "mass extinction" is something of a statistical judgement; it could be any short time period during which the number of species dying out is elevated above a normal rate. After all, nobody denies that animals and plants become extinct in the regular course of things, otherwise the fauna and flora would scarcely change through long periods of geological time, and we know that this is not the case from thousands of rock sections around the globe. To recognize the times of elevated extinction rates requires mathematical analysis of the geological ranges of extinct animals, a subtle business that demands computing as well as much compilation of data. This is a job for J. J. Sepkoski Jr. Jack Sepkoski is one of the brilliant exponents of mathematical palaeontology at the University of Chicago, and has published much of his work in conjunction with David Raup. With enormous patience, he has listed all the thousand upon thousand geological ranges of families and genera of animals since the Permian. Years of tedium and routine lie behind such a compilation. Then the database has been tested, manipulated, modelled and mangled for all it is worth, to determine whether there are cycles in extinction, of which the K–T event was merely the greatest of many.

Raup and Sepkoski famously declared that there was a *cyclical* extinction every 26 million years from the beginning of the Permian. Although the idea was being publicly aired in 1983 at a conference in Berlin, it hit the headlines in 1984 by way of various articles in *Nature*. Other scientists, especially astronomers, moved very fast to give reasons for the pattern. It was one of those times when imagination, fact and reasoning seemed perfectly in step. Analysis of the ages of ancient meteorite craters also apparently matched the same cycles: thus, it was concluded, regular inundations of meteorites bombarded the Earth, causing unusual fatalities on a periodic basis, and catastrophe occasionally. This became known as the theory of periodic mass extinction. Regularity always has an astronomical ring to it; revolutions of the Earth, the visits of Halley's Comet, all measured by the indifferent clockwork of Isaac Newton's Universe. So the astronomers sought a source of death which might turn and turn again to the right period.

The Nemesis explanation was an imaginative attempt to explain all. It proposed that the Sun had a small, dark companion star—the eponymous Nemesis—which, during the course of its *pas de deux* with the Sun, passed every 26 million years through the Oort Cloud (named after its discoverer, Professor Oort). The Oort Cloud is a gathering of countless comets with their own orbits around the Sun—lying beyond the outer planets. The gravitational field of Nemesis as it passed through the Cloud altered the course of some of the comets, a few of which were deflected towards the Earth with lethal effect. Thus we return again to the "watery comet foretold by the worthy Jérôme Lalande," except that 12 million years are expected to elapse before we must be sent hurrying to the confessional, which is time enough to generate plenty to confess about.

Life would be in thrall to the intensity of the visitation, a dark, even invisible pendulum would swing over the future of every organism; no day of judgement, just the random, fatal deflection of a heavenly body, which would one day announce doomsday. No adaptation would help, so the old dinosaur cliché with which this chapter started would not apply. The world would indeed end with a bang rather than a whimper.

Since the early days of the theory some of the glitter has worn off the idea of periodic mass extinction. Several of my colleagues at the Natural History Museum in London have looked hard at Sepkoski's data and found it inadequate. Rather unfairly, they seem to have been branded as spoilsports, as if to cast suspicion on eventual extermination was rather ungenerous of them. The dates of mass extinctions do not match the 26-million-year cycle as well as was once thought, nor are some of them as massive as originally claimed. I have tried to extrapolate backwards towards the Cambrian, and find that some of the bigger events are not on the expected curve at all. It seems a shame to abandon a theory that was both simple and original, but, at least to this observer, that is the way the argument is going. None of this is to deny the arrival in the geological past of large astral bodies on the Earth's surface; it is merely to take away the inexorable clock that allegedly propelled them.

Fossils are the messengers of time. But there are a few examples which almost defy time itself in the perfection of their preservation. Amber is the time capsule *par excellence*: truly forever amber. It formed initially as a resinous blob oozing from a wound or cut on one of several kinds of trees.

You can see and smell comparable resinous secretions in pine forests today, and if you search among the pine needles you will pick up tacky globules around the bases of trees. Their aromatic scent will linger on your fingers for hours. Insects and other creatures get caught in the ooze. Time darkens it and hardens it. After a few hundred years it becomes copal (often yellow and slick); then after a million or so it acquires the indefinably deep, golden-brown colour of amber. Amber is a resistant material that eventually finds its way into sedimentary rocks. But amber beads still carry within them the trapped insects, sealed until the end of time itself. Cretaceous amber is known from several localities—Lebanon, for example. A mosquito trapped in the age of the dinosaurs, and still with blood in its tiny maw, might well reside in an amber jewel. If the DNA of the dinosaur were preserved, could the monsters yet be regenerated and reawakened, as was proposed in the novel *Jurassic Park*? This would defy all theories of the end: a blueprint for life could be handed down directly from the age of reptiles.

The survival of the DNA molecule is not at all like the survival of bone, for it is fragile: it falls to pieces, it loses its signature. The reports in 1991 of bee DNA fragments recovered from amber specimens from Dominica was as astounding in its way as the discovery of the tomb of King Tutankhamun; but unlike archaeological treasures it could not be examined in the light of a quivering torch, nor displayed in glass cases for the delight of the multitudes. The only way to "see" it at all was by means of a sophisticated technique that multiplies and copies the fragments countless times, until there is enough to identify. The problem is that the technique is so sensitive that it will multiply the molecules of anything alive: be it human DNA, the scale from the wing of a butterfly, or a fungal spore brought on the whim of a breeze. The surprise was that the ancient DNA was so much older than anything known before—a thousand times older. Bone DNA had survived (and then just as fragments) for a few thousand years, and only under special conditions, such as the inside of a pharaoh's tomb. The possibility of contamination in ancient amber was difficult to eliminate.

I won a research grant in 1993, together with several colleagues, to attempt to repeat the discovery of DNA in amber. We had specimens in our collections of the same species of fossil bee as that which had been used in the original discovery. A laboratory was set up far from any other experimental lab, out of bounds to everyone except the experimenter, Jeremy Austin, an extrovert Tasmanian who had to scrub himself more often than a brain surgeon. His knees shone, his hands were spotless. The amber was polished to exclude the adherence of even the breath of a fruit fly. Everything was sterilized. The fossil tissues from the inside of the amber were rendered into solution and the clever magnification techniques applied. So far, we have discovered only tiny pieces of DNA, and they could well belong to the ubiquitous fungi, contaminant or not. It is that most dreaded of ends: *the negative result*. It does not disprove the earlier positive results; it merely makes them not readily repeatable, and rather debatable. I am reminded of Lewis Carroll's poem *The Hunting of the Snark*, in which the elusive Snark is hunted through many mock-heroic verses by a battery of nonsensical means until

> In the midst of his laughter and glee,
> He had softly and suddenly vanished away—
> For the Snark *was* a Boojum, you see.

We seem to be left Snarkless for the moment, and any idea of creating Frankenstein's monster from tiny, fossilized blueprints has faded.

In reality, it was already impossible, for there are no known fossil mosquitoes as old as the dinosaurs. Revivification of the ancient monsters has been a dream for many years. Sir Arthur Conan Doyle had them surviving in an unexplored enclave, defiant of time, in *The Lost World* (1912). Professor Challenger was a successor to Rider Haggard's brave but scholarly heroes, and not so different from the academic in *Jurassic Park*. Behind all the fiction lies a dream, a fantasy to reanimate what has passed from the Earth, in defiance of finality, even a denial of the End of the World.

But it has all gone. Extinction is for ever, like amber. And, regardless of the outcome of the meteorite debate, what has been lost is the innocence of the world. It does not matter whether the atmosphere was penetrated only once, or more than a thousand times by a destructive body. We are bound to the rest of the solar system and to the wider wastes beyond, and it has always been so. This fact has never been so clear as in the photographs that space probes transmit back to us with a new accuracy and definition: they show us that life is no more than a glaze upon the surface, or something as delicate as the bloom on a peach. Wrapped in the cocoon of atmosphere we may have shuddered at the arid vastness outside, sought to enclose the world from outside interference. But we cannot escape our origin, for it still intercedes in our fate.

Chapter 11

Suckling Success

MAMMALS ARE ANIMALS whose females suckle their young upon modified sweat glands. They thus embody milk and maternity: they are nurture encapsulated. Most mammals are hairy, and even those that are not had hairy ancestors. And such creatures bearing fur or pelt, and feeding their young upon a specially secreted, nutritious fluid, had spread over all the lands of the Earth in Tertiary times. They were accompanied by the birds, which have thrived alongside mammals in dynamic counterpoint ever since.

After the demise of the ruling reptiles, warm-blooded mammals were released from more than 100 million years of inconspicuousness. The group had appeared at almost the same time as the dinosaurian reptiles, having been derived from one of the numerous mammal-like reptiles, probably at some time in the Triassic. Like poor cousins of a great family, they had to live on the edge of the estate, making ends meet, until a failure of succession allowed them the freedom of the grounds. We can imagine them quivering with bewhiskered sensitivity under the cycad groves as thunderous sauropods stamped past; they were awaiting nightfall, then to emerge to scuttle among the leavings from the reptilian tycoon's repast. There were good livings to be made at small size, especially in the roles of scavengers and insectivores. Insects were abundant in the Mesozoic forests, as amber beads confirm.

Small, insectivorous mammals such as shrews are still common in many places, which is evidence that insects are a predictable food source as well as a nutritious one. I have lain very still in the rough grass on the edge of a field and watched a tiny pygmy shrew dispatch a moth with speed and precision. The shrew emerged from obscure concealment among dense herbiage, its long nose twitching and cautious. It seemed to catch the moth in its paws, severed off the wings in a trice with a couple of nips, and then proceeded to eat it with indulgence. I swear it even made a succulent, chomping noise, like a greedy schoolboy making the most of an apple. Then the creature was gone—bent upon eating its body weight in insects within a twelve-hour period, which is the price that must be exacted for having warm blood. It was not difficult to imagine a similar scene being enacted in the Jurassic, out of sight of large carnivores, and beneath the notice of mighty herbivores. One such creature is *Megazostrodon*, a delicate Jurassic mammal from South Africa that looks like a shrew. Not surprisingly, the early fossil record of these small and fragile insectivores is patchy in the extreme: the discovery of an entire skeleton of a new early mammal is always big news, almost in inverse proportion to the size of the fossil. The commonest evidence of the early history of mammals available to the palaeontologist is tiny teeth alone. It is likely that mammals were commoner from the first than their sparse fossil record suggests.

When the terrestrial dinosaurs had died away all the ecological niches that they had occupied fell vacant; they had been at the top of the food chain for a long time. The Palaeocene period, the earliest division of the Tertiary, was a unique time. It was as if a selection of vast feasts were laid out—not merely fatter grasshoppers or tastier grubs. If Nature truly abhors a vacuum, this is when there was a rush to fill each vacancy in the living world with a practitioner skilled in an appropriate trade: grazers and hunters and every other type of vertebrate activity imaginable. Even if several of the fundamental mammal designs had *first* appeared during the Cretaceous, it was after the demise of the dinosaurs that they went crazy, in a kind of stampede of creation of new designs that turned small, even literally downtrodden animals into the greatest range of shapes that has ever graced the Earth. For it was the mammals and the birds that took advantage of the opportunities that presented themselves. The other survivors that crossed through the great death at the end of the Cretaceous did not—neither the lizards, nor the crocodiles, nor the turtles. What birds and mammals share is a homeothermic metabolism (that is, they are both warm-blooded). Both are accordingly insulated—by feathers and fur, respectively. It must have been this property that favoured them in the repopulation of the biological firmament. This may be because the climate had become generally more seasonal, which warm-blooded animals can cope with successfully. There is good evidence for the growth of cool polar regions during the Tertiary. It is curious to reflect that humans, the bald apes that now rule or misrule over vulnerable continents, have lost the very feature that once gave its kind the edge over the rest of creation.

The rapidity of what is usually called the "radiation" of the mammals can be seen written clearly in rock sections. What were previously very rare fossils become common ones; large species appear, which are usually detected in the field by tapping out their robust limb bones. Shrew-sized animals have dog-sized companions within a mere 3 million years of the extinction of the dinosaurs. Professor S. J. Stanley has shown that in the early Tertiary the rate of evolutionary change in the mammals was many times faster than it was in clams and snails, which were prospering and changing in the sea at exactly the same time. For example, during comparable time intervals many more kinds of new species, genera and even families of mammals appeared as fossils when compared with the molluscs. This may have been because the seashells were not diversifying into "empty" ecological space like the mammals. The marine world was not a *tabula rasa* upon which any efficient design could be successfully graven. As we have seen many times, the sea has always been a haven of continuity, a place where revolutions are muffled.

To the student of the geological past, mammals have one advantage over their reptilian predecessors—they have particularly distinctive teeth. My mammal colleagues can wax eloquent on the discovery of a solitary tooth. Teeth are composed of various cusps, which are the bits that break off when you eat the wrong kind of muesli. I have known mammal workers to get excited by a single cusp perched on the tip of a pin! Since much of the evidence for the presence of mammals in Antarctica hung on such a cusp for several years, this is perhaps understandable. Mammologists can identify animals closely from even a single tooth. My colleague Andrew Currant, an expert on Ice Age mammals, once had the misfortune to swallow the very small tooth of a very important arctic vole. Quite what he was doing with the tooth in his mouth he is not keen to explain, although apparently it had something to do with getting rid of some old glue. He was able to recover the tooth after the passage of several hours, which is the best proof I can muster of the regard in which mammologists hold dental characteristics.

* * *

THERE ARE MANY KINDS of mammals living at the present day, each with a history worth consideration. I can describe in detail only a handful—there are just too many—but I shall attempt to outline some of the mechanisms that underlay their differentiation. Consider their variety. Herbivores vary from ponderous cows, faster horses, through gazelles as elegant and supple as young ballerinas, to massive pachyderms—elephant, rhinoceros and hippopotamus, which are the ecological doubles of the brontosaurs. Odder herbivores must not be forgotten: camels, giraffes and kangaroos, tree sloths, the giant South American rodents—capybaras—and the improbable giant panda, connoisseur of bamboos. Then there is the smaller tribe of rabbits and hares, capable of living everywhere from high Arctic to subtropical desert. Small rodents of a hundred species gorge on seeds and nuts and much besides, nibblers and gnawers all. Pigs are rooters, a talent turned to advantage by the French peasants who use them to snuffle out the world's most delicious and expensive foodstuff, the truffle. Vegetarians are preyed upon by those that are not. Some carnivores are small and vicious, like weasels and stoats—which, it will be recalled, were described by gruff Mr. Badger in *The Wind in the Willows* as definitely not the right kind of people. (Mole, of course, was, and is, a very efficient subterranean carnivore, while Ratty is vegetarian, but not above taking the odd caddis fly. Ratty is no rat at all, but a water vole, at least in zoological terms, although linguistically he might have prior claim to be the true and only rat, since "rat" is an Anglo-Saxon word, predating the arrival of the brown rat and ship rat in Britain.) The lion (*Panthera leo*) is indubitably king of the African predators, and no nomenclatural quibbling, even if the tiger is the more ferocious beast, and both are aristocrats among numerous cats. Dogs, hyenas, foxes, bears, gluttons add to the roll call of the Order Carnivora. Weird mammals have forsaken succulent meat for hordes of ants: aardvarks and pangolins and anteaters with elongate snouts and sticky tongues. The more conventional insectivores are the shrews, but there are also hedgehogs—with hairs modified into spines. The most curious insectivores of all are bats—pipistrelles and barbastrelles and horseshoes—animals that can fold their wings like vellum umbrellas and see in the dark by emitting high-pitched squeaks. They include fruit-eaters, too, and a bloody vampire; the smallest one (*Craseonycteris*) hardly weighs more than a bumble bee. As a cast of characters, do not the mammals already strain credibility?

But there are still more. There are those that returned to the water. Lazy dugongs and manatees are the herbivores; seals and sea lions the carnivores, with the faces of dogs but bodies slick and dappled and rounded, and breath worse than a camel's. The crown of them all are the cetaceans—porpoises and dolphins and whales—who acknowledge their mammalian roots by teat and milk alone, while their bodies scull, glide or wallow through the ocean more comfortably than any submarine, and as gracefully as any salmon. Yet they include giants that would outweigh the largest of the dinosaurs, and put to shame the most outrageous inventions of the reptiles. I wonder if the blue whale were known only from fossils whether any palaeontologist would believe his arithmetic: 170 tons of blubber, and all built on the nourishment gained by filtering planktonic shrimps through slats of baleen. A second evolutionary line led to the toothed whales, the giant of which is the sperm whale, a vast, blunt-headed animal capable of diving to enormous depths in pursuit of squid. Whales have become the symbol of conservationists because they defy probability, and they symbolize the fecundity and inventiveness of life. With whales, you do not need to argue the case that the loss of a species diminishes us all, something which seems more debatable in the case of a subtly different species of vole. The gross carnage of whales is historically familiar, and celebrated in the novel that most closely associates the physical and the metaphysical, *Moby Dick*. On my first visit to Spitsbergen I saw the debris of ancient butchery near Smeerenburg—drifts of ribs, strewn around the boreal waste like so many white poles stranded from vanished encampments. What 65 million years of evolution had accomplished a harpoon could undo in a bloody afternoon.

Yet whales, too, began with legs. If one needed proof that bones never lie, the fossils that record the history of the whale probably give the best example of evolutionary transition, and are as good an example of the contribution of palaeontology to mammalian history as any. The Eocene *Basilosaurus* has a name that sounds as if it should grace a reptile—*saurus*es almost to a beast—and it was originally thought to be one. Yet when its bones were fully known it proved to be a mammal, and a primitive whale twenty metres long to boot. Its head is relatively short, and its body correspondingly long, when compared with living whales. It still retained a relic of the hind limbs, and they have fingers and arms, but shrunken and useless, a kind of evolutionary signpost which pointed the way to an ancestral animal that could walk on all fours. Other recent discoveries from the Eocene rocks further filled in the apparently insurmountable gap between the great whale and a normal quadruped. *Ambulocetus* (the name attempts to describe its salient feature—"walking whale") is known so far from a partial skeleton—and it has legs, arranged after the fashion of a sea lion, so they were still functional. Probably the oldest member of the whale group is *Pakicetus*, known only from the back of the skull, and part of its lower jaw. But even this fragment shows that the ear region was poorly modified for deep diving, compared with the specialized structures in other whales, which can tolerate considerable water pressure. Extraordinarily, it was found with freshwater snails and other fossils which preferred to be away from the sea, which may indicate that the whales were born among mammals that gambolled after fish and shellfish in the shallow rivers and estuaries of the early Tertiary, and only later moved into the ocean, of which they became the masters. There is a modern equivalent among the carnivores: *Lutra lutra*, the otter, which can bridge the marine and freshwater worlds like no other creature except the salmon—which is possibly its favourite food. I have spotted them in trout-rich rivers in South Wales; I have seen their close relatives exuberantly chasing through the kelp groves of California. The story of the genesis of whales is particularly eloquent, and it may stand for so much of mammal evolution. An animal with bizarre shape really can be shown to have had ancient relatives with combinations of characteristics that link them with more conventional animals. It also shows how fragmentary our knowledge is of some of the crucial linking species, and this is something that recurs with many kinds of animals, not least our own species. I do not propose to describe many other stories of the kind: "first this species, then that, then the other," as they soon degenerate into mere catalogues of Latin names. The most reiterated of these genealogies is probably that of the horse (*Equus*), and I have become thoroughly bewildered over the years by how complicated that story has grown since the first, small, five-toed Eocene horse-like animal (*Hyracotherium*) appeared in the fossil record. At the last count there were more than twenty generic names dotted about on the tree of relationships, and to describe it would resemble the beginning of one of those Old Testament books wherein Zebediah begat Obadiah, and Obadiah begat Numquat and Aliquot, and Aliquot Mizpah and Mephaniah etc., etc., and so on for many generations. But there are a dozen narratives of this kind that could be told, and many palaeontologists spend their working lives with teeth and a few bony scraps filling in missing paragraphs or, on occasion, writing a new chapter. Very rarely, the almost impossible is discovered—like a walking whale. It requires a singular kind of devotion, at times almost religious in its self-effacing fervour, to keep plugging away at ungrateful bones.

THE EARLIEST TIMES of many kinds of mammals are poorly known. Placental mammals, those having a womb in which to nourish their babies to some degree of maturity before birth, are considered to be comparatively advanced. The most primitive living mammals are the monotremes, which retain the characteristic of laying eggs from a reptilian ancestor. As with other examples of survivors mentioned in this book, they are Australasian natives. The duck-billed platypus is an animal so extraordinary that when it was first described it was assumed to be a hoax. When it was finally proved that an aquatic mammal with a duck-like bill that laid eggs really existed, it merely turned a sceptic's dream into a zoologist's nightmare. Its commoner cousin, the echidna, looks rather like a hedgehog. Although the monotremes must have existed all the way back to the early days of the Jurassic, there are very few fossils to prove it. Maybe they were always rather rare. Marsupials, on the other hand, give birth to tiny little babies that crawl their way through moistened fur to a special pouch (the marsupium), in which they can suckle and grow. They, too, are abundant in Australia, but not entirely confined to that continent; there are South American opossums which have spread northwards into what is now the United States in comparatively recent times. Much of marsupial history is also poorly known. Their early fossils occur well down in the Cretaceous period, but at this time they appear to have been exclusively North American. Somewhat later in the Cretaceous they are quite well known from *South* America, and one scenario has them populating Australia from there. In the early Tertiary they were practically global. But by whatever route they reached the antipodes, they prospered when they got there.

The history of the mammals is intimately linked to that of the continents. It will be remembered that during the Permian the continents were gathered together into the great supercontinent, Pangaea. For the first and last time in history it was possible for an animal (if it was tolerant of hardship) to wander at will over continental parts of the globe. There would have been barriers—deserts, rivers, mountains—but no great oceans to halt migration. The subsequent story of terrestrial life is entwined with the break-up of that mighty continent, driven by the great engine of plate tectonics. Narrow seas grew into wider oceans. The Atlantic Ocean, north and south, cleaved apart the Americas from Eurasia and Africa. This process was well underway in the Cretaceous, but there were still possible connections, for it was neither a simple, nor a clean split. North and South America then had a kind of sparring history in the Tertiary, sometimes engaging, but more often apart, separated through what is now Panama. Africa had similar interactions with southern Europe, and thence to the rest of Asia. The Tethys, that ancient sea, continually redefined its shape and extent. In the Mediterranean region the shifting relationships of land, continents and seas are as complex and tortuous as the relationships in the dynasty of the Borgias. Antarctica set on its own course southwards, a course of doom for some, but the making of many species of penguins and seals. India left its billet against Africa and drifted away, to collide with Asia, and to ruckle up the high Himalaya, leaving Madagascar behind, stranded near Africa. The tumid west coast of Madagascar is surely the most obvious testament to continental movement; one almost feels compelled to cut it out of an atlas and push it back into the hole from whence it was launched.

Each continent carried with it a cargo of animals and plants, and when they were isolated from their common origin they evolved in isolation. This was a wonderful contribution to the richness of the natural world, for in this way five or six times as many species (or higher ranks of life as time passed) could be supported. Separation breeds diversity. This is why there are so many different kinds of lemurs on Madagascar, or why there are no native cats in Australia, and why llamas look different from pronghorns. Because the number of livelihoods that an animal may have are limited, there were different animals doing similar things on the several nascent continents. Grazers, carnivores, insectivores, canopy-dwellers, all evolved more than once from different ancestral species. And for every continent there were a hundred oceanic islands on which life could also be seeded. So the passing of Pangaea was the creation of the modern world, and a requiem for the vanished supercontinent soon became a celebration mass for the modern world.

The logical way to describe what happened to mammals, therefore, is continent by continent, and I have picked just a handful of examples from each.

The creative effect of isolation is best illustrated by the most solitary continent. Australia launched itself early from disintegrating Pangaea, taking a lonely path which would doubtless be described as rugged independence by the present incumbents of that continent. It took with it a cargo of marsupials; and they remained marsupials while in most of the rest of the world the placental mammals eventually became dominant. There was a huge continent for the marsupials to experiment upon. They are often considered rather a dim kind of mammal, and it is true that philosophical discussion of any kind is wasted upon a koala. But they are by no means the evolutionary dullards that common prejudice might imply. The kangaroo and wallaby, for example, are wonderfully adapted creatures. The female kangaroo is always pregnant, but during hard times development of the foetus is suppressed; during good times the joey develops and takes up residence in the pouch, which is possibly the last word in maternal care. Similar contraptions are sold in childcare shops, made of washable fabric. The lolloping, bounding bipedal gait of the kangaroo is energetically efficient; kangaroos can make good flesh from hard tucker. When they have to live alongside sheep or cattle, both advanced placental herbivores, it is the kangaroo that survives hard times in better shape. Their distinctive design is unique among large plant-eaters. In central Australia you will be grinding along in a Toyota a hundred and fifty kilometres from a town and hours from standing water when suddenly a huge grey or red kangaroo will bound across your path with a kind of earnest nonchalance which is matched by no other mammal. On the contrary, other marsupial designs are extraordinarily similar to those found in mammals with wombs instead of pouches, and this is invariably because they are performing similar ecological roles. There are marsupial mice that look like any other mice, and there was until very recently the Tasmanian wolf (*Thylacinus*), which looked for all the world like a dog. The last one died in a zoo in 1936. I have a photograph of it apparently hugely yawning, but I am told that this was a threatening posture similar to that used by its relative, the Tasmanian Devil. Such a wide gape reminds one of the agonized screams in the paintings of Francis Bacon, and seems oddly appropriate for an animal which is the last of its kind. Without stretching the imagination too far one can find marsupial raccoons, rats, moles and squirrels. Only a million years ago there was a much wider range of marsupials that roamed the Australian plains. They included a large herbivorous quadruped, *Diprotodon*, a curious animal that has a passing resemblance, but hardly more, to an African pachyderm. It was hunted by *Thylacoleo*, which had a plausible likeness to the king of the beasts, though undoubtedly another marsupial. These remarkable animals lived on into comparatively recent times, and some workers claim that man himself may have hastened their end.

Until quite recently the earlier history of marsupials in Australia was largely a mystery. Kangaroos and their like, koalas, *Thylacoleo*, and the devils and wolves all appeared to have had only a sparse history from the rocks. Then some wonderful fossils were discovered at Riversleigh in northern Queensland that suddenly filled in a story that had hitherto been obscure. These fossils were preserved in limestone, and could be etched out to reveal whole jaws. One of the glories of palaeontology is that there are always surprises. An intelligent guess might have anticipated a set of ancestral kangaroos or possums from Riversleigh. Instead, Riversleigh abounded with many kinds of completely new marsupials, the like of which had never been dreamed of. There were mystery marsupials so arcane that they were known to their discoverers as "thingodonts"; they are now scientifically christened with the more prosaic name of *Yolkaparidon*, although still nobody knows what other marsupial they are related to, nor how they lived. There were even giant relatives of the *Platypus*.

Riversleigh is a remote station which lies close to the Gregory river. This is a narrow waterway which winds through the outback, supporting a thin, green ribbon of lush vegetation among the vast plains of gum trees and *Spinifex* bushes that cover so much of the interior of the continent. In Oligocene and Miocene times the whole landscape was as rich as Amazonia, green and moist, with trees and ferns in profusion. This is the vanished landscape that is recalled in the Riversleigh fossil beds. The forest grew upon an ancient limestone terrain, and as so often happens in such country there were pools and fissures etched by the abundant rains. It was in such pools that animal skulls and bones were preserved, covered with a protective layer of tufa deposited from the limy waters; they are still preserved in the same way in limestone karst country today. Even now, the waters of the Gregory river are so charged with lime that tufa builds natural dams and terraces, over which clear water cascades.

Australia was isolated in the Miocene, having by then been long severed from Gondwana, and so marsupials provided the raw material for a forest full of mammals. Only bats flew in from foreign parts, and it is as well that they did, for it was the identification of a fossil bat genus with one from as far away as France which allowed the amazing fossil finds to be dated. Year after year from the first discoveries in 1976 successive expeditions found new locations for fossils in this spacious land. They proved to be of more than one age. Many of the fossil finds have not yet received a full scientific description—and Latin name—to mark their official existence: they are in a kind of creative limbo, awaiting definition. Almost the whole mammal story of Australia proved to be recorded in the rocks as the investigators moved from one side to another: the fossils demonstrated a gradual drying-out of the climate, which pushed rain forest to a few peripheral sites. There was a growth in variety, and a subsequent decline of large, four-footed herbivores like *Diprotodon*; and not one, but several marsupial "lions" and "dogs" fought over the flesh of these large animals. In the streams that ran through the Riversleigh forests turtles were more abundant and varied than anywhere else in the world, a belated reptilian blossoming; and crocodiles like *Baru* were lurking in the swamps, just as they do today in the north around the Gulf of Carpentaria. Kangaroos were a whole tribe of different animals, not just the familiar large hoppers, but also relatives of the rat kangaroos, and potoroos, and, most extraordinary of all, there were meat-eating kangaroos that hunted their peaceable neighbours. We must never assume that what we know from the living world is what pertained in the past, for history can ambush our perceptions. It may be asked how the astonished scientists knew that the carnivorous animal in front of them was a kangaroo. They knew the bones of all kangaroos so well that the features of the skull betrayed their common ancestry— but equally the specialized teeth of a carnivore are unambiguous. The trees of the Riversleigh forests were full of pygmy possums, and many other curious possums besides, some odd and enigmatic (and given odd and enigmatic names like ektopodontids) and known from nowhere else. What emerges is a picture of a forest that was livelier and more varied than anything that remains in the drier climate of Australia today. It was the triumph of marsupials. Far from being poor relations of the rest of the mammals, they were gloriously diverse. Finding the Riversleigh fossils was equivalent to the discovery of the Inca ruins in the Andes, for assumptions about the possibilities of the past were challenged by physical evidence. Just as the hardy peoples that cultivate potatoes in the shadow of the teetering Andean palaces seem to be a pale reflection of a glorious South American civilization, so the marsupials that survive now appear to be but a sample from a vanished cornucopia of life. As the newspapers in Queensland reporting the Riversleigh discoveries put it, "History was rewritten," and, for once, the cliché was nothing less than the truth. The caution is that it may be rewritten again, that some other corner of Australia will disclose yet another lost episode in the history of life. The sad thing is that some of the most extraordinary marsupials survived until comparatively recently. Would life not have been richer if a walk in the remaining rain forest patches in northern New South Wales still carried the risk of an encounter with a marsupial lion?

The Riversleigh case is the most startling recent example that I know of the discovery of an unexpected fauna. There have been similar cases before. In the last century, the effects of the Tertiary isolation of South America became known to an inquisitive public. We have become accustomed to the strangeness of the animals now, because reconstructions of life on ancient pampas have long been one of the obligatory dioramas which accompany standard histories of our planet. But when they were first known the Tertiary and Pleistocene fossil animals of Argentina caused a sensation. For long periods after the dissolution of Pangaea, South America was almost as much an island as Australia. As a consequence, the animals that evolved there in isolation were as various and striking as those I have just described from our present island continent. Charles Darwin encountered their remains on the voyage of HMS *Beagle*. He considered *Toxodon*, which he described on his visit to Patagonia, as "perhaps one of the strangest animals ever discovered."

Well, so he might—after all, it was a massively built creature three metres in length, which may have had a trunk like an elephant but retained something of the rodent about it, and is thought to have grazed upon water plants like a hippopotamus, using its spatula-like snout. Just as in Australia, there were a whole series of extraordinary South American natives. Many of them belong to two groups called notoungulates and litopterns, which include horse-like, camel-like, rabbit-like, deer-like and even rhino-like animals. Several hundred different kinds of these animals are known—they were no flash-in-the-evolutionary-pans. These were womb-bearing mammals, which bore their young not only alive, but nourished by the bountiful placenta within the body. Most of the South Americans were thus more closely related to the mammals that dominate the rest of the world—and to ourselves—than are the majority of Australians. But among them there were other, curious animals that seem to confirm the notion that isolated animals become slow, bumpkin-esque creatures. The sloths still slothfully dangle from rain-forest trees, stuffing leaves and favouring the observer with a stare which is quite evidently dim. These animals even have greenish algae growing in their fur, an appropriate tribute to their vegetating existence. The giant sloth, *Megatherium*, was a huge relative of these creatures—large enough to stand up against a tree and pull down edible fruits and leaves. The skeleton of one of these magnificent sloths stands outside the entrance to the Department of Palaeontology in the Natural History Museum—not, I trust, an ironical comment on the work that goes on there. It is commonly confused with a dinosaur by those who mistake size for zoological affinity. A relative of this sloth evidently survived into historical times, for its dried pelt, still with a covering of hair, has been recovered from a vast cave eaten out from a conglomerate formation, the Cueva del Mylodon, near Puerto Natales in Patagonia. The animal evidently lived in such caves, for its dung occurs there too, in heaps, and it can still be burned. Would it not be wonderful to drink a cup of tea brewed on giant sloth's dung! The same species of algae have been recovered from the fossil fur as still live upon the tree sloths. The last of these giants were probably alive only 10,000 years ago. As with the extinct marsupials, there are those who believe that man himself may have hunted the lugubrious giants to extinction.

There were other curiosities in Tertiary South America. Rodents as large as bears were the evolutionary acme of the great tribe of rats, dormice and lemmings. They were named dinomyids—literally "terrible mice," a wonderful contradiction in terms. On the ancient pampas they were serious grazers. There was a carnivorous marsupial "dog" called *Borhyaena*. The glyptodont was hardly less extraordinary, and represents the mammals' closest imitation of the heavily armoured vegetarian dinosaurs such as *Ankylosaurus*. Nothing closely similar survives, unless it is its living relative, the armadillo. For *Glyptodon* is a three-metre tank of an animal, with a bony covering forming a great domed shell on its back as massive as that of a giant tortoise, and probably quite as impenetrable. It could scarcely have been a swift mover, but then I doubt that it needed to concern itself with running away from mammalian enemies. When the isthmus of Panama formed, finally connecting the two Americas, it was among the animals that trundled northwards, and survived and prospered for some considerable time. The armadillo was one of its companions in the northward trek, and that animal still flourishes among the *mesquite* bushes of the southern United States. The glyptodonts died out as the result of a climate change, perhaps, or maybe it was another victim of that uniquely human combination of ingenuity in hunting and killing, and a propensity to destroy beyond recall.

The Panama link ended the isolation of South America; there was complete connection between the northern and southern continents about 3 million years ago. The thin isthmus served as a corridor, and animals moved both ways along it. At least as many mammals successfully moved north as moved south. The horses and deer and sundry carnivores (pumas, bears, dogs) that soon occupied the forests and plains of South America were recruited from the northern hemisphere, where they had had a long and very complicated previous history. Elephants, too, were present among the southward invaders. Armadillos, porcupines and cavies were part of the trade the other way. Many of the other South Americans survived well for a while in Central America and some moved further north, but, as with the glyptodonts, they perished under mysterious circumstances.

There was an altogether opposite effect in the sea. Seashell and coral species were common to the Pacific and Atlantic Oceans through much of the Tertiary across what is now Panama. They were inhabitants of a single sea. When the isthmus prised the two oceans apart, so the animals gradually acquired a personality appropriate to one side of the divide or the other; the gene pool was divided in two. Different and new species appeared that were confined to the Pacific or to the Caribbean, respectively. Specialists can often tell at a glance from which side of the divide a shell originated. This is only now being thrown into reverse by the connection afforded by the Panama Canal, doubtless aided by the steamers that pass through it. So the paradox is that the very land bridge that served to unite the mammals served equally to part the marine fauna.

*   *   *

IT WAS COMPARATIVELY EASY to describe salient features of the occupancy of Australia and South America by mammals after the dinosaurs' demise. The histories of these continents have a coherence guaranteed by their comparative insularity. Those migrations just described that brought South America's isolation to an end are bewilderingly and continuously part of the mammal story over the rest of the world. North America and Europe, Africa and Asia, India and Arabia—all have histories which interact time after time. The story is like that of a military campaign in which the various forces surge this way and that; there are flanking movements, and peace treaties, which are then broken on a whim. The movements of each battalion are fascinating to the military historian, even as they are extraordinarily difficult to reconstruct with certainty. The non-historian might be intrigued by a human detail, or the outcome of one or another battle, but cannot be expected to hold in his head the hundred thousand events which are a true description of history. So it is with Tertiary biological history. There were hundreds of different mammals in Eurasia and Africa, and each one has some story to tell. An inventory of these animals would take up the rest of this book, and I am not going to provide it. Almost any summary is a simplification which would arouse the ire of a specialist.

There are some animals so extraordinary that they cannot be omitted without losing some of the most improbable and unpredictable episodes. For example, in the early Tertiary of North America there were giant birds called *Diatryma*. These birds had lost the capacity to fly, and, because they had taken up the trade of hunter, they had massive bills capable of shredding many of the early mammal herbivores. Looking at these animals it is easy to believe that the dinosaurs did not really breathe their last at the K–T boundary. They lived on as voracious ground eagles striding about on massive, muscular legs. Had these monstrous birds survived, how different children's stories would have been. Red Riding Hood would have dreaded a monstrous ostrich, and the king of the beasts would have hatched from a royal egg.

Then there are some generalizations which are worth making. Mammal history divides into two halves: the earlier half is the more exotic, populated by many strange beasts that no longer survive; the latter half includes, if not actual members of the living fauna, a selection of animals that are clearly related to species we can now go and examine in a zoo, if not in our back yard. The first half embraces the Palaeocene, Eocene and Oligocene; the latter half much of the Miocene, Pliocene and Pleistocene periods. There were a number of ancient exotics which survived at least until the last, Pleistocene ice age, and even overlapped with man himself, as did the giant sloth. Even so, there are sufficient species related to the living faunas and floras in the earlier phase to allow a different kind of reasoning for interpreting vanished environments from any we have met before (I exclude algal mats): this is a *direct* comparison with living equivalents. So a fossil palm can be easily assigned to a living family of palms, and their ecological preferences can be directly investigated where they still live today. This makes the assumption—not necessarily a sound one—that their habits have remained unchanged for maybe 50 million years. None the less, this comparative method gives a new precision to scenarios of Tertiary life.

Perhaps the best way to approach European mammal history is to describe another site where the record of life is truly exceptional. This site is in Germany, near Frankfurt, at Messel. Set in the midst of a tract of secondary woodland there lies a vast pit, a kilometre across, now partly filled with water, in which there are a series of dark oil shales so rich in the remains of animals and plants that the deposit has been characterized as a fossil ecosystem, preserved in its entirety. Oil and paraffin were formerly extracted by distilling the shales, and this fossil fuel was itself derived from decay of the profusion of life that lived in and around Messel. For this was the site of an Eocene lake, 50 million years ago, in which soft sediments accumulated to preserve the fossils. The lake was surrounded by a subtropical forest humming with life. From time to time slurries of muddy sediment swept across the lake, entombing animals, plant fragments and insects alike. The rapid burial preserved them all in wonderful detail. Imagine a delicate bat, *Palaeochiropteryx*, as fragile as a paper kite, with every bone laid out upon a dark slab, as if it had been waiting its turn as an extra in a Dracula movie. Then there are Jewel beetles (Buprestidae), which in the living fauna shine with iridescent greens and blues as precious as emerald: and so they do in the Messel specimens, a dance of colours preserved so perfectly as to mock time. There are giant ants and termites, drowned in the lake during their nuptial flight, which was probably very short—a single day in remote history ensnared for ever upon a rock surface; even their wings and antennae are perfectly preserved. There may have been blooms of algal plankton in the lake which removed oxygen that would otherwise have promoted decay; the anaerobic conditions which resulted safely carried the preserved carcasses of these soft-bodied animals through to the present day. A spider, born aloft over the water on a silken thread, met the same fate. Normally, these creatures would have been snapped up by fierce

fish like garpikes and bowfins, which abounded in the Messel lake. Similar
fish still survive in the waterways of North America, and ichthyologists are
convinced that they are among the most primitive of the living bony (teleost)
fishes; they were evidently commoner and more widespread in the past.
There is one fossil eel.

Many amphibians turn up in the dark oil shales at Messel: there are toads
and salamanders and frogs. The frogs appear to be preserved in mid hop, legs
tucked in and bowed ready for action—there is even a tadpole or two to show
that the lake could nurture the young just like lakes today. Freshwater turtles
abounded, especially *Trinoyx*, and there were no less than six kinds of crocodile, of which the commonest is *Diplocynodon* (so much for the setback for *these*
reptiles at the K–T boundary!). There are some preparations of these animals
that have been extracted entirely from their rocky matrix, bone by bone, so
that they look as if they might be freshly imported from the upper reaches of
the Parana river. Lizards and snakes were terrestrial animals, then, as now,
and by no means as common as turtles in the freshwater sediments. They are
of great importance to herpetologists as they are so well preserved. The lizard
fossils include early representatives of many living groups: skinks, monitors
and iguanas, as well as the legless varieties that do their best to pretend that
they are snakes. Not that there is a lack of true snakes; the Messel evidence
confirms that this is one group of reptiles that really did have its main evolutionary burst after the extinction of the dinosaurs, although, like the marsupials, their origins were earlier. The fossils preserve every rib, as numerous
as the legs of millipedes: there are boas several metres long, fossil stranglers
and squeezers, but none of the fanged, venomous snakes which are most
numerous in species today; they seem to have been a comparatively modern,
Miocene invention.

There were birds. It may be easily imagined that birds are hard to fossilize, having an unpropitious combination of characteristics: they were both
delicate and tasty. None the less, some survived as skeletons, even with feathers, and a curious and interesting assemblage they were, too, absolutely vital
to our knowledge of the history of the most popular item of contemporary
wildlife (there are more pairs of binoculars in Great Britain than there are
dogs). There were a few types of birds which can now be matched only by
South American seriemas, odd survivors from a formerly common group. But
there are also recognizable flamingos, owls, plovers, nightjars, swifts and
rollers. Some of the fossils are so delicate that they have to be studied by
X-rays. The anatomy of the fossil flamingo solved a controversy that had
riven the ornithological community for years. It proved that the flamingo

relatives lay with the avocet rather than with the ducks and storks; the fossil had not yet acquired the weird bill of living flamingos, although the rest of the anatomy, including the legs, was already typical. The avocet is the most elegant of wading birds, and has acquired familiarity in Britain as the emblem of the Royal Society for the Protection of Birds. Apart from its distant relative, the majority of Messel birds probably lived in the bushes and trees surrounding the lake. We can be sure that the still, humid air was punctured by the shrill or sonorous cries of birds.

As for the mammals, there was a variety of delicate species that would not have been preserved anywhere else. The marsupials were there in the shape of an inconspicuous opossum; in the Eocene they were still widely distributed. Insectivores included an extraordinary long-tailed, shrew-like animal called *Leptictidium*, which probably ran half upright on its long hind legs. There were relatives of the hedgehog, complete with spines; but one of them probably (or should it be improbably?) hopped like a rabbit. I have mentioned the bats, preserved complete with their wing membranes. There are three main kinds, and they are commoner as fossils here than anywhere else in the world. They still have their stomach contents preserved intact, which indicates that they died suddenly. It has even been suggested that they were overcome by poisonous fumes emanating from the lake. You can occasionally see vile-smelling and toxic black bubbles glooping up from stagnant pools in wet jungles today, so perhaps it is not stretching probabilities to imagine a stricken bat hitting the water and unable to relaunch itself into the air, struggling fitfully, and then drowning and sinking downwards, to be preserved in the rank mud.

There was a lemur-like animal lurking in the bushes, a small creature with forward-facing eyes, which might not be thought too remarkable, except that the lemurs are primates, and thus primitive members of the group that includes both apes and humankind. The face of this diminutive lemur is the face of the future. A naturalist descendant of this animal hiding among the bushes would have had little trouble in recognizing several species of rodents, for the Messel animals had the characteristic incisors in upper and lower jaws, teeth that need to nibble continuously to hone their gnawing edges. The Eocene species were rather large, as rodents go; some of them were as big as a small cat. *Ailuravus* has stomach contents preserved, which are invariably full of macerated leaves, so it was probably arboreal, scuttling and weaving rapidly through the branches, nibbling frantically. It might have had cause to leap as fast as it could to escape the attention of predators, for there were inevitably some of these in attendance, although they are very rare as fossils.

They have been studied by Dr. Springhorn, who provides another example of the marriage of name and specialization—what better name for a mammal worker than a cross between springbok and pronghorn? One of the hunters was a very early example of the true carnivore line, somewhere close to the common ancestor of lion, bear and seal. Another is an altogether stranger creature, a creodont, belonging to an extinct group that has a different arrangement of its cutting and rending (carnassial) teeth, a dental feature which at once betrays carnivorous habits. Neither probably had much success in chasing pangolins across the forest floor. The Messel pangolin is hardly different from its living counterpart, with its curious but impregnable armour of overlapping, triangular scales, and a mouth specially adapted for a diet of ants and termites. It can curl into a ball to defy even the most determined attacker. There can be no more striking demonstration of the differences in rates of change which happened during the long history of mammals—50 million years ago rodents were near the beginning of their history, but the pangolin had already done much of its evolutionary work. The other anteater is equally suggestive, because it was undoubtedly related to the South American anteater that still roams the pampas from one ants' nest to another. It is an edentate—a mammal with a reduced dentition—a group which also includes sloths; edentates have always been regarded as quintessential to the South American continent in its island phase, and this muddling migrant was an enigma. It surely proves that there was more contact between South America and the rest of the world than had been thought.

Finally, there were hoofed mammals (ungulates), the great group of grazers that now provide us with food and milk, and steeds and wool—the very nub of most tribal societies. The Messel deposits have yielded the most perfect small horses, *Propalaeotherium*, with two species, the size of a terrier and an alsatian respectively. That these exquisite creatures are fossil horses is clearly shown by the anatomy of the head bones, yet they still retain several toes in their "hooves," an ancient feature retained from still more distant ancestors. There were even pregnant mares, preserving the foetus inside the body, down to its last fine bone, surely the ultimate proof of the ascendancy of the womb. These small horses shyly picked their way through the undergrowth, plucking at leaves, only stirred by fear into sudden flight. Later species, of which there are many, became dwellers on the plains, and were progressively built for speed, culminating in forms with one toe—the hoof. The tapir and rhino are distant relatives of horses, and they, too, have early relatives in the Messel fauna. The other major types of grazers are those with cloven hoofs, like cows, goats, camels, sheep, pigs and deer, which between them provide most of the human race with nourishment, and a smaller tribe with sport. Two rather undistinguished-looking animals that foraged on the forest floor were close to the ancestral deer. The most splendid ungulates appeared elsewhere, and in other faunas. This account of Messel inevitably begins to sound like an inventory—but what an inventory! This was a world we could recognize. Rich and varied in its ecology, it was populated by many animals which were only starting their evolutionary history, but which had already acquired the habits that their descendants have today.

Amid this diversity, there were glimmerings of difference in intelligence. The hunter must outsmart the hunted. Warm-blooded physiology hotted up the stakes. More food had to be consumed to feed the metabolic fires; reptilian opportunism had to be supplanted by stratagem, which is part instinct, part experience. Intelligent hunters must be outwitted by delicate nerves, subtle senses. The twitch of a leaf would set a bird into its alarm call and provoke it to flight, or send a small mammal scuttling away down a specially constructed tunnel. Sense built upon sense. Consider the cat—how acute its vision, precise its attack, discreet its stalking, persistent its lust for blood. I have seen feral cats in the middle of the outback in Australia destroying the marsupials created by 60 million years of isolation with their unsheathed claws and nocturnal virtuosity. In the archives of the Natural History Museum in London there are drawings of marsupials which no longer survive in the wild. The books were made in the early days of Australian naturalists, who must have come across the animals in their surveys. Their only record is now bound in leather in the care of the museum archivist. I have examined some of these drawings intently, as one might scan the face of a lost relative, and could not see in their wide-eyed expressions any obvious deficiency; but a cat or a fox evidently knew their vulnerability for what it was. Some cats (and many dogs) established social systems, the better to ensnare their victims and nourish their offspring. Social communication demanded mutual understanding, and that in turn probably placed a premium on greater intelligence. Since intelligence is reflected in elaboration of neural pathways in the brain, this requires more cortical tissue—more of what Hercule Poirot always irritatingly described as the "little grey cells." Larger brains (relative to total body size) are reflected in brain cases, and this, too, can be discovered from fossils.

We attribute virtue to intelligence, and tacitly condemn many species to a role of stupid, bit-part players. Sheep are legendarily near the bottom of the league, quivering dullards animated by nervousness alone, dunces of the mammal class, dolts and dimwits. It is grudgingly acknowledged that these allegedly obtuse animals can survive in places and conditions where sparkling wits are useless, but somehow the poor sheep acquires no credit for this performance. I suppose, on the contrary, that the sheep is quite as intelligent as it needs to be, and if the wolf is brighter (which it is), this is only in proportion to the ingenuity it must exercise to catch the sheep. If, like those Australian feral cats, it were in an altogether different class, the poor sheep would not long withstand the brilliance of the onslaught—but, within a few generations, unrestrained appetite would have destroyed the larder, and malnourished wolves would then quickly die out. Intelligence is also a matter of context.

Carnivores today lack one of their most distinctive historic designs: the sabre-tooths. They survived until comparatively recently. The last of them, *Smilodon*, is known from the Pleistocene tar pits of Rancho La Brea. The tar pits are in the middle of downtown Los Angeles in California. Within a few blocks there are the granite-skinned, gleaming facades of banks. It has been a long time since there was a ranch there. The natural tar seeps were once patchy water pools, and any grazing animal that lost its footing might be trapped in the bitumen, never to escape. It is likely that *Smilodon* was drawn there by easy pickings, and then itself became a victim. Its exhumed skeleton, perfect in every detail, can be examined in the adjacent museum, which is easily recognized among the thicket of high blocks by its modest proportions. There is something ironic about the relics of vanished carnivores secreted in the midst of this chrome-and-glass shrine to business competition, and when the great San Andreas quake finally brings it all tumbling down, I like to imagine the shades of *Smilodon* prowling among the carcases of the greediest animal of all, and then wildness will return again to the Pacific coast. *Smilodon*'s sabre-like canine teeth were fifteen centimetres long, and the cat was capable of opening its mouth into a great gape, a precursor to stabbing. The teeth were probably able to penetrate the tough hides of the rhinoceroses and elephants that also inhabited North America (and Europe) at the time. The sabre-toothed cat is reputed to have been able to bite out a chunk of flesh before leaving the victim to bleed to death. It is an extraordinary fact that similar fangs were evolved not once, but on several occasions in Tertiary times, and among very distantly related animals—they even appeared among the South American marsupials. This kind of parallel (correctly termed "convergent") evolution is familiar to palaeontologists. Nature is not profligate with useful designs.

In those television series in which actors in spaceships explore paradoxical corners of the space–time continuum it is odd that most of the aliens they encounter look remarkably like *Homo sapiens*, apart from their often being greenish. I have occasionally wondered whether these episodes might have been scripted by a palaeontologist, who has deduced that the optimum shape for an intelligent being is upright with arms, legs, eyes and clean teeth. Otherwise, there really seems no explanation for the high degree of convergence, which is arguably a similar case to that of the sabre-tooths. A little more reflection reveals the improbability of my intergalactic scenario. History is not just a matter of this chapter, but of the whole book. The sabre-tooth arrived at its design as much because of events in the Devonian, when land tetrapods first acquired legs and fingers, as because of events since the extinction of the dinosaurs; design is a consequence of a thousand prior circumstances. The chances of the worthies of *Star Trek* encountering a matching history on some distant planet—especially a history which could produce a blonde with lips—is statistically remote. Even the simplest planetary difference would redirect every detail. To take a simple case, a larger planet with higher gravitation would have implications for the size and musculature of any animals that evolved—and who knows if the answer to intelligent locomotion might not be the evolution of a cerebral worm with a hundred wheels? I conclude that the predominance of humanoids in space fiction is because it is difficult to generate drama with life forms resembling custard.

A less flippant question is whether the possibilities of design have been fully explored on our seas and continents. For example, are there ways of being a predator which *could* be produced by our mammalian anatomy, but which have never been tried? Is it easier to imitate than to innovate? The answer to this might be, as they say on *Star Trek*, "affirmative." To consider one example, the possibilities of venom have not been fully explored in mammals, although so effective in one order of reptiles—the fanged snakes (a lizard, the Gila monster, is also venomous). One could imagine that a poison-fanged cat would be invulnerable. There is no reason, in principle, why a venomous hunting mammal should not have evolved, given the fact that poison glands have evolved on so many occasions in the animal kingdom as a whole. Instead, the cat's adaptations were towards refinement of the senses, combined with intelligence and claws. Doubtless that is enough. The converse question is whether there is some ecological *role* that has been unexplored, some wholesale opportunity missed. I stated early on in this book that there is but a finite range of ecological roles, although there are equally many living players willing to act them out. This seems to be as true in the age of mammals as in the age of trilobites. After some thought I have identified one niche which never seems to have been occupied. High in the atmosphere there is a stream of air which transports insects and spiders, like some plankton of the ether. Could an aerial "whale" have evolved to harvest this stratospheric protein: a light, flying animal with a wide feeding gape, an animal that could cast a shadow across the sky? There may be several respectable, mechanical reasons why such a creature could never exist, or maybe it is that, like Icarus, creatures of the earth were not meant to soar too close to the Sun.

On the contrary, it is difficult to think of anything of which an insect might be capable which some species or another has not succeeded in doing. Feeding on living flesh? Mimicking a bird dropping? Or a dead leaf? Or another, nastier insect? Living on nothing but paper in the complete absence of water? There are insect species which are dab hands at all of these activities. Insects steadily continued to proliferate through the Tertiary as never before, even as the flowering plants prospered and diversified. Butterflies added their gaudiness to the forest glades. Bees perfected the art of pollination. The sights, smells and sounds of a summer afternoon grew from the growing interdependence of plant and pollinator. Honey appeared. So the biblical paradise-on-Earth, the promised land "flowing with milk and honey" could not have flourished prior to the Tertiary, since milk and honey only became abundant in the Tertiary. In millions of rotting logs, or dungheaps, or dark places in caves, the phalanx of beetles was doubling and redoubling. We know little of the process, but much of the effects of the success of the Coleoptera. Beetles are nowadays so rich in species that we will never know or name them all. When asked what characteristic God might display, J. B. S. Haldane famously remarked, "An inordinate fondness for beetles"—a phrase which might well serve as a motto for all coleopterists.

WE KNOW THAT as the continents moved slowly to their present positions, and while the oceans opened, volcanic islands welled up from the mantle and broke the ocean's surface. They, too, would have been populated by insects, and there is no doubt that special beetles would have evolved on each island. The island may then have sunk beneath the waves, and, no question, carried their endemic beetle species with them to extinction. Island endemic species are famously vulnerable. The most timely example is a death on 31 January 1996, when the last specimen of *Partula turgida*, which was kept in London Zoo, was declared "demised." This is (I should say was) a tree snail, rather than a beetle, one of more than 100 species endemic to the volcanic Society Islands, near Hawaii, which arose as a response to their remoteness and isolation. Introductions of other slicker, foreign snails are held to be responsible for their subsequent fate. There was a French collector, M. Thirioux, who acquired and "preserved" the last two specimens of an exotic lizard species from Rodriguez Island, in the Pacific Ocean, only to die himself of a heart attack later the same day, thus neatly ensuring the simultaneous demise both of a species and his own person.

It is not surprising that the fossil record of ephemeral islands is poor. In the Mediterranean region half a dozen ancient islands have been recognized, dating from Miocene times. They, too, produced bizarre endemics. My favourite is a giant hedgehog, *Deinogalerix*, five times longer than the hedgehogs in my garden, which was probably the terror of all the smaller mammals on its island home. Once the island was no longer isolated from the mainland, and thus the biological history of Europe, the inelegant monster no longer filled the bill, and it became extinct. There was a giant dormouse on Malta in the Pleistocene which was the size of a squirrel. There are some who link the myth of the one-eyed giant, Cyclops, with an extinct Mediterranean island elephant, fossils of which show the large, median nasal opening (surely a gaping "eye") which is a distinguishing character of these pachyderms. Odysseus and Theseus voyaged among the Aegean islands, which were the abode of several incomparable monsters. One can imagine an astonished mariner coming across a fossil in a cavern, or weathered out of a limestone bluff, and deducing the form of the creature that left it behind. The hero then scurried off, wide-eyed with hyperbole, and a legend was in the process of being born. In all truth, there really *were* islands which must only be imagined where unmourned monsters once thrived, now consumed beneath plate margins, or plunged beneath the sea into an early oblivion. We may people these islands with what phantasmagoria we please.

THERE IS NOTHING more superficially ordinary than grass. It is there to be lain upon, to be fed to our animals and cursed roundly every Sunday in summer. It is a flowering plant, although its wind-pollinated flowers are scarcely conspicuous—notwithstanding that hay fever is a conspicuous human reaction to its prolific pollen. However, grass is a special herb. The appearance of grasses in the Tertiary was of crucial importance to the modern mammal fauna, for many of the animals that figure prominently in human history feed, predominantly, upon grass. It has a remarkable property: its leaves grow from concealed bases—not from the tips of shoots, as is the case with most plants. So grass can be cropped—its leaves endlessly nibbled—without compromising its generative heart. Grass makes meadows, which virtually nothing else does.

"All flesh is grass," so the Book of Isaiah tells us, and indeed much of it is. Even the flesh of wolves is (in fairy tales at least) the flesh of sheep, and hence grass transformed. Its place in the economy of things can be compared with the endless "soup" of algal plankton that forms the basis of many food-chains in the oceans. Grasslands finally took over large tracts of the world during the Miocene, a time which was so often a watershed at the origin of the modern world—the great grasslands, savannah, prairie and pampas, date back to this period. The ultimate control on the spread of grass may have been climatic, especially the amelioration of the tropical belt produced by the growth of the Antarctic ice sheets. These became established when the Antarctic continent had drifted fully to its southern polar position. In tandem with the expansion of grasses, ruminant animals—those that chew the cud—were best able to exploit the new feeding opportunities. The several "stomachs" of ruminants like cows, camels and deer means that the process of gathering food can be separated from the longer process of digesting it. The pulp is brought back up into the mouth for a second chomp at leisure, and this creates an efficient method of converting grass to flesh. The rise of such grazers may even have discouraged plants without the special, regenerative growth habit of grasses. In rich herbivore communities, like that of the African savannah, grass-grazers lived (and still live) side by side with animals that nibble shoots from trees and shrubs, or even process whole branches, as do elephants.

These magisterial mammals were once much more various and widespread than they are today. The earliest Eocene elephants were about the size of a small pig. As their tusks grew in length from one species to another, so, presumably, did the trunk, which took over the job of stuffing the mouth with the great quantity of food an elephant needs. It really is the most extraordinary structure—without parallel for muscular flexibility in the animal kingdom—apart from the octopus's arm. By the Miocene, there were many different kinds of elephant, of which my favourite has to be the splendidly named *Gomphotherium*, a massive elephant with four tusks rather than two, which were carried on the head like two pairs of ungainly tongs. The tusks originated from both lower and upper jaws. There was another elephant, *Deinotherium*, in which only the lower jaw was so favoured, its one pair of tusks curving down like the tines of some kind of primitive agricultural instrument; it looks improbable, like one of Dr. Dolittle's inventions. The rest of the anatomy of both Gompho and Deino looks conventionally elephantine. Modern elephants, of course, have only the *upper* pair of tusks developed. Their extinct relatives roamed the world in the Pleistocene Ice Age. The woolly mammoth is the one always pictured in reconstructions of life in the Ice Age. Vast, with tusks as elegant as Brancusi sculptures, the shaggy animals fed in the tundra and forests which fringed the great ice sheet that covered so much of the northern hemisphere 40,000 years ago. We know its anatomy in extraordinary detail because deep-frozen specimens have been recovered from several sites in Siberia. They were engulfed by bogs and then frozen into the permafrost. Their meat is so fresh that it has been eaten by modern dogs; the fossil hair is coarse and brown, and concealed within it there are even parasites. Surely, if any animal could be revivified it should be this one, for even its DNA is preserved—in pieces. A straight-tusked, mastodont relative of this creature was one of the victims caught in tar pits at Rancho La Brea, but this species is now bones alone. The great, grinding teeth of elephants turn up quite regularly from the terrace deposits of the river Thames, even in the middle of London. Each tooth is ribbed like a bony washboard, but as massive as a kerbstone. When I was a child I had an encyclopaedia which pictured London not so many thousands of years ago, somehow cleverly superimposing Nelson's Column in Trafalgar Square with faded elephants (and somewhere in the background rhinoceros and hippo, too) so that in the artist's eye the extinct animals were as real as the Column, and dwarfed the famous lions at its foot. This picture has stayed with me as a memento of how even the solidity of paving stones and the pomp of monuments are little more than a phase in our history, and may yet vanish and be forgotten as so many other scenes have passed away before.

The Ice Age of the Pleistocene period is the third one I have described in this history; in geological terms, we have only just emerged from it. The waxing and waning of ice sheets during the last 1.5 million years forced great migrations upon animals. A number of species died out for ever at the onset of global refrigeration. But it made opportunities for others. Cold-lovers, like mammoths, cave bears and woolly rhinoceros, prospered at times when the ice advanced, while warmth-loving animals, like hippos and elephants, supplanted them in the northern hemisphere during warm phases (interglacials) between glacial pulses; some of the interglacials were even warmer than the climate is today. The vegetation changed in harmony, and obvious changes in the types of pollen preserved in sediments supply a kind of thermometer for the past. The Pleistocene was so close to the present that we can be quite certain that the plants will not have changed their habits since they contributed their fossils. There were several major cold phases, and more have been recognized in the last few decades. Even the four major advances of the ice sheets on land in the last million years are now known to have been interrupted by many more minor pulses, short-lived ameliorations of the climate known as "interstadials." At its greatest, an ice cap extended through the centre of North America beyond the Great Lakes, and in Europe covered much of England and Germany and Russia. The ice left its signature on scratched rocks, and deposited moraines, just as I described from Oman as the legacy of the ancient Gondwana glaciation of the Carboniferous and Permian periods. The signs are more blatant in Scotland or Wisconsin, less glossed over by the passage of tens of millions of years. You can still see glaciated valleys, or pluck scratched pebbles from the glacial drift, as if the ice had retreated only yesterday.

It is not difficult to visualize herds of reindeer and mammoths moving in around the tundra as warmth-loving mammals drifted southwards, prompted by their subtle instincts of a change for the worse. Some giant mammals evolved specifically to cope with cold, because large size is efficient in a cold climate for conserving heat. There were giant bears, and Irish elk, with antlers to match. There is a whole room in the basement of the Natural History Museum in London hung with these elk antlers, like the abandoned trophies of some megalomaniac big game hunter. Not all the interesting animals are giants; the arctic vole also has its story to tell, as its appearance signifies a cold pulse in the climatic history. The history of ice advance and retreat has become very complex, and ice age climate modelling is becoming a major field of investigation in its own right. The least ambiguous record of climatic events is probably preserved in sediment cores recovered from the deep sea, where the gentle rain of microscopic fossils continued unabated even as ice sheets grew and shrunk on land. Various species of planktonic animals moved back and forth, north or south, in sympathy with the fluctuations in the climate. Hence a core taken from the sea floor shows a diary of climate change that is more easily read than the shifting glacial deposits left on land, where a younger glacial pulse may have polished away the record of an older one. Limy fossil shells have even retained a climatic signature in the elements that make up their skeletons. Modern equipment can measure isotopes of oxygen precisely enough to read the fluctuations in temperature directly. This is how the new, and very complex curves for waxing and waning ice sheets were computed. Nick Shackleton at Cambridge University was one of the first to do these calculations. He is one of a dynasty of geological Shackletons. His father, Robert, is famed for his indestructibility as well as his acumen. At the age of eighty-five he led an expedition across the high plateau of Tibet, which would test the stamina of anyone not brought up on rancid yak's butter. Both Shackletons are related to Shackleton the explorer, and it is oddly appropriate that Nick should be doing his research into the vagaries of climate which tested his great-great-uncle.

Our fellow mammals are so closely bound to human history that it is especially hard to consider their evolution in a detached way. Mammals have been our companions and our food, and in many societies they have also been the barometer of wealth; a man may be judged by the number of cattle he has, in the same way that corporate man takes note of the model of his neighbour's company car. We talk to dogs and cats as to uncomplaining friends. It is scarcely surprising that any account of mammals has a tendency to anthropomorphize. Beatrix Potter's animals still look like the creatures they are supposed to be, give or take a few clothes. But the descendant of her mice is called Wiffly the Mouse and lives in a chintzy drawing room and never seems to do mousy things like eat undesirable creatures: home baking's more the thing. Mickey Mouse takes the process a stage further, having lost most of the protuberant nose, and with the face shortened to make him— well, virtually human. The same thing has happened to bears; the early teddy bears had long noses like the dogs to which they are related in the mammal Order Carnivora. They have got shorter and shorter, until now the teddy bear is, essentially, an orangey, furry human baby. Whatever the rights and wrongs of hunting it, the baby seal is such a popular icon because of its big eyes and round face, and general babyness. We breed the snouts out of dogs and treasure cats with huge, moon-like faces, like babies freshly placated. Koalas are the marsupials that small girls want to take home from the zoo. Ugly animals are those who either resemble unappealing—indeed very ugly—humans, or those that do not resemble us at all. Some of the baboons appal by their uncanny resemblance to Uncle George after a few drinks. Of all the animal kingdom, the mole rat is probably the least appealing aesthetically, however interesting its social life, looking as it does like a bloated, toothed maggot.

The tendency to anthropomorphize has produced varieties of pseudoevolution. Rudyard Kipling did it best with his *Just So Stories*, amusingly mythic but emotionally plausible accounts of how peculiar animals got their peculiarities. It is genuinely difficult to find an entirely adaptive explanation of how the leopard got its spots; nor are fossils likely to help us, since the one thing that is never preserved is spots. There *are* some narratives which are revealed by the truth of bones: I have mentioned whales with limbs, and growing elephant trunks, and I might have explained how much is now known of the early history of giraffes prior to development of the neck. Tribal societies often intertwined the stories of the origin of their favourite or revered mammals or birds with stories of their own gods. These tales serve to bind the people with the animals they hunt. The kind of ritual dancing that many North American native peoples employed is as much an expression of reverence for the hunted animal as it is part of a sympathetic magic to ensure the success of the hunt. Explanations can do more than simply square with the facts of descent. I am sure that my blandishments about the breast being a modified sweat gland, or about the history of the limbed whales, would be treated with amused astonishment by peoples who know how the gods made the animals. And surely the story I have related in this chapter is as astonishing as any tale of creation told to young hunters around the fire, even as yelping mammals called to one another in the distance, and the undergrowth rustled with the scurrying of tiny, furry creatures in search of the night's supply of insects.

67. One of the last of the South American endemics. The superbly preserved pelt of a sloth, *Mylodon*, which was probably still alive 13,000 years ago

68. An Eocene bat from Messel—a perfect preservation of a delicate mammal

69. An X-ray photograph of the same species of Messel bat *Palaeochiropteryx*, showing the finest details of the skeleton

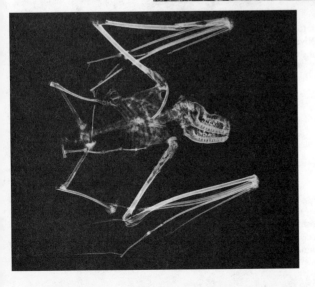

70. The extinct elephant *Zygolphodon* from Greece, with its Cyclops-like nasal opening

71. The mammoth found in 1793 in the delta of the Lena River, Siberia, in the company of a woolly curator

72. Leg of mammoth complete with wool, mummified by permafrost, from the Indigirka River (found 1972)

73. Giant single-celled foraminifera, *Nummulites gizehensis*, abounded in the warm seas of the early Tertiary. They are the size of small coins. This limestone is famous for providing the building material of the Egyptian pyramids.

74. Lady Smith Woodward's tablecloth. The wife of the Keeper of Palaeontology in the British Museum embroidered the signatures of all her teatime guests upon her tablecloth.

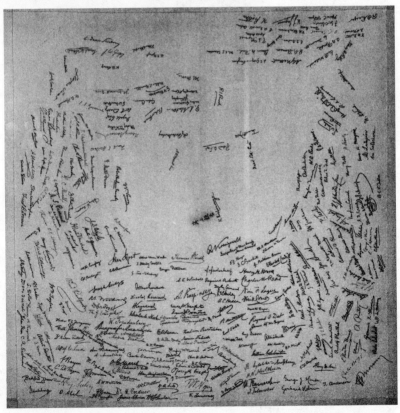

75. The "father" of continental drift, Alfred Wegener, in field gear

76. ABOVE RIGHT: Dr. Louis B. Leakey— a pioneer in the quest for human origins in Africa— studying *Australopithecus boisei*, one of the "robust" australopithecines

77. Footprints proving upright gait in hominids more than three and a half million years ago—the famous Laetoli (Tanzania) discovery of 1976

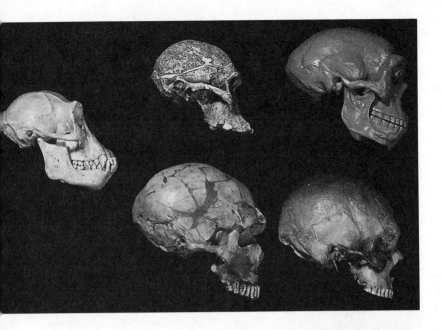

78. The story in the skulls. The conventional
evolutionary alignment from primitive to advanced
(and, more or less, old to young); clockwise from left:
chimpanzee; *Australopithecus africanus; H. erectus*
Zhoukoudian, China; *H. sapiens* "Cro-magnon";
H. neanderthalensis

79. Above, a comparatively
crude and ancient early
Palaeolithic stone tool
from Olduvai Gorge;
left, flint hand axes of
Upper Palaeolithic type
(Thames Valley, England),
showing more sophis-
ticated manufacture

80. Neanderthal man—an image of the "ape-man" in atavistic, Yahoo-like form published in the *Illustrated London News* in 1909. He recalls the evil Mr. Hyde in the silent film versions of R. L. Stevenson's *The Strange Case of Dr. Jekyll and Mr. Hyde*.

81. An illustration of Yahoos from a nineteenth-century edition of *Gulliver's Travels* by Jonathan Swift, c. 1840, illustrated by J. J. Grandville

82. A Neanderthal skeleton, buried on death, indicating their reverence for the dead (Kebara, Israel; about 60,000 years old)

83. House of mammoth bones, about 20,000 years old. The construction had been completely buried beneath a covering of loess and was reconstructed by archaeologists. Mezhirich River, Ukraine

84. Excavating a cave for remains of the history of *Homo neanderthalensis*. Palaeolithic Vanguard Cave in Gibraltar being excavated in 1996 by Tracey Elliott (left) and Lucy Gibbons, who rest on one time plane while successively older ones show as bands in the rock layers beneath their knees

85. Artefact as art: an exquisitely wrought late Palaeolithic stone tool in its original haft of antler—surely the product of advanced sensibilities. Late Palaeolithic, Switzerland

86. Temples to the laws of chance: ranks of one-armed bandits in the gambling casinos of Las Vega

87. LEFT: Emmer wheat, one of the ancestors of modern cereals

88. ABOVE: Isolation produced a list of endemic insect species on the Hawaiian Islands in a few million years. This example is an Antlion, *Eidoleaon*

Chapter 12

Humanity

I AM WATCHING MY WIFE threading a needle. In one hand she has the needle, precisely grasped between forefinger and thumb, raised in the gesture that the French employ to indicate good cooking. In the other hand, the thread, held similarly, has been delicately moistened on the tongue to smooth down the whiskers that might otherwise stop it passing through the eye of the needle. The eyes are directed forward, intent, precisely focused upon the manoeuvre, which requires exact manipulation, coordination and timing of hand and eye. Her lips are slightly apart, and the tip of the tongue projects as a measure of her rapt attention. With a deft touch, it is finished.

The most routine of skills reveal what it is to be human, and also what we share with our ape-like relatives. When Jane Goodall recorded the use of tools by *Pan*, the chimpanzee, there was film footage of these animals using straws to pick edible termites safely from their well-protected colonies. The straws were chosen in the same way as a needle might be selected. The attention paid to the task by the chimpanzee was as intense as my wife's to the task of threading, and even the expression on the face of the African primate was not very different. One of the early definitions of what it is to be human was coined as a book title by Dr. Kenneth P. Oakley, an anthropologist who was still working in London twenty-five years ago when I started my career:

this was *Man the Toolmaker*. Clearly, upon this definition the chimpanzee was a human.

The features that we share with apes are legion. I can use "feature" in a physical sense or a behavioural sense: there is the shape of the nose, the form of the face, the good eyesight, the dexterous hands, and a whole vocabulary of physical postures and facial expressions, several of which were noticed by Charles Darwin. Several more have helped Dr. Desmond Morris to become the promulgator of a view that the human is simply an unclothed primate. Once upon a time this opinion prompted outrage; indeed, it still does in some quarters. It may have been this prejudice that, unconsciously perhaps, formerly influenced mainstream anthropologists to believe that apes were not *closely* related to man himself. If common ancestry of human and ape was widely acknowledged, the separation of mankind from his uncouth cousins could at least be placed 20 or 30 million years ago, thus giving ample time for the evolution of our higher faculties, and banishing those shared grimaces to a remote prehistory. On this view, the higher apes—gorilla, chimpanzee and orangutan—comprised a separate group of animals with a long and distinct evolutionary history. The anthropologists emphasized the inhuman features of apes, especially the gorilla/chimp habit of walking on the knuckles, a feature which is instinctively parodied by small boys "aping around." Prominent, dog-like canine teeth, strikingly different stature of males and females, to say nothing of general hairiness, were all deemed consistent with the long separation of apes from humans. This view is almost certainly wrong.

Investigations of the genetic legacy of our ancestry have shown how much genetic material we share with the chimpanzee. Our DNA molecules are more alike than those of two species of the same genus of clams! A cladogram based upon several different molecules shows that humans are more closely related to chimpanzees than they are to gorillas, and all three may be more closely related one to another than to *Pongo*, the orangutan. But maybe all the apes—and man—should be placed together in the same zoological family, since they descended from a common ancestor.

When I was a boy there was a popular ritual at the London Zoo: the chimps' tea party. It was held, of course, at four o'clock. At this time, children and their parents would rush towards an open space near the restaurant, where two or three chimpanzees would be seated around a table. There would be teapots, and cups and saucers, and bananas and sandwiches. It was a kind of ghastly parody of the (then) family ritual of afternoon tea, which was ironical if we and the chimps really do belong together in the same family. After a semi-civilized start, disorder would break out. The animals would display their prowess at dismembering a banana. Tea would be slopped, cups approximately aimed at the table; with luck there would be one or two breakages. One of the chimps might climb on the table and try to steal something from his friends. We children loved it—I suppose we responded to an atavistic longing for days before tables and teapots and don't-talk-with-your-mouth-full. This entertainment has been discontinued in the modern zoo, and quite right, too. But in a grisly way it showed both how similar we are to what is almost certainly our closest living relative—and how different. We could recognize many of the facial expressions, although we could never emulate the wide, toothy gape. The grasping of the cups worked well enough—but it was mostly whole hand grasping, and had nothing of the delicacy of tea at the vicarage. The configuration of the feet—especially the curved big toe—and of the knees was most unlike our own. Even when they were walking on both hind legs there was a kind of crouching stance, a long-armed swagger, like a sailor too long on the sea and on the rum. The arms could have swung into action up and away and into a tree if there had been one around, and then we would have realized our own limitations. For the chimpanzee is no half-baked human but a beautifully versatile animal, no ersatz person ripe for mocking, no humorous homunculus. If it is as close to us as the genetic similarity suggests, it is evident that some rather important changes have happened with what small genetic changes there were since the time of our common ancestor. Recent work on chimps has shown that they are capable of understanding symbolic "language"—but of course are quite unable to speak. Since they are also capable of using tools (even if they are mostly ready-made) Kenneth Oakley's definition of humankind seems more than a little inadequate. Man the chattering ape might be better, perhaps?

In the early 1990s molecular biologists have estimated that the divergence between man and his closest relative would have happened about 5 million years ago, the time necessary to account for the differences in genes that they observed—the 20 or 30 million years of previous estimates was considered far too long. But were the fossils consistent with this estimate?

It is easy to forget that until quite recently the fossil evidence of humans was woefully inadequate. Even now, the total evidence for our early skeletal history would fit into a large packing case. Whole field seasons pass, and there is unfeigned delight if a single jaw bone is discovered as a result of searching and digging. For many years, following upon Huxley and Darwin's contention of our common ancestry with the apes, there was a search for a "missing link" that would forge the connection. It was not generally successful. There were red herrings, like the Piltdown Man hoax "discovered" by

Charles Dawson in 1911, an artificial confection of human and ape; such was the desire to believe in the "link" that more reputable scientists believed it than should have done, and for longer. It was the final exposure of the forgery in 1953 that confirmed Kenneth Oakley's reputation.

Now, the situation is almost the opposite. Since the notion of African genesis has achieved universal acceptance, more and more discoveries of hominid fossils have been made there—especially in South Africa, Kenya, Tanzania and Ethiopia. The link first became a chain, and then the chain developed branches. The result of the early finds was—at first—an absolute certainty about descent. Then, still more finds showed that there were extinct apes that were definitely *not* on the line to mankind. Were they men or not? They were assuredly not *Homo sapiens*. The only recourse was more and more critical examination of skeletal material: after all, *bones never lie*. As this duly happened, the names applied to the fossils have changed and changed again in the most bewildering way. Each name is attached to a species, and the problem has been that there have been grave disagreements about how many species there are and what they should be called. The poor student has had to learn a new name, only to unlearn it again as opinions have changed. Nor has the process stopped. Recently, I read an article by Bernard Wood of Liverpool University claiming that there were probably rather more species around in the early days of mankind's evolution than are currently recognized—once again, more names will no doubt result if this view prevails.

The early discoveries of linking forms were not in Africa, but in Java and China. Eugène Dubois, a military doctor posted to the Dutch East Indies, reported the discovery of *Pithecanthropus erectus*, Java Man, at a meeting in Leiden, Holland, in 1895. It had been discovered at Trinil, on the Solo river, by one of those wondrous strokes of luck with which the history of palae- ontology is punctuated. *Pithecanthropus* is the first name in the inventory— but he has disappeared. Now, Java Man has become *Homo erectus*, thereby establishing an adjacency to ourselves, but the early name is worth record- ing because *Pithecanthropus* means, literally, "erect ape-man," which neatly encapsulates Dubois' view of the fossil's place in the scheme of things. He had noticed that its brain size (or cranial capacity) seemed to be about halfway between that of an ape and that of a human—but its limb bones indicated an upright posture. The teeth showed loss of the prominent ape-like canines. This was—no doubt about it—the "missing link" and, as if to seal its status in a name, *Pithecanthropus* it became. Although the name is no longer accepted,

Dubois' views are now regarded as correct in their essentials. But he had grave troubles with his contemporaries. This is the first example of what has happened so often in the history of our history. The story of anthropology is peppered with examples that illustrate the deficiencies of our species: argumentativeness and bloody-mindedness first among them. "Man men" fight and squabble over bones in a way which is reminiscent of the sandwich wars at the chimpanzees' tea party. They are a very contentious crowd. Poor Dubois became so dispirited by the carping and aspersions of his critics that he eventually hid his bones from public view. Thirty years after his original report a young German aristocrat, Ralph von Königswald, found more remains of *Homo erectus*, and showed them to be closely similar to specimens from China, which were famously discovered in an apothecary's shop, where they had been part of the lavish pharmacopoeia of traditional Chinese medicine: "dragon's teeth," no less. In 1925 Raymond Dart reported the discovery of another and different "missing link" at Taung in South Africa, an altogether more ape-like "link" this time, which was named *Australopithecus africanus*. This name still stands.

The Keeper of Palaeontology at the Natural History Museum through most of these exciting times was Sir Arthur Smith Woodward. He was in the habit of inviting the leading palaeontologists of the day to afternoon tea—at four o'clock, naturally—where I am sure that the behaviour was unremittingly decorous. Fingers were no doubt delicately crooked, while cucumber sandwiches filleted of crusts were consumed, and thin slices of Dundee cake were served afterwards. At the end of the ritual Maud, Lady Smith Woodward, requested that the guest sign the white, lawn tablecloth. Fountain pens were produced and the cloth was adorned with that most direct symbol of human consciousness: the signature. After the guest had left, Lady Smith Woodward carefully embroidered each signature, so as to preserve every nuance of handwriting. Both Raymond Dart and von Königswald signed the cloth. Their irreconcilable differences about the linking status of the fossils to which they had attached their reputations were beautifully reconciled in the democracy of the tea table. The palaeontologist's tea party might, after all, illustrate what the nineteenth-century anthropologists were wont to call the "higher faculties" of the human being. The tablecloth has been carefully preserved, and now hangs in a frame outside the Keeper's office in the museum.

In another corner of the cloth, the name L. B. Leakey can be found. Louis and Mary Leakey (and, in turn, their son Richard) were Dart's most redoubtable successors in the pursuit of African hominids. From the late 1920s onwards they discovered and classified countless stone tools, some of them chipped from volcanic glass, as at Kariandusi, in Kenya. Tools are far commoner than bones for the obvious reason that they are much tougher. Stone tools are especially common in two sites in Ethiopia, at Omo and Afar. The Leakeys became convinced that another site, Olduvai Gorge, in Tanzania, would reveal a full history of the hominids, because it had a full succession of rocks spanning the last few million years. Their patience was eventually rewarded by the finding of *Zinjanthropus* in 1959, and "handy man," *Homo habilis*, on several digs between 1960 and 1963.

It is difficult adequately to convey the patience required for this kind of work. Most hunters after the past can go to a productive rock formation with a fairly good chance of finding the objects of their desire, even obscure ones. It is not so with hominids. Years can pass with nothing but hints, splinters that entice but do not satisfy. The skull is the grail, providing evidence of brain size and teeth, and hence diet and intelligence. In the story of African genesis there have been numerous unsung heroes among the workforce, mostly local black tribesmen, who helped the Leakeys, and many who followed in the pursuit of the booty of the big discovery. They may have been the ones who spotted the tell-tale fragment, their eyes sharpened by the promise of glory. It can be hot—very hot—in Olduvai, and only those accustomed to the heat, or moved by the lust for fame or truth, can bear to work on through the day. The Leakeys were sustained by such faith for years. But systematic searching reveals not only precious remains of human fossils, but also equally precious, if less newsworthy, information about past climates. This reveals the *milieu* in which humans first prospered. In their own way the plants and animals that are associated with the hominid remains are as important to the story as the human bones themselves. Different kinds of grazing animals—gazelles and their relatives—were found in different ancient ecologies. Climate change may have been the prime mover in the changes that happened along the links in the chain that connects modern man with his first ape-like relatives. Some would say that the change from dense forest to sparse forest and savannah 4 to 5 million years ago triggered the most important steps in human evolution. These changes are revealed by patient excavation which does not make the front pages of the *Daily Telegraph* or the *New York Times*. Tools are almost as revealing, because they broadly track the skills that were learned, even as the brain enlarged from ape to human capacity.

There is a final ingredient in modern studies of human origins. In many sites in Tanzania, and further along the Great Rift Valley that almost cleaves eastern Africa in two, there were active volcanoes, which from time to time blew out hot ashes and lavas. These eruptions happened at the same time as crucial changes in climate, and in hominid anatomy. Some remnants of the volcanoes still fume, and Kilimanjaro remains today as the most perfectly-shaped volcanic cone, reaching so high as to be capped with snow, for all that it lies deeply in the tropics. Volcanic deposits—ashes and flows—are interspersed with lake sediments bearing fossils in some sites. They yield a special bonus for our narrative, because they can be dated quite accurately by means of the naturally radioactive minerals they contain. They provide a "clock" which was not available to Raymond Dart. I was a research student in Cambridge when John van Couvering was doing some of the early work attempting to get accurate dates from difficult rocks. It was my first experience of the American work ethic, under which work must not only be done, but must be seen to be done. No matter what time I passed the department, there was John, doing something or other very enthusiastically. It instilled not a little guilt in this Englishman, brought up in the equally dubious tradition of apparently doing no work at all while pulling results out of hats by sheer brilliance (a deception, naturally).

What lies at the base of what it is to be human may be no more than a footprint. On a September evening in 1976, at Laetoli in Tanzania, Andrew Hill discovered two footprints that looked human, among several others belonging to rhinoceros or elephants. All had left their imprints upon one of the volcanic ashes, much as a holidaymaker might mark his passage over the sands near the sea's edge. Two years later Mary Leakey found two sets of prints walking, as it were, side by side. The beauty of these discoveries was that the volcanic rocks could be dated by radiometric methods at 3.6 million years. The prints proved that hominids at this time walked upright, like any family on a Sunday afternoon stroll. The discovery of this enduringly intimate record calls to mind Longfellow's verse, itself jogging along at a regular pace:

> We can make our lives sublime,
> And, departing, leave behind us
> Footprints on the sands of time.

It was estimated that the larger of the two walkers was 1.4 metres tall. Thus, long, long ago, before any skulls with anything like the brain capacity of modern humans are known, there were upright, bipedal hominids. The date was closer to that suggested by the molecular evidence noticed previously. Can it be that the crucial breakthrough on the line to humankind was this bipedal habit: Man the Walker? Walking has been shown to be an efficient way of travelling over long distances. The easy gait of the experienced walker can cover a lot of ground on little food. To be unpermissibly teleological about subsequent events: upright legs evolved first and freed the arms; thus liberated, hands learned to manipulate and to manoeuvre precisely; this required the kind of coordination with which this chapter began, and then, and only then, brain size became important: and thus the collaboration of hand, brain and eye allowed my wife to thread the needle. It is appropriate that when our species reached the Moon, and thus broke the bounds of Earth for ever, it was the first footsteps on that alien surface that we all remember. To walk upon it somehow symbolized possession of it. But, equally, 4 million years ago, one small step for early proto-man became a giant step for mankind.

We still regard baby's first steps as a defining point in the development of our children. It does not happen without difficulty. Parents support the body, hold out their hands, make encouraging noises. The new toddler knows that it has done something important when it is able to totter three steps from a chair leg to its mother's outstretched arms: it beams in delight with an expression that any primatologist could recognize. I wonder if the deep importance we attach to this developmental Rubicon is not truly an unconscious recognition of the place that bipedal behaviour has in our history, a kind of replaying of a vital evolutionary breakthrough. The baby's delight coincides with the moment that it becomes human.

Although upright walking has its benefits, and is almost unique in the animal kingdom, it carries hidden costs. More days are lost from work because of back problems than through any other single cause. This is because our bodies have still not entirely forgotten the quadripedal past, and sometimes our bones ache to return to a time when we scrambled as happily in trees as on the ground. And is there not something in the simple, joyful triumph of a small boy announcing his presence from high in the branches of a tree which speaks of a time that our mind has forgotten but our body has not?

Without fossils, there was no way that we could possibly have known that upright gait—and walking—*preceded* large brain size, not to mention many of the other skeletal and mental attributes that together make up *Homo*. A change in climate which led from thicker forest to more open habitats has been associated with the change in stance. Patchy nutrition favoured those who could amble across long distances. There were other changes that are much more difficult to speculate upon. Man the Talker is the most nebulous of all, for, although some minor secondary skeletal changes have been associ-

ated with the acquisition of our advanced vocal cords, nothing is wholly convincing. We do not precisely know when the first shouts spread from mouth
to mouth across a nomadic band, or when an elder gave the benefit of his
experience to an admiring youngster.

Even the upright gait may not have happened "all at once"—I have just
heard the South African anthropologist Philip Tobias describe the bones of a
fossil foot which indicates that at least early *Australopithecus* species were
likely to have retained partly arboreal habits. Our earliest relatives might
have frequently needed to scramble up trees in a hurry to escape numerous
contemporary predators. . . .

As for who made the footprints in the sands of time, the culprit was in all
probability the earliest species of *Australopithecus* yet known, *A. afarensis*. The
most complete of its kind is Lucy. She (for the bones reveal the gender) is
known from something like 40 per cent of her whole skeleton, a remarkable
percentage in view of her antiquity. The skeleton was collected piece by piece
by Donald Johanson and Tom Gray at Hadar in Ethiopia in 1974. Johanson
has described how he woke up in the morning on 30 November with what we
would have referred to in the early seventies as "good vibes." He felt lucky.
Yet the find was not made until he and Tom were just about to pack up for the
night, after a day wandering up and down arid slopes picking up a selection of
mammal bones—everything but hominids. They agreed to a last walk at the
end of the day—and, almost immediately, there was a limb bone. Within a
few moments Johanson had the back of a small skull. More pieces were scattered around—and, incredibly, there only seemed to be bones from one individual. Even as the light faded, the anthropologist's dream was growing in
solidity—an articulated skeleton of an early hominid! There was no sleep in
the camp that night as the discoverers drank beer and attempted to convert
the pure adrenaline of discovery into discussion and dreams. Johanson tells us
that "Lucy in the Sky with Diamonds" was played over and over on a gramo-
phone in the background; and by the morning the discovery recorded as
AL 288–1 in their notebooks had become Lucy.

Lucy was an adolescent girl. Her legs were comparatively short compared
with those of later hominids, a legacy of ape ancestry. However, her canine
teeth were already reduced compared with her first cousin, the chimpanzee.
There is no question that she was already on the path that led eventually to
man. The best date extracted from lavas nearby in Hadar is 3.1 million years,
and thus somewhat younger than the footprints. Quite a number of fossil
fragments from other localities were placed by Johanson and his friends into
Australopithecus afarensis, and they regard the case that an early hominid like

Lucy made the footprints as well founded. As always happens when human origins are concerned, there has been much dispute about how many different species might be involved with these early remains, and Richard Leakey was among those who thought there were probably several different early ape-men. Since all animals do vary from one individual to another, there is always the possibility of confusing this kind of variation with differences *between* species. The consensus has moved towards regarding all the early remains as belonging to a single species, *A. afarensis*, and placing that species at the bottom of the tree of human descent. But it would take a bold scientist to predict that no further discoveries will be made which will change this account.

A richer cast of extinct species in the story of the rise of mankind is now known than when Raymond Dart and Louis Leakey sat down to tea with the Smith Woodwards. They prove that, after Lucy and her kind, there were several man-like apes (or, if you prefer, ape-like men) living as contemporaries in the Africa of 1 to 3 million years ago. The early ones are assigned to the same genus, *Australopithecus*, as Lucy. I have already mentioned Dart's *Australopithecus africanus*, the first one to be discovered. The other names must be introduced if this account is to approach the full story. What follows is an attempt to record evolutionary changes species by species, something I have not attempted for any other species mentioned in this book. Maybe man deserves it, and maybe he does not; but his example will serve to demonstrate what may have happened in the history of any other species on our deliciously complicated planet.

One line of ape-men has proved to be a red herring, if that is not too shameless a mixture of zoological metaphors. When Mary Leakey discovered *Zinjanthropus boisei* on 17 July 1959 it was another of those moments in which long patience was rewarded by joy. This was a skull, which showed a very low forehead and prominent ridges above the brow. Curiously, a crest ran along the centre of the cranium, rather like a Mohican haircut. There were huge concave areas on the flanks of the skull available for the attachment of cheek muscles. The reason for these features can be found by examining— as always—the teeth. They are huge, grinding molars, and they must have been powered by bulky muscles. These teeth were capable of processing seeds and roots and other tough vegetable material; "Nutcracker Man" was a resulting vernacular name. He would probably have enjoyed the muesli they sell in my local supermarket. Nowadays this animal is considered closely related to Dart's *africanus* and hence is placed with it in the genus *Australopithecus*, along with another stout vegetarian, *Australopithecus robustus*, whose name speaks for itself. Recently a further, somewhat similar, but older species has been named from Lake Turkana in Ethiopia as *Australopithecus aethiopicus*. These comparatively harmless creatures lived alongside early *Homo* species in Africa—two-legged, of course, but using their massive jaws to cope with vegetable food that other animals might spurn. The human family would have been much more interesting at the time than it is now. It is odd to reflect that had this evolutionary sideline prospered rather than our own there might have been nothing but vegetables, cereals, nuts and roots in the local stores,* and we might have judged pulchritude by the bulginess of cheek muscles. Such a hypothesis of survival of a species of this kind has been suggested as an explanation for the mysterious Yeti, allegedly haunting the high Himalaya: sadly, there is no evidence that *Australopithecus* ever left Africa before it disappeared about 1 million years ago. Nobody knows why.

If these robust *Australopithecus* species were side branches in human evolution, the same was not true of *Australopithecus africanus*. Most theories of human evolution place this comparatively slender, so-called "gracile" species close to the *fons et origo* of the line that leads ultimately to mankind. In this sense Dart was right about its "link" status; it is just that the link was rather far down the chain. *A. africanus* has so far been found in South Africa in rocks dating from 3 to 2.3 million years ago, so it is younger than many, but not all, specimens of *afarensis*. It differs from the robust forms in lacking the prominent crest on the skull and having smaller cheek teeth, which makes it much more like *Homo*. The brain size was not greatly different from that of its robust compatriots. There is evidence that *africanus* lived in a less arid environment than the latter, and it may have been more of a fruit-eater, doubtless enjoying the odd grub or bird that came its way as a welcome dietary supplement. The rest of its body (so far as it is known) may not have been very different from that of other *Australopithecus* species.

Finally, in this gallery of extinct species, there is *Australopithecus ramidus*, discovered as recently as 1992 in northern Ethiopia, at a site about seventy-five kilometres from Hadar. This animal is even older than Lucy and her kin, at 4.4 million years old. According to its discoverer, Tim White of the University of California, the seventeen bone fragments are sufficient to prove that this oldest species is even closer than *afarensis* to the divergence between the apes and the hominid lineage. They suggested that even this early ape-like animal may have been dominantly bipedal and thus provided another

---

*I should note that not all scholars are in agreement that the robust species were vegetarian; there is an argument based on the chemistry of the bones that implies some meat in the diet; maybe, like chimpanzees, they got meat when they could.

"link" in our burgeoning prehistory. Perhaps there will be another evening when a fortunate stroll in some arid corner of Ethiopia will reveal more of this enigmatic creature to a vigilant pair of eyes.

Out of the evolutionary "bush" of the several *Australopithecus* species scrambled *Homo* himself. It was very probably the earliest members of our own genus that made the first tools—those tools that the Leakeys had found so easily, even as the bones of the animal that made them proved so elusive. Stone tools date back to about 2.5 million years ago, while the first bones attributed to *Homo* are about 2 million years old. There is a geometric law about the bones of *Homo*—the nearer to the present day they are, the disproportionately more abundant they become. Near to the time of origin of our distant relatives bones are excessively rare, and still exclusively African. The earliest species, discovered at Olduvai, was named *Homo habilis*, and was probably the original tool-maker. It will come as no surprise that the status of *H. habilis* has been hotly disputed. There are those, like Philip Tobias, who regard *H. habilis* as the true link between the more ape-like *Australopithecus* and the species of *Homo* that are clearly on the route to modern humans. Others wonder whether it is a good species at all, disputing whether some individual specimens are distinct enough from *Australopithecus africanus*, on the one hand, or *Homo erectus*, on the other, to be put in a species of their own. But however you read the specimens belonging to this early member of our own biological clan, it is clear that they do sit between undoubted cousins of our own species and the more ape-like australopithecines.* The brain capacity of *H. habilis*, for example, lay somewhere between that of *Australopithecus africanus* and *Homo erectus*. It is, if you wish, another link.

Every discovery of a new hominid fossil makes the news. The reports that have appeared in the newspapers over the years are interesting for the light they cast upon the psychology of both scientist and reporter. I have never seen a new discovery reported as SMALL TOEBONE ADDS DETAIL TO AFRICAN HOMINID. It is always something like NEW FIND OF FOSSIL MAN OVERTURNS GENESIS, and the accounts nearly always include phrases claiming that the textbooks will now have to be rewritten. It always conjures up a wonderful picture of scribes with pens poised to scrub out the old text and write in the new.

Early tool industries have recently been reinterpreted thanks to experiments by the anthropologist Nicholas Toth. It seems rather a good idea to test how tools worked by trying them out, and how easy they were to make by

---

*Very recent studies recognize an additional species for African *H. erectus* called *Homo ergaster*.

making them. Nicholas did this. Virtually all stone tools—early and late—were made from hard and homogeneous stones; volcanic glass was often favoured in African sites, flint in European sites. This is because, when struck in the right fashion, these tough rocks will fracture conchoidally (literally, like a shell), and controlled working can then turn these fractures into sharp edges along the line where they meet. The tools that result were hard enough to cut most natural materials. I can vouch for this by the gash in my thumb that resulted from my own feeble attempts to emulate palaeolithic techniques on the flinty beaches in Suffolk. The flash of scarlet blood was shocking—but, I told myself, this was only a trial with flint; but it proved its efficacy. Professor Bordes, a pioneer in experimental creation of flint tools, can manufacture a beautifully efficient stone tool in an hour or two. The conventional interpretation of early stone tools was that a lump of the appropriate rock had flakes chipped off it until what remained (the "core") was a useful artefact for clubbing things on the head or skinning them. Nicholas Toth's elegant work on the earliest stone tools came to the somewhat surprising conclusion that it was not the "cores" that did the business—but the flakes. He found that he could skin, dismember and prepare a carcase more efficiently with the flakes than with the core, although the latter might have had some part—for example, in clubbing bones to extract bone marrow. What is also striking is that a simple style of stone tools persisted with only a few modifications for an enormously long period of time through the earlier part of the Old Stone Age, or Palaeolithic. They are known from rock successions dated at 2.5 million years old and they last for more than a million years with no change. Such conservatism is staggeringly uninventive by modern human standards, and has led to claims that these early "men" gave no thought to the construction of tools. Rather, it was as automatic as the nest-building of a weaver bird.

Homo erectus is known from African fossils collected around Lake Turkana in Ethiopia, dating at about 1.7 million years. We have met him already when I reported on the pioneering hominid discoveries in Java and China. The latter lead to the conclusion that after the earlier history of mankind in Africa there must have been a phase of invasion from Africa into Asia, and possibly beyond. *H. erectus* is known from a nearly complete skeleton discovered in 1985 by Richard Leakey: the "Turkana Boy." He does not seem to have acquired a more familiar epithet *à la* Lucy. What he reveals is something of a mix between modern human and australopithecine design. His brain capacity lies between the two (850 cubic centimetres is typical), and the teeth are well on the way to modern human form, though still robust. There were prominent ridges about the brow. It is likely that the difference in stature between males and females was only about 25 per cent, whereas male *Australopithecus* were almost twice the size of the females. The closely similar stature of men and women is more than just a convenience in dancing cheek to cheek—it is an indication of a thoroughly different social system from that which pertains in apes. *Homo erectus* may have been edging towards greater sexual equality. Where (s)he appears, so also do characteristic stone tools with a "teardrop" shape, which have been found very widely beyond Olduvai Gorge. These tools were probably used as hand axes, and thus signify the more sophisticated use of the "cores" of the worked stones. Many of the tools were worked along both faces of an edge, to produce a sharp cutter. Very similar tools from Europe have been described as Acheulian after the town of St. Acheul in northern France where this kind of Old Stone Age industry was first characterized by nineteenth-century prehistorians. The French love classification, and soon pigeonholed these early artefacts into various categories: scrapers and trimmers as well as axes. Whatever the truth about their function, it is certainly true that the appearance of *H. erectus* was marked by a greater variety of tools, even if it does require something of a *connoisseur* to appreciate their finer points of difference. It may not be hyperbole to describe this as a technological breakthrough.

So far, so subhuman. Nobody has claimed that *Homo erectus* might be only a variety of *H. sapiens*, although several students would like to divide the specimens on which *erectus* was based into more than one species, of differing proximity to modern man. But there are surprises. If the newest dates are right, Dubois' specimens of Java Man found so long ago on the Solo river, at Trinil, are only about 100,000 years old—thus proving that the primitive species lived on while elsewhere in the world humans of modern aspect were already thriving. The kinds of tools *erectus* made suggest that hunting and scavenging were already important to their diet. There were sites where carcases were processed using stone tools—the cut marks scribed on the bones remain as intimate reminders of a meal enjoyed long ago. In a few places in Olduvai broken bones have been reconstructed from their constituent pieces, thereby proving that they were split open for their marrowbone on that very spot. Whatever the protests my vegetarian friends might make, it is beyond question that meat-eating became very important early in human history (doubtless supplemented with roots, leaves and berries); meat is nutritious stuff, and a good meal may have left time for lying around and developing innovative activities. At about the same time, the remains of firesites show that fire was domesticated. Some would claim that *Homo erectus* had already used fire late in his history: its uses for cooking were probably almost incidental compared with its use as a threat to ward off lions and tigers. Prometheus may have been a smallish individual with a bony brow and a large nose.

The conventional story of the next stage in mankind's evolution goes on with *Homo erectus* melding, as it were, into modern mankind—our own species—possibly about half a million years ago. This was supposed to have happened over its whole geographical range, advantageous changes being propagated through the population by interbreeding. The tendencies towards large brain size were carried further, while the social habits, tool-making, and all the paraphernalia attached to hunting and gathering tribes were added piece by piece until you could say of the creature standing before you: *ecce homo*. Once again, the story is pieced together from precious few fossils, mostly skulls. My colleague Chris Stringer lugubriously describes the business of trying to reconstruct the evolutionary narrative from these tantalizing fragments as like being in a darkened theatre, in which some saga of the complexity of *The Monkey King* is being played out on stage, lit only by occasional and random flashes of illumination. What is seen is dramatic enough, but how the glimpses fit together is open to interpretation. Many of the crucial fossils are African, and they are referred to by anthropologists with the familiarity that you might reserve for an old acquaintance. They will talk about Broken Hill, Jebel Irhoud or Awash—all skulls—much as one might refer to old Percy at Number 49, or Mrs. Jones across the street. However, familiarity in this case breeds not contempt, but contention.

The argument currently preoccupying Chris Stringer, so much so that I can occasionally hear him growling the names of some of his detractors under his breath, is the antiquity of ourselves, *Homo sapiens*. The orthodoxy of gradual transformation from *H. erectus* is being challenged. New analyses of skull characteristics have persuaded Chris and many of his colleagues that a species which can genuinely be identified with ourselves originated as recently as 40,000 years ago. This figure neatly separates two former chronologies: being one-tenth of the previously identified time of transformation of *H. erectus* to *H. sapiens*, and almost exactly ten times the antiquity of the Earth suggested by Bishop Ussher on the basis of the Holy Bible—thus, I imagine, offending the maximum number of parties. Like the major hominid events I have already mentioned, the alleged origin was in Africa. Modern *H. sapiens* spread in this geological twinkling of an eye from Africa to colonize the Arctic ice caps, the Brazilian rain forest, and Wall Street.

At this point I have to consider the Neanderthal men. I have also to recall the great Ice Age of the Pleistocene mentioned in the last chapter—especially the period 110,000–35,000 years ago. For, lurking beyond the edge of the ice sheets, contemporaries of cave bears and woolly pachyderms, there were stocky human figures, short-legged and with prominent brows and big noses, hanging together in tight little groups. They were equipped with stone tools of the most complex and exquisite workmanship; their brain capacity was as big, or even bigger than modern man's, although the cranium was low slung. These were the ancient men originally recovered from the Neander Valley, near Düsseldorf in Germany. The Abbé Breuil found carefully buried bodies at La Ferrassie in the Dordogne, France, in 1909. Other comparable examples of reverence for the dead ensured that these early Europeans have a better fossil record than most hominids. They have been investigated over a longer period than other fossils in the human lineage. One touching example has cornflowers placed in the grave. Surely these were humans of a sort, who so demonstrate affection for their kind.

When I wrote a more conventional predecessor to this book, more than ten years ago, I sought the advice of anthropologists about the status of Neanderthals. At that time they were considered to be a subspecies of *Homo sapiens*, one that through the best part of 100,000 years was especially adapted to cold conditions in Europe and the Near East. Now, these tough men are often considered as a separate species. They had a hard life; few men or women lived much beyond thirty; the healed breakages that are commonly seen on their skeletons have been compared unfavourably with those carried by seasoned rodeo riders. Perhaps their lives were "nasty, brutish and short," but one thinks of those cornflowers, the little ritual tools, and surmises that around the camp fire stories were told and magic invoked in a way we would all recognize. Their compact build was no doubt an adaptation to the exigencies of a glacial age; even their large noses may have helped to conserve heat as well as capture all manner of smells. Perhaps they existed in proud independence well to the north of African, southern European and Arabian humans, who looked more like ourselves.

The origin of *H. sapiens* was from among a series of populations spanning a time interval between 700,000 and 125,000 years ago. Some of these really do show a mixture, or mosaic, of features between *H. erectus* and *H. sapiens*. One of the most famous of such fossil specimens is the Petralona skull, discovered in 1959 in a cave close to the village that gives it its name, which is not far from Thessaloniki, in Greece. Its brain capacity is large, but its brow ridges are like those of a typical *H. erectus*. It is something like 220,000 years old. These early populations gave rise to what was formerly referred to as "archaic *Homo sapiens*," whose fossil remains are known from a dozen or so sites, mostly through Africa, but also in Europe and at least as far as the Middle East. Archaic humans, too, had spread beyond the mother continent. The tendency at the moment is to distinguish the same specimens as a different and distinct species (some would even recognize more than one species). This nomenclatural quibbling is a product of the theory which has all modern humans—*H. sapiens* by anybody's measure—originating only 40,000 years ago. If this "Out of Africa" theory is correct, modern man *by definition* cannot be the same *either* as the older "archaics" *or* the Neanders. On this reasoning, therefore, they were separate species. Thus it was that our direct ancestors pushed aside the Neanderthal men, and displaced the archaics. This may or may not have been accomplished by violence—but it would be a mistake to assume that warfare was necessarily part of the diaspora, for shifting climatic conditions alone may have sufficed to make some of the specializations of Neanderthals redundant. As our species spread, so new cultures developed. This is shown by spectacular advances in the technology of stone tools; within the compass of a few thousand years more innovation had been achieved than in the previous million years by *H. erectus* and his "archaic *sapiens*" successor—whatever they should be called. These "industries" show variation from region to region and, surely, the signature of the craftsman taking pride in his work. The best examples of the advanced Solutrean or Magdalenian tools are satisfying to the eye in just the same way as a well-turned pot. Some of the polished hand axes are perfectly and symmetrically formed.

The coincidence in timing of the accelerated change in the sophistication of tools with the "Out of Africa" scenario is impressive. To this there has recently been added genetic evidence. It is a fact that human genetic variation is greater between the native peoples of Africa than between all the peoples of the rest of the world. This might seem surprising, given the obvious physical differences between, say, an Australian aborigine, a Han Chinese and an Irish redhead. None the less, so it is, and such variation is evidence that the African populations that show it have been established for the longest time. In some respects the people of the Kalahari are closest to what might have been the human "rootstock," at least as measured by genetic change. All the differences attributable to different racial types are but a comparatively recent gloss hung upon a fundamentally similar skeleton. Melt away that "all too solid flesh" and we are truly brothers under the skin. Fundamental similarities in the genetic structure of human mitochondrial DNA among living peoples—which is handed down through the female line alone—has suggested that we *all* include genetic material derived from one woman, and that the base of the evolutionary tree lay in Africa.* She became widely known as "Mitochondrial Eve." This name is misleading, for it connotes descent from a single "parent," as in the biblical account. It would be more correct to say that "Eve's" genes survived through several thousand generations, through the female line, by a succession of lucky chances; the population among which she and her successors lived contributed to the fitness of all her survivors. There were, successively, many "Adams."

These different lines of evidence combine to make a rather convincing case for the comparatively recent origin of our species, at least as precisely defined. But we have seen the notions—even the names—of mankind's descent change so many times in the past; there are so many different versions of what is the important missing link, that it would be foolish to assume that no further changes will happen: indeed, we may depend upon it. There is no reason to suppose that we have now arrived at a more definitive version than was the case in Darwin's time, or Dart's, or Leakey's. In some ways, history has come full circle. In the early days of physical anthropology almost every new discovery was given a new name: *Pithecanthropus, Sinanthropus, Zinjanthropus;* only to fade into comparative obscurity as they were relegated to species of *Australopithecus* or *Homo.* This is what biologists usually refer to as "lumping." Now, once again, new species seem to be emerging from the flux of human creation, some of them old names revived, others coined afresh.

It is salutary to recall that early arguments upon the antiquity of man depended largely upon his artefacts, such as the stone tools found in Maccagnone in Sicily, or the Somme Valley, and published with such exquisite drawings in the 1850s. Close association of artefacts with extinct species of mammal such as mammoths and cave bears was crucial to appreciating that the antiquity of our species extended widely into prehistory. The first fossils of Neanderthals were found in the valley that gave them their name in 1857; the great Sir Charles Lyell confirmed their authenticity to his satisfaction in 1858. Hugh Falconer wrote to Lyell that year: "Had anyone else mentioned the loess human skeleton, I would have put it in the bundle with mummy wheat, the Sea Serpent, and live frogs of existing species hopping out of Palaeozoic rocks." Clearly, there was a certain scepticism in the air. The techniques of precise excavation beneath the stalagmite on cave floors were refined by such devotees as William Pengelly, working in Brixham Cave in Devonshire, at virtually the same time as the Neander Valley discovery. The

*I should note that there have been criticisms of the technique used to analyse these results, and we await the final answer.

association of tool industries with assemblages of bones delicately recorded the dual track of time and climate change. Poor Pengelly was not to get the credit he deserved, as others sought to promote their own discoveries at his expense. Charles Lyell was apprised of this. In 1863 he wrote to Pengelly: "the history of the obstacles put in the way of publishing the Brixham results makes me somewhat indignant." As well he might be at this, probably the first of many examples of inhuman behaviour associated with the history of humanity. But Pengelly's careful work was given full credit by Lyell in *The Geological Evidences of the Antiquity of Man*, his book published early in 1863. It was an immediate success with the public, and its first edition sold out within a week. The same year saw the publication of Thomas Henry Huxley's *Evidence as to Man's Place in Nature*. Thus it was in one momentous year that two of the most trenchant minds of the age asserted both that man was alive at the time of the Ice Age—and that his ancestry was shared with the apes. The history I have recounted has been a vindication of those views. Neither Lyell nor Huxley could have predicted that the antiquity would be that much older, and the descent so much more labyrinthine; nor that Africa would have played so important a part in the story. It was at least conceivable at the time they wrote that the birth of our species happened close to where Paris, London or Berlin now stand. And who is to say that we will not have to revise our ideas as radically again over the next 100 years?

The first finds of ancient, but anatomically modern humans were found five years after the publication of Lyell's book, at Cro-Magnon, a shelter in the cliffs near Les Eyzies in France. If you dressed Cro-Magnon man in a suit and tidied him up a little he might have been a suitable tea-table guest for Lady Smith Woodward. His language would have been unfamiliar, but there would have been no doubt about his species. His appearance would have been acceptable in society. He was proficient in another manifestation of humanity: he could paint. Man the Artist. In dark corners of caves—as at Lascaux, in the Dordogne region of France, or near Altamira in northern Spain—Cro-Magnons drew exquisite icons of the animals they held sacred, or those that they hunted. These outline drawings, in ochre, in charcoal, or in natural pigment, portray mammoths, antelopes, bison, oxen, horses. Man himself appears as an emblem, rather than a portrait, a spindly figure, a dark, attenuated sprite, less characterized than the animals around him. In some caves, handprints, also in ochre, have been added later, as if to assert individual identity. The drawings combine economy of means with precision of characterization in a way which leaves no doubt that there was joy or reverence in the skill of their rendition. The spirits would not be placated by a botched job. These are not the jagged approximations of childhood sketches. They are deft. If they recall anything later in art history it is the masterly likenesses that Picasso could achieve with a few pencil lines. Some of the animals that these early Europeans portrayed, like the larger mammoths, passed away with the Pleistocene Ice Age. Others, such as Przewalski's horse or the musk ox, have lingered on. I wonder how many art students today could capture the essentials of these animals with the assurance of early humans, working between 15,000 and 30,000 years ago, crouched uncomfortably in the reverential silence and uncertain light of a cave etched deep into limestone cliffs.

The separation of the line leading from African *sapiens* to modern humans, and the relegation of Neanderthal man to a separate species and extinct sidebranch, has one strange consequence. Consciousness is no longer the sole prerogative of one species—ourselves.

The history of life can be thought of as a crossing of thresholds—each threshold allowing more freedom for further biological growth and change. The first, replicating molecules producing living cells—these cells then collaborating in tissues and organisms—and, later, sexual differentiation promoting enhanced rates of change; the colonization first of the land, and then of the air, which life itself had manufactured. But then the final threshold is consciousness, freeing the mind from the confines of mere cells, allowing imagination to probe situations not yet encountered: a sense of self—and reason—are those properties we like to consider uniquely human. Was this a species character, like the plumage of a pheasant or bird of paradise, or the spots on a leopard? It is the uniqueness of this threshold which allows this one species a disproportionate place in my narrative: Man the Thinker, the sapient one.

Noam Chomsky has revealed how the root structure of language is shared between tongues—the building blocks of Babel. Language is not merely a learned ability, like dancing the foxtrot. Conversation around the tea table *is* what it is to be human. That language and thought developed in tandem is as reasonable as assuming that symphonic music developed alongside the orchestra. Parts of the frontal lobes of the brain concerned with language are prominently developed in humans, as would be expected. But nothing would be possible without the "plumbing" of the larynx and vocal cords being in place. Brain, voice and consciousness itself are linked together as a *Gestalt*— what in business would be called a "package." I leave to the theologian whether an immortal soul was part of the same innovative collaboration. But we cannot question that the sense of "I-ness" is one of the defining possessions of our species. So, with the separation of the sentient ancient inhabitants of the Neander Valley into another species, we have also dethroned *H. sapiens* as the sole possessor of consciousness. If "archaic" *H. sapiens* (now often recognized as one or more species) also had the same abilities, then consciousness becomes just one characteristic among others—its development a matter of degree, like the opulence of the tail feathers of Paradisidae, the birds of paradise. This might well be the final abnegation of a special place in Nature for humanity.

In the bleakest of the travels of Lemuel Gulliver, Dean Swift separated human qualities into two embodiments. The higher faculties, especially Reason, were possessed by the Houyhnhnms, horses of quite exceptional nobility, governed in all things by the faculty of logic. To the modern reader they seem rather too good to be true. The other half, the grubby side of humanity, was embodied in the Yahoos—hairy, dwarfish, scrabbling creatures of unbridled appetites. Swift's disgust at these horrible and untrustworthy animals is palpable, and you are left in little doubt that, in his view, they embrace the greater part of humanity. They bear a curious resemblance to some reconstructions of Neanderthals, for all that *Gulliver's Travels* appeared well over a century before their discovery. "Yahoo" has been employed as a pejorative description of uncivilized behaviour, and I have heard "Neanderthal" employed in much the same way. Swift's bifocal view of humanity is a separation of the altruistic side, supposedly supported by blessed reason, from a darker, instinctual side that is too often seen to dominate. A nineteenth-century view might be that the Yahoo side is a legacy of our "apeness"; and our Houyhnhnm side the more recent innovation, which pulls us away from dark entanglements. Studies of chimpanzee and gorilla social structure are sometimes quoted as evidence for the foundations of our own behaviour, and particularly our bad behaviour, as if these were not really different species at all, but a Yahoo version of our own. I suspect that the giggles at those zoo tea parties were rooted in such a perception. Now that we know something of the complexities of our origins we probably have more reason than ever to be cautious about such easy comparisons. Even the loss of the considerable size difference between males and females which has happened on the human line may be sufficient to generate a radically different social structure from those of our living ape cousins, in which males are physically dominant. While we cannot deny the similarity of our gestures, this is no guarantee of similarity of meaning.

Whatever the truth about his origins, *H. sapiens* spread around the world over the last 30,000 years, tracking changes in climate, and adapting as he went. The use of animal skins for clothes gave humans greater tolerance for a range of ambient temperatures than any animal that had ever lived. From Africa, humans moved through the Arabian Peninsula into the Middle East and thence into Asia. The superficial differences of colour and proportion, and subtler, invisible differences in the genes—distinctions which are the basis of the several races of mankind—were acquired during this diaspora, and linger today. During one of the glacial phases, probably about 15,000 years ago, so much water was locked up in ice sheets that global sea levels fell sufficiently to expose a bridge across the Bering Strait, an area which remained free of ice. Ancient Asians ambled across after antelope. And thence these people moved further, southwards, to found the Plains "Indians"; and further again across the Isthmus of Panama, and still further through Amazonia, and ultimately to Tierra del Fuego, where the remotest humans stayed tough and alone, until they astonished the young Charles Darwin upon the voyage of the *Beagle*. Other groups of humans threaded their way through what is now Indonesia, and New Guinea, and ultimately to Australia. They have been called Aborigines, but how inappropriate a name for people that have come so far and learned so much.

The properties of early tribal societies are inferred from limited archaeological evidence, combined with comparisons from surviving tribes as studied by social anthropologists. However, when earnest documentaries describe a tribe in Amazonia as "hardly changed since the Stone Age," this is surmise. How, after all, could we know? It is only reasonable to claim that tribal peoples who still live a wandering life present some kind of model for the state of mankind after the first diaspora. Hence it can be confidently asserted that humans were originally hunters and gatherers, each tribe attuned exactly to place. The members of a tribe were mutually supportive, every individual having a prescribed function within a society frequently on the move. Knowledge of the edibility of plants, and the uses of their poisons, were acquired by trial, the error of which we shall never know: somebody, somewhere, tasted a *Camellia* species and pronounced it tea. The hunter-gatherers moved through the land and were part of it. Myths bound people together, and animals and plants were incorporated in those myths. In some societies the land itself became an extension of consciousness, as Bruce Chatwin has most memorably described among the inhabitants of outback Australia in his book *The Songlines*.

When the sea rose again—as the great ice caps melted—the new peoples were isolated. For example, South America developed along its own lines until the time of Columbus and Vasco da Gama. Differences in cultural lives became hardened through the peculiarities of rites and language. The cultural diversity of humankind arose during 10,000 years or so of tribal differentiation. Different hunting tools set a signature on local carnivory: blowpipes, boomerangs, clubs or harpoons according to available prey species. In all peoples, music and dance enhanced rituals, and such celebration of rituals served to unite societies before the complexity of written law. Rites preceded rights. Rituals connected with initiation into manhood, with hunting and with feast days appear to be universal among living tribes, and often acquired religious significance. Curiously, the neural "wiring" for music apparently runs independently through the brain from other cognitive functions (the literary psychologist Oliver Sacks has described how a man whose damaged brain would not allow him to distinguish his wife from a hat could still play Chopin and Scarlatti). It seems rather wonderful that a corollary of humanity is the urge to sing.

Skirmishes between adjacent tribes were probably a routine part of life from the first. The warrior class continue to be the best dressed, the most magnificent, least Yahoo-like members of the tribe. I fear that the wonderful inventiveness of humans was not mostly directed towards the discussion of abstract virtue. Where mankind spread, several large animal species became extinct. It is the subject of passionate debate whether an intelligent, two-legged hominid caused these extinctions, with a kind of casual voracity. A huge, flightless bird—the moa—strolled in thousands across New Zealand; it was a kind of feast on legs. It seems that the arrival of the ancestral Maoris coincided closely with the decline and extirpation of this elephantine emu. Fossil feasts have been found. This conjures up a disagreeable picture of bloated humans feasted beyond satiety, still grabbing yet another bird until they could gorge no more. But what of the giant sloth, the mammoth, or the cave bear? These animals were also vulnerable to climate change. Such changes more subtly render the species redundant, forcing it into ever more peripheral reserves, until a vulnerable population succumbs to some aberrant virus and cannot recover. It is at least as plausible that climate change—the very change that accompanied Neanderthal man into extinction—rendered these wondrous creatures vulnerable. There are numerous cases where tribal man both venerates and eats his prey. North American Indians had to apologize to the spirit of the beasts they slaughtered. At the hunter-gatherer stage, mankind has lived alongside other large species in equilibrium, if not harmony. It seems to have been the arrival of the gun that has upset a balance. The aurochs, *Bos primigenius*, the wild ancestor of the cow, is painted on the walls at Lascaux, but survived until 1627, when the last specimen is said to have been killed in Poland. Consider, too, that on Wrangell Island a small species of mammoth survived until 5,000 years ago—even to the time that men were erecting pyramids elsewhere.

Some tribes dogged the herds of wild animals in their annual migrations, much as the Lapps do today. The animals provided food, their skins clothing, even shelter. But, in some richer pasture, husbandry was conceived. Why follow herds when they can be contained and protected? This happened independently in several societies, for some Far Eastern regions domesticated pigs, others in the seminal Middle Eastern area bred sheep and goats, and possibly, cows, while llamas were recruited in Central America. The ancestry of the pig and the cow is a single wild ancestral species, of which the wild boar beloved of Obélix the Gaul still survives. The pedigree of sheep and goats is more complicated: quieter beasts were selected—the more tractable favoured over the temperamental—until gradually a domesticated variety evolved by continued human selection. This is why gazelles have never been domesticated; their fine and nervous sensibilities will not bow down to husbandry. Those animals that produced abundant milk or sweeter meat were favoured over their wild counterparts. The cow, however, was originally used for work rather than milk over much of its range. Most of the Han peoples of China cannot digest milk products even today, and find the notion of eating cheese (ugh . . . *rotten milk*!) singularly repulsive. It was not more than a thousand years before different breeds suited to different purposes or grazing had been bred. Livestock must have been one of the first topics of inexhaustible interest, and I imagine that certain Neolithic men would have understood rather well the gestures, and patting of animal flanks, taking place in an English market town on a Friday afternoon.

Even the Houyhnhnm was taken from his wild pasture, and the bit inserted through his supple mouth. These animals are still "broken in" by trainers, the word implying both subjugation and a sense of lost nobility. Scholars do not agree who first tamed one of the last of the "odd-toed" herbivores, although no one doubts that when they swept in from Mongolia in the fifth century carrying warriors it was to new and devastating effect.

Domestication entailed enclosure—fields—and the idea of bringing food to the animal rather than allowing it to roam free in a hostile world. Enclosure equally implied possession, and thence a measure of wealth in numbers of cattle or sheep. The source of materialism may have been the domestica-

Telegrams : "NATHISMUS, SOUTHKENS, LONDON."
Telephone : KENSINGTON 6323.

BRITISH MUSEUM (NATURAL HISTORY),

CROMWELL ROAD. LONDON, S.W.7.

Dear Sir/Madam,

The specimen(s) you have submitted for examination is/are a tooth/teeth/bone(s) of horse, cow, pig, sheep or goat, dog.

The specimen(s) is/are of no great age and is/are not fossil

Keeper of Geology.

tion of pigs and goats. Today worth and wealth are still often measured in cattle among African tribesmen. Sheep and goats became as abundant as people. No longer true fossils, they turn up when sewage pipes are laid or gardens deeply dug. They are so common that the Natural History Museum in London had a special card printed to fend off the enquiries that still come in through the door. This card is reproduced above in its faded eloquence.

Plants were domesticated into crops. This was a release from the drudgery of finding a berry here, a root there, constantly foraging. While there is debate as to whether or not the transformation of man from his antecedents happened in one place or over a wider area, it is not in question that different cereals were brought into cultivation in defiantly separate parts of the world, over a period about 5,000 to 10,000 years ago. The domestication of maize (along with the guinea pig) in South America 5,000 years ago was not connected with the selection of cultivatable varieties of wheat, barley and peas some 5,000 years earlier in the "fertile crescent" embraced by the rivers Tigris and Euphrates. The idea of cultivation happened independently in different societies. In many places fossil evidence of cultivation of animals and plants—registered in bones or spores—appears at the same time in archaeological sites. In Jericho the remains of cultivated grains predate those of domestic animals. There are grains of emmer wheat and two-rowed barley. There is evidence that hunter-gathering continued into the Neolithic period (7,000 years ago) even as small, irrigated fields started under cultivation. Genetic studies have proved the complex ancestry of our modern cereals: hybridization and doubling up of the chromosome number in each cell is implicated. In the Middle East wild varieties still grow. They are scruffy little plants; you can rub their ears in your fist to get a few coarse, edible grains. You marvel at the skill of the early agriculturalists, who employed artificial selection in a world hitherto ruled by the natural variety. For the fattest and tastiest grain might not be the fittest under wild conditions. Cultivation of grains yielded surplus and plenty in human communities.

Humanity shares bread. The grinding of grain, preparation and baking of breads is a first act of civilization. Claude Lévi-Strauss has shown how rituals of food preparation have a seminal role in social cohesion. How often in the Bible is the breaking and sharing of bread both the simplest and most profound expression of altruism and fellowship. I like to think that the thin cucumber sandwiches served by Lady Smith Woodward to Louis B. Leakey were in cultural continuity with the gritty, unleavened loaves eaten by the early cultivators. I have tried a loaf of emmer wheat, a primitive grain prepared by an experimental archaeologist. It was not particularly tasty, rather stiff and chewy, though probably better than the spongy blocks sold in supermarkets. Maybe, after all, this is what repels us from that chimpanzee's tea party: they don't share the bread.

Cultivation led to settlement. Dwellings made of mammoth bones were discovered at Mezhirich in the Ukraine, a complex, carefully interwoven mass of ribs and thigh bones and vertebrae. They were 15,000 years old, and may well have predated cultivation in that region. Presumably, unconventional building materials ensured their durability, for early habitations of the younger Stone Age (Neolithic) are otherwise rare. The origin of permanent villages is still largely speculation. We can see how surplus food might lead to a structured system, those myriad societies studied by the social anthropologists, united only in having chiefs and shamans. The shaman guards the secrets of Nature by holding esoteric knowledge; careful rituals under his supervision ensure sunrise or rain, and the abundance of crops.

There were further discoveries dependent upon settlement. Clays mixed with straw made bricks, and bricks made houses. River clay, baked in the sun, was used to manufacture utensils. Firing the clays made stronger pots; storage jars were fashioned, which cushioned the bad seasons by saving from the good. Now there was a cornucopia of new artefacts. In many areas domestication of animals and the appearance of pottery closely coincide. The population doubled, and redoubled. More people meant more inventiveness: the rush of invention tracked the population and has never ceased. Writing began in the most functional way—the tally of lentils or the baking of a dozen loaves recorded on a clay tablet. Villages joined together. As Alexander Pope described it in his *Essay on Man*:

Great Nature spoke; observant Men obey'd;
Cities were built, Societies were made:
Here rose one little state; another near
Grew by like means, and join'd, through love or fear.

Let history begin.

Chapter 13

Wheels of Chance

When all that story's finished, what's the news?
In luck or out the toil has left its mark:
That old perplexity an empty purse,
Or the day's vanity, the night's remorse.

                                    W. B. YEATS

IN THE CAVERNOUS INTERIOR of the Excalibur Hotel in Las Vegas, Nevada, there are ranks of gambling machines lined up like gaudy soldiers on parade, each one to a common design, though subtly different. They invite you to make a close inspection. The machines are designed to be close to human height. You prop yourself before them on skimpy stools, or stand to attention inspecting the arcane rules that govern how you may be paid back richly as you gamble. The winner takes all.

It is a temple to chance. Between the machines there are roulette wheels, spinning symbols of chance itself. Punters cluster intently around the wheels, seeking by willpower alone to guide the little ball into the numbered haven of their choice. There are various ways in which to place your bets. The long shot is to bet on a single, winning number, with the faint possibility of colossal returns. Or, alternatively and more conservatively, you can bet on red or black. Whatever you do, the table has the edge. The odds can be evaded temporarily—enough to fund an occasional small fortune. But the odds will reassert themselves in the end with the inexorability of a mathematical certainty.

The bleeping, winking, sleepless cavern at Excalibur knows neither day nor night. It lives under perpetual, but dimmish, electric light. It is maintained at the temperature of a warm spring day. The temperature outside is about 100 degrees Fahrenheit—not a particularly hot day for this part of the world. All around there is the Nevada Desert, and if the water dried up, or the electricity were switched off, the desert would soon reassert itself. The people would depart in their Chevys and suburbans, and the whole absurd farrago of neon and paste would crumble away; and soon hot winds would excoriate the boulevards, and whistle among the abandoned ionic capitals of Caesar's Palace, around the glistening Nilotic pyramid of the Luxor, and beat against the castellated concrete of Excalibur. Maybe, at that time, a shy pronghorn antelope would pick delicately among the debris, coming down from the hills in search of rewards from remnants left behind by departing men and women. Lizards would crawl out to bask upon abandoned stretch limousines. Those creatures lacking in presumption, but rich in the qualities required for survival against the odds, would inherit the Strip. For the meek survive, even if they don't inherit the Earth.

Life—as everyone knows—is a gamble. Chance promotes or damns according to the whim of history. The analogy of permutations in a game of chance has long been applied to genetic mutations, spontaneous changes in the genetic code. Like the vast ranks of gamblers in Las Vegas who come away with nothing, most mutations also lead to nothing. They do not result in an increase in fitness. Some are lethal—as might be the production of a wingless butterfly. Others might put the mutant at a disadvantage—a new colour pattern might not be favoured by a potential mate, for example. But those rare mutations that hit upon an advantageous combination produce the big payoff—the jackpot. They are rewarded not in the gross currency of quarters or dollars but in the irresistible coin of many successful offspring. Unlike the lucky gambler in Excalibur or Caesar's Palace the luck of the successful gene is passed on to make luck for future generations.

Through the 4,000 million years of life's history I have trodden but one of countless possible paths. One recurring motif has been luck of a different order. Those animals and plants that survived the death of the dinosaurs did not prosper because of a small change in genes at the critical time. They already *had* whatever was needed to survive—by another sort of luck. Small, warm-blooded mammals and birds, together with insects and magnolias, survived the crisis at the end of the Cretaceous that extinguished dinosaurs and

ammonites because they were already equipped to survive, not because they
invented some wheeze at the last minute. The whole course of life hinged on
the fact that some of the animals that appeared in the Cambrian were unlucky
enough *not* to survive, and thus failed to propagate their designs in their
descendants. Recall, too, the constant splitting and rearrangement of the
continents as the great plates that make up the Earth's surface cruise in their
leisurely fashion over the surface of the mutable globe. It is luck—it must be
luck—which determines which animals, when and where, are attached to
a particular continent. As Antarctica was carried towards the South Pole after
the break-up of Pangaea its cargo of terrestrial animals was doomed to
die. Rare fossils tell us that there were mammals there before the ice caps
grew. Doubtless there were mutations galore, some of which favoured cold-
tolerant species over less tolerant ones, as the climate deteriorated. But that
was useless in the face of implacable ice and everlasting winters, a cold that
freezes blood. This was far, far worse than Las Vegas with the water and electricity turned off. It was bad luck.

Geography can be just as creative. When the island of Hawaii was born
from the eruption of oceanic basalt lavas it was isolated within the young
Pacific Ocean, far from the nearest continent. It was a *tabula rasa* upon which
Nature might write. A tropical climate almost guaranteed that it would be
fruitful. Volcanic soil is rich and capable of nourishing any plants that can
reach it. Some did. Seeds floated, brought by storms; other seeds blew in,
light as thistledown, borne in upon delicate parachutes carried by the breeze.
Distance was selective—very few species of animal landed safely. For birds it
may have been less difficult to reach the freshly green island, but even so the
founding species were small in number. There may have been as few as seven
different insects that won the lottery. But good luck here provoked a fever
of creative evolution—the opposite of the Antarctic blight. From the few
species that arrived in this evolutionary Eden dozens of species arose. Since
change could only work upon what material was to hand, wonders happened.
Harmless pollen-eaters evolved into predators; small insects became large;
many specialists arose to eat the newly native flora. None of these unique
species was found outside Hawaii. Good luck was fenced off from the rest of
the world.

But it has become bad luck, now that contact has been established with
the world, thanks to human colonization. For, in a tragic mirror-image of
that original creativity, introduced animals such as the rat are forcing out the
erstwhile lucky ones from their private Eden: it is a reversal of fortunes.

Australia also carried its cargo of marsupials southwards, and I have described how they, too, evolved richly on their own. Man's arrival as part of the diaspora of *Homo sapiens* also coincided with the decline of some large marsupial herbivores, but more damage has been done by the greedy ways of modern humans and their cats. None the less, not all Australia's endemics were losers when faced with competition from the wider world. The eucalypts have been Australia's contribution to the diversity of arboreal life. There are great forest trees like ironbarks, or drought-tolerant, dwarf mallees, or the white-barked ghost gums—no more varied range of designs for every purpose exists in the vegetable kingdom. They have prospered around the world against all comers. I have sat under *Eucalyptus* groves of great height in western Argentina; the local acacias are dwarfs by comparison. *Eucalyptus* leaves dangle in the sun, fending off its full blast. Their fragrant oils help to slow transpiration. They can even stand fire whipping through their glades, hot and fierce, but leaving the cool heart of the trees untouched to sprout again. They are world-beaters. Now, they line roads in Andalucia, thrive on university campuses in California, and decorate the town squares in Portugal. A strong-minded woman of my acquaintance once tried to argue with me that eucalypts were natives of Portugal—taken by early mariners to Australia—such was her conviction about how naturally they fitted into the Iberian landscape. Evidently, the isolation that ultimately proved bad luck for many Australian marsupials when they faced competition from placental mammals was good luck for gums. The skills acquired by trees along creek and by billabong proved to be useful elsewhere. We cannot simply pin labels on products of the intricacies of 4,000 million years of the history of life and specify doom for some and unlimited prospects for others. There will be surprises.

For life, unlike any gambler in Las Vegas, has made its own luck. Every one of the innovations on which my story turns—from the inception of photosynthesis, which modified the primitive atmosphere in a way suitable for "higher" life, to the colonization of land and, eventually, the skies—both altered the odds and reset the tables. A world so enriched might better endure the slings, not to mention the arrows, of outrageous fortune; life has survived the worst catastrophes that could be dealt by the cosmos: bolides have been beaten by beetles and barnacles. These mass extinctions were times when, briefly, all bets were off. At other times, life survived ice ages and changes in sea level, or fluctuations in the atmospheric concentration of carbon dioxide.

In short, life has gained an edge.

Nor is it trivial to draw a parallel between the biography of life and an autobiography. The interweaving of episodes from my own scientific life into the narrative of the biosphere does more than provide the occasional diversion. From my early days in Spitsbergen, luck—fortune, if you prefer—has played a part in deciding which direction I have taken. On occasion, a course has been decided by decisions almost as arbitrary as the spin of a coin. Our lives are not merely the inexorable unwinding of the consequences of our own genome, for all that it would be foolish to deny that genes are extremely important. One might compare genes with a good (or bad) poker hand— which can be modified during the course of play, but which can never be ignored. Luck, the environment, nurture, all contribute to the personal biography: "in luck or out the toil has left its mark." But the individual also has use of his free will to make choices—to make his own luck. Hence I see more than a facile resemblance between our own biographies, with all their quirks and minor setbacks and inconsequences, and the grander story of Life itself. Stephen Jay Gould's notes on the contingencies which have shaped the history of life amount to the observation that no two human biographies are identical, and no life would unwind twice in the same fashion. It is, in a sense, a trivial point, since clearly no biography can *ever* be relived. There is only one narrative. This was an observation made by Sir Karl Popper in *The Poverty of Historicism*. But a life remains a compromise between what is dealt, and what is experienced, mediated by the effort of will. The greater story of life on our planet is partly a story of luck, of changes imposed upon the world by earthly and universal forces, and partly a story of genes, and finally a product of the changes life itself has wrought to modify the odds.

We cannot easily imagine a planet stripped of its forests or with an atmosphere rancid with sulphur, or thick and suffocating with carbon dioxide. The changes made upon the globe by life have altered the way in which the very rocks are eroded. Green-clothed hills are sponges that absorb energy which would have scoured away slopes in the Palaeozoic. Even now, we are only beginning to see the true complexity of the tangled skein of life, let alone understand its workings. I like Goethe's description of "the eternal Weaver's masterpiece: look how one press of her foot sets a thousand threads in motion—how the shuttles dart to and fro, the flowing strands intertwine and a thousand connexions are made at one stroke!" As I write this there is a debate about whether the biosphere extends deep into the Earth, because living bacteria have allegedly been recovered from within rocks recovered from boreholes. I have no idea how the debate will be resolved, but "at one stroke" it could double the biological sphere and add yet further to the intimate complexity that links life with its geological matrix.

The last component of this complexity is consciousness. The rules of the evolutionary game were changed at such a critical point in the evolution of *Homo*. Bishop Berkeley's observation may not be right that, were it not for the contemplation of a conscious mind, the world might cease to exist, but it is assuredly true that perception has altered all the rules that once held indifferent sway over organic evolution: because choice is the companion of consciousness. In consciousness the narrative of history and that of the individual finally meet. The choices that modify the personal legacy of our genes, or the exigencies of our environment, are paralleled by the influence that human consciousness *as a whole* may exercise upon the narrative of the future. Our future entails that of the world; we have control of it, if we could only have control of ourselves. Human conscience and consciousness are closely allied.

It is interesting how military metaphors have often been used to describe what I prefer to think of as thresholds which life has crossed. A dozen popular illustrated books refer to "the conquest of the land." Who, I wonder, was the enemy in this conquest, and who the vanquished? It recalls the combative metaphors employed in disease that Susan Sontag described: "the struggle against cancer, the patient fighting for life, the battle against AIDS." This has a heroic subtext, like Theseus grappling with the Minotaur. Yet one of the remarkable things about the important thresholds in the biography of life was that they were explorations, innovations. The "enemy" was the challenge of a new physiology, no cryptic Minotaur. The metaphor seems instinctively to be appropriate, probably because the struggle for success is part of the legacy of classical Darwinism. The implication is: no change without challenge. This may be the stuff of regular evolutionary change (some scholars do not believe so), but when thresholds were crossed life gloriously broke free of combat, at least for a while, and lost itself in creative innovation, like those pioneer insects on Hawaii.

The shape and tempo of the story of life as a whole resembles that of the history of mankind, a single species. I think of Maurice Ravel's dance, *Boléro*, which starts slowly, uneventfully, a long series of slight variations upon a recurrent theme, gradually gathering pace, shifting from one instrument to another, while an underlying pulse goes on and on. From time to time there are shifts in key, then more instruments join in, and the pace and excitement build, until, at the end, it is a scurrying, swirling mass of interwoven instrumental activity. The slow growth of brain size in humans took several millions of years. To be sure, the innovation of tool-making was a threshold, but how similar those tools remained for more than a million years, slowly turning changes on familiar themes. But technology built itself up, and interactions between a growing population increased, until there was a sudden flowering of stone tool types, building techniques, domestication of animals and plants. In a geological instant society mushroomed in all its uncountable complexity. Similarly, for thousands of millions of years the bacterial and algal world slowly, almost imperceptibly changed—a long, leisurely theme repeated and slowly modified—while within the last 500 million years the richness of life has increased a thousand-fold again. The history of one species, the one to which the writer and reader belong, took a chapter of this book to sketch in only the broadest outline. Imagine if a narrative as complex were known for all the other millions of species living in this tangled and prolific biosphere; why, the paper used to print their histories would strip the trees from the world! Yet the chances are that every species has a story worth telling. In your mind's eye imagine you are an eagle easily gliding above the rain forest canopy, and as far as you can see in every direction, there are billowing canopies of trees reaching upwards into the light. Each one of these trees might represent the branching history of a living species, and the forest itself might be a crude representation of the density of the historical past. We shall never know every detail of every tree, but we can understand the vitality of the whole, and thus see the forest for the trees.

My tale stops where civilization starts, and where prehistory blurs into history. This is the moment when records begin, when humble steles or grandiose monuments tell of mankind's inhumanity, or his aspirations to godliness. Appropriately enough, some of the earliest writing is biographical, inscriptions trumpeting the achievements of kings. The truth of a life also depends on the selection of its incidents, no less than does an account of a civilization. As such unnatural selection proceeds there is also the possibility of lies and unwitting deception; although Nature is full of camouflage, we are the first animal ever to deceive ourselves.

A review of the history of life should provoke awe, above all else. As Goethe said, *"Zum Erstaunen bin ich da"*—I am here to wonder. There are no trite moral lessons, nor are homilies desirable about cycles of history which are destined to come around once more. It is only certain that there will be change, and change again. Man will doubtless be an additional cause of it. The difference from any of the hundred incidents I have described in this biography is that we should be able to anticipate effects. Let us hope that we act wisely. Spinning tumblers, geared by chance, will still intercede in our fate. There may be bolides, there will certainly be climate change, there may be incidents with no precedents.

Life will probably cope.

Glossary

THIS IS NOT a textbook, and I have avoided technical terms for the most part, or I have explained them when they were introduced. However, it may be helpful to have a brief list of simple and simplified definitions to serve as succinct reminders.

amnion—a membrane which surrounds the developing embryo

ammonites—extinct mollusc group with coiled shells and tentacles, distantly related to living squid and nautiloids

anhydrous—lacking any trace of water

Archaea—or Archaebacteria, primitive bacteria and probably the earliest forms of life on Earth (see fig. 8)

Archaean—the most ancient period of geological time, from the assembly of the Earth at about 4,600 million years ago to the Proterozoic 2,400 million years ago

archaeocyathids—a group of sponge-like organisms of Cambrian age, notable for forming some early reef-like structures

arthropod—member of the great phylum Arthropoda—animals with jointed legs—and including insects, spiders, scorpions, crabs, shrimps and extinct trilobites. The skeleton in these animals is on the *outside*.

australopithecine—"ape-men" of the genus *Australopithecus*, hominids undoubtedly related to *Homo* himself

basalt—a dark, fine-grained volcanic rock that (in various forms) makes up much of the oceanic crust and the oceanic islands, such as Hawaii

biota—the sum total of all living organisms

bolide—general term for large, extraterrestrial body

brachiopod—a group (phylum) of animals encased in two valves, superficially resembling a clam, to which they are biologically unrelated

brittlestars—a type of starfish with flexible arms, frequently abundant on the floor of the deep sea

bryozoa—a group of colonial animals commonly found encrusting stones or seaweeds; sometimes, and erroneously, called "moss animals"; individuals of the colony are tiny and filter-feed edible particles from sea water

calcite (adj. calcareous)—the mineral form of calcium carbonate, common in nature as limestone rock—but also as the shells (often fossilized) of molluscs, corals, brachiopods, trilobites, etc.

cartilage—"gristle"; makes up the vertebrae of sharks

cellulose—strengthening material of cell walls of plants; molecules of cellulose are long, unbranched chains (polysaccharides), which are allied in bundles, accounting for their toughness

chalk—a pure white limestone made largely of coccoliths. The Chalk is a Cretaceous rock formation, strictly speaking, and referred to in this way should be capitalized.

chert—a tough sedimentary rock made of silica (q.v.)—used in the manufacture of stone tools, and preserving remarkable fossils from the Precambrian onwards. Flint is a pure variety of chert from the Chalk.

chlorophyll—pigments (the "green" of plants) responsible for photosynthesis

cladistics—technique for estimating how organisms are related one to another using the features of their bodies (or genetic structure) to compile branching *cladograms*. The cladograms identify the important branching events in their history at which new characteristics were acquired.

coccoliths—minute plates produced by oceanic, single-celled algae, which are a significant component of many marine sediments

coral—member of the jellyfish group (phylum Cnidaria) which lays down a skeleton of calcium carbonate, and hence is readily fossilized *crinoid*—a "sea lily," and no plant at all but an echinoderm, related distantly to sea urchins and starfish

cycads—a group of superficially palm-like shrubs and trees, especially important during the Mesozoic, but still quite numerous today, especially in the southern hemisphere

DNA—abbreviation for Deoxyribonucleic acid—the basic genetic (hereditary) molecule of the cell, constructed of nucleotides arranged in the famous double helix discovered by Crick and Watson

dioecious—describes a plant species in which male and female are carried on different individuals

echinoderms—literally "spiny skins." The phylum Echinodermata includes a range of marine animals covered in calcite plates: starfish, brittlestars, sea urchins and sea lilies (crinoids) are the most well known.

endemic—confined to one particular area (the opposite is pandemic)

enzyme—a cell or body protein that turns on vital functions in the cell—a chemical messenger, or catalyst

Euglena—one of the more familiar single-celled protists bearing a whip-like flagellum

eukaryotes—those organisms with cells having a discrete nucleus containing the genetic material

eustatic—a global sea-level rise or drop is described as eustatic

family (of organisms)—a classification category usually embracing several genera (singular genus). Humankind—the genus *Homo*—for example, belongs to the family Hominidae, or hominids, which also includes the ancient genus *Australopithecus*.

flint—see chert

foraminiferan—single-celled (protist) organism often with skeleton, some kinds of which can grow to large size

fumaroles—vents associated with volcanoes, from which gases (frequently toxic) escape

genus (of organisms)—a classification category often embracing several species, which share characteristics showing that they are likely to have arisen from a common ancestor. Our own genus, *Homo*, also includes the extinct species *H. erectus*, *H. habilis* and *H. neanderthalensis*. The generic name is customarily italicized, and abbreviated to an initial.

ginkgos—at present represented by one species of tree, a "living fossil," but in the Mesozoic a diverse group of shrubs and trees

Golgi bodies—organelles of plant and animal cells (discovered by Camillo Golgi in 1898—hence their name) concerned with many vital chemical agents of the cell such as enzymes and hormones, and production of poly-saccharides such as cellulose

herbivory—plant-eating

hominid—member of the family of humankind, Hominidae

hydroid—a small, and often branched, attached jellyfish

hydrothermal—literally "hot water"—associated with volcanic activity, pro-ducing minerals, hot springs and the like

inlier—an area of older rocks entirely surrounded by geologically younger strata

invertebrate—literally, lacking a backbone—a term broadly applied to all animals other than vertebrates

lancelet—the amphioxus, a standard laboratory animal for biology students to study the primitive features of animals with notochords—in the most general terms ancestral to vertebrates

liverwort—primitive, creeping land plant growing in damp places

loess—a fine sedimentary deposit covering much of central Asia, including China and the steppes—it is an accumulation of fine dust, much of it of glacial origin, and may be only a few thousand to a few hundred years old

lycopods—a great group of plants, including trees in the Carboniferous, and today represented by club mosses

lysosomes—organelles within the cells of animals and plants being particu-larly associated with digestive enzymes; these are involved with numerous vital functions, from food-processing to consumption of alien microbes (in white cells)

Magdalenian—one of several styles of stone axe "industries"; these are usually named for a type locality

magma—the melt from which igneous and volcanic rocks crystallize on cooling

magnetization—when volcanic rocks are extruded at the mid-ocean ridges they acquire the magnetization prevalent at the time. This can be either "normal" (i.e. the same as today) or "reversed" (North Pole becomes South and *vice versa*). Reversals of the Earth's magnetic field have happened many times, and thus provide another kind of chronometer. Regular "stripes" of normal and reversed magnetism to either side of mid-ocean ridges provided a crucial confirmation of the reality of the spreading apart of plates from these ridges.

marsupials—mammals that nurture their young in a special pouch, the marsupium. The babies are born in a very immature condition compared with those of the placental mammals.

metazoa—animals composed of numerous cells, which are organized into tissues with different functions

mitochondria—generally sausage-shaped organelles found in eukaryote (plant and animal) cells, and containing the enzymes responsible for aerobic (oxygen) respiration

mollusc—member of the great group of invertebrates, phylum Mollusca, including snails, clams, squid, octupus, nautiloids and the extinct ammonites

nautiloid—swimming, marine molluscs with a coiled, chambered shell represented in the living fauna by the pearly *Nautilus* (it will be recalled that Jules Verne gave this name to Captain Nemo's submarine in *20,000 Leagues under the Sea*)

nematodes—members of the phylum Nematoda (roundworms)—unsegmented worms, mostly small, and some of them parasitical

Neolithic—see Palaeolithic

nucleus—the central body of the eukaryotic cell, where resides the genetic information on the chromosomes

organelle—one of the membrane-bounded bodies contained inside cells; there are several kinds of organelle, each with a particular function; for example, chloroplasts are the critical organelles in photosynthesizing organisms

Palaeolithic—"older Stone Age" applied to styles of flint and other tools, and beginning approximately 1.5 million years ago; Neolithic—"new Stone Age" is typified by advanced stone tools, generally 10,000 years ago to the beginning of the Bronze Age; Mesolithic—between the two—is often distinguished

Pangaea—the Permian–Triassic supercontinent

phosphates—compounds of the elements phosphorus and oxygen, and implicated in many vital chemical reactions in the cell

photosynthesis—the vital process whereby atmospheric carbon dioxide is broken into its constituent elements, the carbon to form cell materials, while the oxygen is released into the atmosphere

polyp—the feeding individual of corals and their relatives—a circlet of tentacles surrounding a mouth (the free-floating "jellyfish" is termed a medusa)

productids—giant kinds of brachiopod which were abundant in the Carboniferous period

prokaryotes—"bacteria" lacking a discrete nucleus in the cell

proteins—the basic building (organic) chemicals for life. Their molecules are long chains of amino acids joined together

protist—general term for a single-celled organism larger and more highly organized than a bacterium. Formerly referred to as "protozoa," which artificially implied that they were animals. Now known to comprise highly distinctive and separate groups of organisms close to the base of the tree of life

protoplasm—the viscous and colourless material comprising much of the general cell contents

pyrites—an iron sulphide—a chemical combination of the elements iron and sulphur—and a very common mineral

RNA—abbreviation for ribonucleic acid—the controlling agent for the synthesis of vital proteins in a cell. Ribosomal RNA is associated with the organelles known as ribosomes

radiolaria—tiny, single-celled organisms with a beautiful and delicate silica shell or "test"

regression—especially marine regression, a "draining off" of the sea from the continents—often as a result of eustatic control

respiration—breathing in its broadest sense, utilizing oxygen (aerobic respiration) in animals

schist—a metamorphic rock, often finely splitting with shiny surfaces because of the presence of the mineral mica

sea lilies—elegant, often stalked echinoderms with food-gathering arms—animals despite their name, which derives from their superficial resemblance to tulips

sedimentary rocks—rocks which were originally laid down as sediment, usually beneath marine or fresh water, but airborne sediments also result in sedimentary rocks. Clay is a sedimentary rock despite its plasticity.

sequencing—the technique of identifying the order of nucleotides on genes

shale—a fine-grained sedimentary rock that splits readily into thin sheets

shield area—e.g. Canadian Shield—ancient "core" of present continents often dating back to the Archaean

silica—silicon dioxide, one of the commonest chemicals in nature, familiar as golden sand; utilized as a skeletal material by sponges and radiolarians

spectrum—light of different wavelengths, characteristic, for example, of the various chemical elements

sponge—primitive marine organism with rather loosely constructed tissues, member of the phylum Porifera. The bath sponge is just the most familiar of a very varied and ancient group of organisms

stromatoporoids—a group of sponges, or sponge-like organisms forming an important part of reefs in the Palaeozoic, where they form low mounds and crusts

stromatolite—finely layered or laminated structure constructed by cyanobacterial mats and typical of later Precambrian limestone sedimentary rocks. Living stromatolites are being formed today in a few protected marine sites.

subduction—in plate tectonics, the process whereby oceanic plates are consumed at the edges of continents (subduction zone); these areas are marked by ocean trenches and, naturally, are closely associated with earthquakes and volcanic activity

symbiosis—literally "living together"—the close association of two organisms for their mutual benefit

tetrapod—"four-footed," a term descriptive of the land vertebrates and their descendants

transgression—especially marine transgression, the flooding of continental areas during a rise in sea level, often under eustatic control

trilobite—an extinct type of arthropod with hard external skeleton of calcite (calcium carbonate)

vertebrate—animal having a backbone—embracing everything from fish to humans

Suggestions for Further Reading

Archer, M., S. J. Hand, and H. Godthelp. *Riversleigh: The Story of Animals in Ancient Rainforests in Inland Australia.* Reed International, 1992.

Bengtson, S., ed. *Early Life on Earth.* New York: Columbia University Press, 1994.

Briggs, D. E. G., D. H. Erwin, and F. J. Collier. *The Fossils of the Burgess Shale.* Washington, DC: Smithsonian Institution, 1994.

Clutton-Brock, J. *A Natural History of Domesticated Mammals.* Austin: University of Texas Press, 1989.

Dawkins, R. *The Blind Watchmaker: Why the Evidence of Evolution Reveals a Universe Without Design.* Reissue of 1987 ed. with new introduction. New York: W. W. Norton, 1996.

Desmond, A., and J. Moore. *Darwin.* New York: Warner Books, 1992.

Eldredge, N. *Time Frames.* New York: Simon and Schuster, 1985.

Gould, S. J. *Full House: The Spread of Excellence from Plato to Darwin.* New York: Harmony Books, 1996.

———. *Wonderful Life: The Burgess Shale and the Nature of History.* New York: W. W. Norton, 1990.

Hallam, A. *Great Geological Controversies.* 2nd ed. New York: Oxford University Press, 1990.

Lehmann, U. *The Ammonites: Their Life and Their World.* New York: Cambridge University Press, 1981.

Lipps, J., H. Boardman, S. Richard, A. Cheetham, and A. J. Rowell. *Fossil Prokaryotes and Protists*. Cambridge, Mass.: Blackwell Scientific Publications, 1992.

Novacek, M. *Dinosaurs of the Flaming Cliffs*. New Haven, Conn.: Yale University Press, 1996.

Index